MICROBIOLOGY of
WELL BIOFOULING

The Sustainable Well Series

MICROBIOLOGY of WELL BIOFOULING

Roy Cullimore

CRC Press

Taylor & Francis Group
Boca Raton London New York

CRC Press is an imprint of the
Taylor & Francis Group, an **informa** business

CRC Press
Taylor & Francis Group
6000 Broken Sound Parkway NW, Suite 300
Boca Raton, FL 33487-2742

First issued in paperback 2019

ISBN-13: 978-1-56670-400-7 (hbk)
ISBN-13: 978-0-367-39946-7 (pbk)

Library of Congress Cataloging-in-Publication Data

Cullimore, Roy.
    Microbilological of well biofouling / Roy Cullimore.
          p.   cm. — (The sustainable well series)
    Includes bibliographical references and index.
    ISBN 1-56670-400-6 (alk paper)
    1. Wells—Fouling. 2. Fouling organisms. I. Title. II. Series.
TD405.C85 1999
628.1′14—dc21
                                            99-35336
                                          CIP

Library of Congress Card Number 99-35336

Visit the Taylor & Francis Web site at
http://www.taylorandfrancis.com

and the CRC Press Web site at
http://www.crcpress.com

The Sustainable Well Series

Series Editor Roy Cullimore

PREFACE

My first involvement in examining the causes of clogging in water wells was when, as the microbial ecologist, I was asked what I knew about iron bacteria. To that question from the director of the Saskatchewan Research Council, I replied "absolutely nothing but prepared to learn." As a result of a subsequent grant application which was successful, I began what has now become almost three decades of traveling down on an often lonely pathway towards learning about the diversity of microorganisms in ground water and in the clogging structures in and around water wells. Perhaps more importantly as a result of conducting research in a little understood field, there was the passage through the early stages of scientific progress.

Today, it would appear that the growth of a particular area of science passes through four stages. In the first stage, there are just a few who appreciate there is a concern that needs to be addressed; the vast majority do not see any and explain the events associated with the concern as being irrelevant or explain them by some unsubstantiated hypothesis. This first stage is one of denial. The persistence of the growing "few" coupled to a gathering of scientific evidence leads to the concerns becoming identifiable through scientific studies. It is at this stage that some of the "majority" now begin to accept the possibility that rationalized scientific evidence explains the concerns. This is the second stage of progress which is a grudging acceptance. Now begins a growing enthusiasm for the scientific concepts and findings that become embraced by the "majority" except for the very conservative elements that have been thriving from the ignorance that previously existed. This third stage is one of overindulgence where the concepts and the findings are embraced and claims exaggerated. In the fourth and final stage in the development of a scientific endeavor, there is a rationalized acceptance of the validated principles and the information enters the "mainstream" of scientific understanding.

This book addresses concerns relating to the microbiology of wells and the relative importance of these indigenous organisms in the biofouling of water wells, ground water systems, hazardous waste sites, remote and on-site bioremediation plants, as well as some reference to the implications in the operation of oil and gas wells. For ground water microbiology and well biofouling, the status of the science is somewhere in the early phases of grudging acceptance. This is seen in the growing acceptance of the need for routine testing and preventative maintenance programs in the industry. Today, the majority believe that a severely biofouled (plugged) well becomes essentially a burden and should be abandoned and a replacement well installed. Such decisions obviously support the well drilling industry and encourage earlier abandonment and less rigorous attempts to rehabilitate the well. In these tighter economic times with more stringent environmental regulations, there is a growing concern that water wells should become more sustainable by better preventative maintenance and preemptive rehabilitation of the wells when there is evidence of plugging (clogging). Rehabilitation should be practiced when a well has only lost between 20 and 40% of its original specific capacity. Once the well loses more than 40%, it is less likely that the well can be recovered to its original production. Today so often, a well is under-pumped so that production is dramatically impacted as the plugging suddenly affects pumping. Canada Agriculture, through the PFRA-TS, have co-operatively been addressing the issues of water well sustainability and this series of technical monograms is partly a product of that initiative.

Human society is very much influenced by the economic costs of undertaking particular projects. When a water well is functioning satisfactorily, the water is considered virtually free and little attention is paid to the causes of the well losing its production through plugging. More thought is given to the time frame before it would have to be replaced with a new well. As the cost of replacement wells climbs and water becomes more valuable, economic studies on the relative value of rehabilitation are being undertaken. In general, the premise that ground waters are often cheaper to treat than surface waters can cause the ground water to have more value by volume than neighboring surface waters. This cost difference can translate into real savings for the water user. From the PFRA-TS initiative demonstration at the city of North Battleford on the rehabilitation of two plugging water wells (40 and 80% losses in specific capacity) came additional experiences on the value potential for the rehabilitation of plugging wells. A recovery to 100% was achieved for the 40% impaired while the other doubled its specific capacity. For the city, the lower treatment costs for the ground water versus the surface water meant a saving, after treatment, of over C$100,000 due to the greater production obtained from the ground water sources.

Over the last decade, various aspects of diagnosis, treatment and management of water wells have been addressed. Some of these issues were discussed in the book "Practical Manual of Groundwater Microbiology" published by Lewis Publishing in 1993. This book, along with the other books in the technical series on sustainable wells, represents an attempt to bring the reader up-to-date with present day field and laboratory experiences relating to the biofouling of wells. While the emphasis is on water wells and the surrounding ground water, many of the observations are relevant to oil and gas wells. In oil wells, the microbes are actually "mining" the most limiting substrate for growth from the oil itself and that is the water! Bioaccumulating that water starts a plugging process that can then reduce the production capacity of the well significantly. In gas wells, the water can become locked into slime formations that have a similar effect in reducing the production of the gas well. Hence, this volume is entitled "Microbiology of Well Biofouling" and, while it focuses on water wells, many of the principles can be applied to oil and gas wells. Similarly, there can be very significant biofouling at hazardous waste and bioremediation sites. These principles are also applicable there.

This book has been written as a comprehensive practical account of the origins, forms and functioning of the biofouling. To detect the biofouling, the biological activity reaction test (BART^{TM}) has been selected since it is simple to apply in field situations, is relatively easy to interpret, and more sensitive by one to three orders of magnitude than the classical agar spread- and pour-plate techniques. There have now been ten years of field experiences with this technique and it was selected as the field test for incorporation in the recently funded American Water Works Association Research Foundation (AWWARF) supported study on water well rehabilitation being undertaken by Leggette, Brashears & Graham, Inc., Connecticut. The tests were also a major part of the Canada Agriculture, PFRA-TS sustainable water wells initiative operated in Alberta and Saskatchewan, Canada.

There is still considerable debate on the importance of microbial biofouling in the operation of water wells which has caused some confusion in the eyes of the managers of water wells. This polarization has generated one group that downplays the microbially driven biofouling to be of a minor significance and easily controlled. That ease of control usually comes from the application of self-selected products being promoted by members of that group! On the other side of the polarized window are those who have seen the extensive forms of infestations that can occur around a well, and the ability of the microbes within that growth to "rebound" after a treatment and regrow with a vengeance. The AWWARF and the PFRA-TS are both to be congratulated on their recent initiative to conduct an evaluation of the present knowledge and science base that has developed around water well rehabilitation.

The latter part of this volume is devoted to practical examples of the diagnosis of well biofouling, application of treatments and long term effectiveness. It has always to be remembered that each well is a unique entity and may not respond to a treatment in the same way as a neighboring well. Each well has to be treated as a separate challenge and the approach to diagnosis, treatment and preventative maintenance may have to be adjusted. It is a very good idea to "walk the wells" and listen to the operator's concerns before beginning the pathway to diagnosis, treatment and preventative maintenance.

This volume is not the be all and end all of the topic of well biofouling but simple a "local library" along the pathway to a matured knowledge base on the biofouling of wells, porous media and essentially the crust of the planet Earth. When you stop to think about it, we really are the planet Oceana and those "oceans" actually reach right under the solid: air interfaces of the planet as ground water. It could be argued that the "sea snow" in the oceans become the slime plugs under the land and the "well snow" down the bore holes. This volume then becomes are very small part of the "big picture."

Numerous are the people who have had an influence in the creation of this volume. While this list may be long, I would like to acknowledge the following: George Alford, ARCC Inc., Florida, co-inventor of the BART™ and the BCHT™ methods. George thought, after our first telephone call, that I was a total idiot wanting to heat water wells. That was then, but this is now! John Lebeden of the PFRA-TS has become an avid believer in the need to make water wells more sustainable. The first biological survey of wells in a municipal district in Alberta, and the setting up of a municipal well field for the evaluation of routine management practices to increase the life span of the wells in that field occurred in the last four years. This was possible due to the cooperation of the following: Stuart Smith of Ada, Ohio who has persistently pursued the iron bacteria as a phenomenon impacting on water wells, and Peter Howsam of Cranfield University in England who, as an engineer, recognized the importance of biofouling and organized two symposia in 1990 on "Microbiology in Civil Engineering" and on "Water Wells, Monitoring, Maintenance, Rehabilitation." Within the Regina Water Research Institute, University of Regina, and Droycon Bioconcepts Inc., there have been many enthusiasts over the years who have helped not only in their work time but also on their own time, to press forward the boundaries of knowledge on biofouling. These include, but are not limited to Natalie Ostryzniuk who over the years has patiently developed my scribblings and bad typings into a form that the reader can (hopefully) appreciate. Also I would like to thank past and present staff including: Brent Keevil, Lori Johnston, Dr. Kripa Singh, Delee Roberts, Wade MacLean, Vincent Ostryzniuk, Derek Ross, Jeff Rheil, Dr. Abimbola Abiola, and Twyla Legault. Sponsorship for some of the activities associated with the biofouling has included: National Research Council of Canada, Industrial Research Assistance Program, Natural Sciences and Engineering Research Council of Canada, AWWARF, IPSCO Steel Company of Regina, Agriculture Canada (PFRA-TS, Agri-Food Innovation Fund); Kneehill Municipal District, Alberta; and the City of North Battleford, Saskatchewan, Canada.

D. Roy Cullimore
Ph.D., Registered Microbiologist

July 1999

ABOUT THE AUTHOR

Roy Cullimore is an applied ecologist trained at the University of Nottingham, U.K. Since 1975, he has been Director of the Regina Water Research Institute at the University of Regina. He along with co-inventor, George Alford, have five patents including the biological activity reaction test (BART™) and the blended chemical heat treatment (BCHT™). Roy has published over one hundred refereed papers, two hundred and seventy technical reports, and has received over 3.5 million dollars in research funding. At present, he is President of Droycon Bioconcepts Inc. of Regina, a biotechnology company involved in research, development and manufacture. He authored *Practical Manual of Groundwater Microbiology* which was published by Lewis Publishers in 1993. Currently, Roy is editor of a series of books on sustainable wells. He is now involved in research on the rusticle growths on *RMS Titanic* and dove to the ship in 1996 and 1998 as a part of the Discovery Channel expeditions. Also, Roy is currently involved in the AWWARF water well rehabilitation project being undertaken by Leggette, Brashears & Graham, Connecticut.

CONTENTS

INTRODUCTORY OVERVIEW

Of all the environments in or upon the surface crust of this planet, the biology of ground waters and the deep subsurface of the crust is one of the least understood. It has been a matter of convenience to consider that the biosphere does not extend down into the crust beyond shallow water wells. It was common even a decade ago to read that water from groundwater was "essentially sterile." As a result, it has traditionally been thought that the potential for biological events was at the most very small and, all of the time, not significant.

In the last decade, applied microbial ecologists have been reporting microorganisms as existing even in the deeper ground waters. It was established that these microorganisms perform a number of very significant functions in changing both the chemistry and hydrological characteristics of the ground waters passing through the porous and fractured media of the aquifers. These changes can affect the production from a water well, the rate at which hazardous substances such as gasoline move through the aquifer and degrade, and the quality of the product water from a well. Today, there is sufficient evidence that recognizes these microbial events in ground water to allow the prediction of biofouling occurrences and control these processes to the advantage of the users.

It may seem strange that, over the last century, so many advances have been made in science and technology at the same time as very little attention has been paid to subsurface groundwater microbiology. However, it is clear that the mandate of "out of sight out of mind" can be applied in this case. For that matter, relatively few direct relationships have been established between the activities of the indigenous microorganisms and the structures and processes designed and managed by civil engineers. Traditionally, the main focus has been on major concerns of economic significance such as the microbially induced corrosive processes which have had serious annual costs measured on a global scale in the billions of dollars for the oil industry alone.

In the last two decades, the ubiquitous and dynamic nature of the intrinsic microbial activities in ground waters is becoming recognized as one of the inherent and essential factors. These are involved in such events as plugging, remediation of pollution events, hydrophobicity in soils and the bioaccumulation and degradation of potentially hazardous chemicals. Some of these events can become manageable once the key controlling factors have become understood.

While the microorganisms are, by and large, very simple in structure, small in size (ranging from 0.5 to 20 microns or 10^{-3} mm in cell diameter) and commonly spherical (coccoid) or rod shaped, the impact that these organisms can create can have dramatic effects on the activities of what is probably the most sophisticated species, *Homo sapiens*. In recent millennia, man has learned to construct and devise many sophisticated systems that not only provide protection, safety and transportation but now also allow rapid communication. During the period of evolution, microorganisms have also developed very sophisticated systems of protection, community structures, modes of transportation and even communication.

In today's human society, engineering has created many advances that have facilitated both the intellectual and material developments within a global society. So rapid have been these developments, particularly in the last century, that the interaction between these systems and structures with other cohabiting species has received but casual secondary consideration. Traditionally, surface dwelling plant and animal species that are deemed to be desirable strains are often "landscaped" into the environment or allowed to flourish within these manipulated structures for the direct or indirect benefit of man. Less progress has been made to understand the extent to which subsurface dwelling organisms occupy, and are active within, the soils and geological "crust" of the earth. Only today are extensive studies being initiated.

CHALLENGE OF DEEP SUBSURFACE MICROBIOLOGY

Today, the recognition of the impact of human societal developments on the variety of the animal and plant species present on this planet is becoming more acutely appreciated. The whole nature of the crust as well as the surface of the planet is going to become subjected to a range of increasingly engineered management structures and practices. The challenge is primarily the responsibility of the professional engineers.

In recognising the need to "engineer" the subsurface biological communities, knowledge of the environments within which these microbial communities thrive needs to be gained. The durable versatility that these microorganisms possess allows them to flourish in environments alien to the surface dwelling organisms. And yet in the deep subsurface environments, these microorganisms are able to form communities that are not only stable and durable but also dynamic. To undertake such comprehensive management practices, a very clear appreciation has to be generated regarding the role of these biological components as contributors to the dynamics of the crust of this planet. For *Homo sapiens,* as a surface dwelling species, it is relatively convenient to comprehend the role of the plants and animals since these are, by and large, either surface dwelling or flourish in the surface-waters which abound on this planet.

Perhaps not so recognized at this time are the roles of the microorganisms. These are the insidious organisms of relatively small dimensions and (apparently) simple abilities that are almost totally ubiquitous within the soils, surface- and ground- waters and even upon and within many of the living organisms that populate this planet. An example of the ubiquitous nature of microorganisms can be found in the fact that 90% of the cells in the human body are, in fact, microbial. The remaining 10% are actually the tissue cells that make up the human body as such. This is possibly symbolic of the role of microorganisms throughout the surface structures of this planet. They are small, insidious, numerous and can also be biochemically very active.

Additionally, these microbes are able to resist, survive, adapt and flourish in some of the very harshest of habitats. Some examples of these extreme environments in which microorganisms can flourish are:

Acidophiles	0.0 to 5.0 pH
Alkalophiles	8.5 to 14.0 pH
Psychrophiles	-36 to +15°C max.
Thermophiles	+45 to +250°C
Aerobes	oxygen concentration from >0.02 ppm to saturated.
Anaerobes	no oxygen required, commonly toxic
Barotolerant	hydrostatic pressures of 400 to 1,100 atmospheres
Halophiles	growth in 2.8 to 6.2 M Sodium chloride

So extreme are some of these environments that the recovery of living (viable) cells and culturing (growing) these organisms represents a set of challenges in each case. The nub of the concept of versatility within the microbial kingdom is that, provided there is liquid water along with a source of energy and basic nutrients, microbial growth will eventually occur once adaptation has taken place.

Important effects of extreme environments include the fact that the number of species able to (1) survive, and (2) flourish will decrease in proportion to the degree of extremity exhibited by the local environment. For example, the range of species of acidophiles would diminish as the pH falls from 5.0 to 0.0. Typically, the number of species would decline from as many as thirty to as few as one or two.

Most of the microorganisms that have been studied originated from the surface waters, soils or the surface dependent biosphere. The widest variety of these microbes grows within a common physical domain which normally involve pressures close to 1 atmosphere, pH regimes from 6.5 to 8.5, temperatures from 8° to 45°C, redox potential (Eh) values down to –50 millivolts, and salt concentrations of less than 6% sodium chloride equivalents. Outside of these domains there is a severe restriction in the range of microbial species that will flourish. A number of survival mechanisms commonly exists amongst microorganisms. These range from forming spores, attaching to surfaces within biofilms, "clustering" in the water as biocolloids, or forming minute suspended animation survival entities (called ultramicrobacteria, or UMB for short).

THE GROUND WATER ENVIRONMENT

In the ground water within the crust, there are a number of physical and chemical factors that would begin to stress any incumbent microorganisms at deeper and deeper depths. These stressors would change the nature of the dominant microbial species. However, at shallow depths where the physical factors are within the normal range for surface dwelling species, these species may continue to remain dominant. As the groundwater environment becomes deeper there are a number of factors that will influence the microbial species dominance. These factors are the increases in temperature, pressure and salt concentration together with decreases in oxygen availability and the movement to a very reductive Eh (redox) potential.

In general, geothermal temperature gradients show average increase in ordinary formations of 2.13°C per 100 meters of depth. Ground water at a depth of 1,000 meters would therefore be expected to have a temperature elevation of 21.3°C. For 2,000 meters, this rise would be on average 42.6°C. That would take the environment up into the range where thermophilic microorganisms would flourish but not the surface dwelling microorganisms (mesophiles). These temperature gradient rises would be influenced by the presence of hot rock masses or unusually high or low hydraulic conductivity.

Hydraulic pressures will also be exerted to an increasing extent at greater depths particularly below the interface between the unsaturated (vadose) zone and the saturated (water table). In general, the soil-water zone will extend 0.9 to 9.0 meters from the surface and will overburden the vadose zone. Generally in igneous rocks, the bottom of the ground water could be at depths of 150 to 275 meters while in sedimentary rocks, depths could reach 15,900 meters. In the latter case, extremely high hydraulic pressures could be exerted upon any incumbent microflora. Hydraulic pressures will be very much influenced by the hydraulic conductivity of the overburden and the degree of inter-connectivity between the various related aquifers within the system. In oceanic conditions, these factors are not significant and pressures can reach 600 atmosphere (8,820 p.s.i. or 1,280 kPa). At these higher pressures, the temperature at which water would convert to steam would also be elevated so that many microorganisms could still exist and flourish at superheated temperatures provided that the water was still liquid. For example, water will not boil until the temperature (°C) is reached for the pressure shown (in parenthesis, p.s.i.): 125°C (33.7); 150°C (69.0); 200°C (225.5); 250°C (576.6); 300°C (1,246); and 350°C (2,398). There would therefore appear to be a strong probability that there may be a series of stratified microbial communities within the crust. These may be separable by their ability to function (with depth) under increasing hydrostatic pressures in more concentrated salt solutions, at higher temperatures under extremely low very reductive (-Eh) redox potentials (e.g., -450 millivolts).

For those microorganisms able to survive and flourish in extremely deep subsurface environments, the geologic formations must provide openings in which the water (and the organisms) can exist. Here the microbes will, in all probability, grow attached to the surfaces presented within the openings. Typical openings include inter-grain pores (in unconsolidated sandstone, gravel, and shale), systematic joints (in metamorphic and igneous rocks, limestone), cooling fractures (in basalt), solution cavities (in limestone), gas-bubble holes and lava tubes (in basalt) and openings in fault zones. All of these openings provide surfaces large enough to support such attachment if the cells can reach the site. Such a restriction would be relatable to the size of the microbial cells in relation to the size of the openings.

Porosity is usually measured as the percentage of the bulk volume of the porous medium that is occupied by interstices and can be occupied with water when the medium is saturated. This volume is sometimes referred to as the void volume. For coarse to fine gravel, the percentage porosity can range from 28 to 34% respectively. Porosities for sand tend to range from 39 to 43%. Fine and medium grain sandstone generally have porosities from 33 to 37% respectively. Tightly cemented sandstones would have a porosity of 5%. However, rocks tend to be porous structures with pore sizes large enough to accommodate bacterial cells that, in the vegetative state, have cell diameters of between 0.5 and 5 microns. When these cells enter a severe stress state due to starvation or environmental anomalies, many bacteria will shrink in size to 0.1 to 0.5 microns and become non-attachable. Such stressed survival cells are referred to as ultramicrobacteria (UMB) and are able to pass through porous structures for considerable distances. Recovery is dependent upon a favorable environment being reached that would allow growth and reproduction. UMB, therefore, have the potential to travel through porous media and remain in a state of suspended animation for very prolonged periods of time. Even normal vegetative cells (e.g., those of *Serratia marcescens*) have been shown to pass through cores of Berea sandstone (36.1 cms) and limestone (7.6 cms) during laboratory studies. There is indeed a growing body of knowledge supporting the ability of bacteria to invade and colonize porous rock formations. Viruses have also been recorded as being able to travel through porous structures.

As microorganisms attach to surfaces within porous media and begin to grow into communal biofilms, these biofilms occupy space in the void volume. This void volume occupancy reduces the amount of the void volume through which water can freely pass. The porosity is therefore reduced by the growth of the biofilms. This can result in reduced hydraulic conductivities and eventual plugging of the porous medium.

Fractures in rocks tend to have much larger openings than pore structures and can therefore form into potential ground water flow paths (conduits) that could act as focal sites for microbial activities. This heightened microbial activity would result in part from the passage of ground water containing potentially valuable dissolved and suspended chemicals (e.g., organics, nutrients, and oxygen). In addition, the flow paths would allow the transportation of the microbial cells and also the dissolution of waste products. Microbial mobility through fracture flow paths in rocks can be very fast. In 1973, bacteria were recorded to travel at rates of up to 28.7 m/day through fractured crystalline bedrock.

Little attention has been given to the scale of microbiological activity present in soils or the crust of the earth. It is only in the last decade that attention is being paid to the potential size of the biomass occurring in the soils, muds and subsurface elements. Present extrapolations indicate that the biomass is indeed very significant in relation to the biomass found on the surface of the planet. Further research will undoubtedly allow a more precise comparison of these various parts of the biosphere.

MICROBIAL MECHANISMS OF COLONIZATION AND GROWTH

There arises the need to understand the mechanisms by which these microbial invasions of porous and fractured structures can occur in both saturated, capillary unsaturated and vadose zones. It has perhaps tradition-ally been thought that, since the ground waters are essentially "sterile", any microbial problem in water wells must have been the result of a contamina-tion of the well during installation. This would involve microorganisms contaminating the well installation from the surface environment. In reality, while such contaminations do occur, there is also the very real probability that microbes may have migrated towards the well as a very "inviting" new environment in which to colonize (or infest, depending upon the viewpoint).

A new well installation offers the advantages of the presence of oxygen (diffusing down both inside and around the well to support the aerobic microorganisms) and a more rapid and turbulent flow of ground water as it approaches and enters the well. This will deliver a greater potential quantity of nutrients to the well environment as opposed to the more laminar slow flow in the bulk aquifer. It is interesting to note that the maximum microbial activity tends to occur at an interface where the oxygen is diffusing one way (i.e., away from the well) and nutrients are moving towards the well (i.e., from the aquifer formations). In general, such focused microbial activities occur under conditions where the environment is shifting from a reductive to an oxidative regime. This is as a result of the movement of water towards the more oxidative conditions in the well. Maximum microbial activity commonly occurs where the environmental conditions are just becoming oxidative (Eh potential, -50 to +150). This is known as the redox front. A series of concentric "cylinders" are formed around the well by this enhanced microbial growth. These may form into distinct biozones that may form concentric cylinders around the well or layered as a series of stratified biozones within a common biofilm.

One major step in developing an understanding of these biozones is to appreciate that the growth of different species of microorganisms often occurs within common consortial structures where the various species co-exist together. These structures are bound with water retaining polymeric matrices (extracellular polymeric substances, EPS) to form biofilms that are attached to inert surface structures. These biofilms, more commonly known as slimes or plugs, act as catalysts which will then generate such events as electrolytic corrosion (through the generation of hydrogen sulfide and/or organic acids), and bioaccumulation of selected inorganic and organic chemicals travelling generally in an aqueous matrix over the biofilms. Biodegradation may also occur under these conditions.

Once a biofilm has become sufficiently enlarged to occupy a significant fraction of the void volume (e.g., 70%) then a total plugging of the infested porous medium can occur. This total plugging (occlusion) holds a tremen-dous potential for the control of the many hazardous wastes that have entered the ground waters, soils and rock strata on this planet by the deliberate management of these natural plugging processes. By coming to understand the methods by which these microbial activities can be utilized, it will become possible to include these manipulations of microbial processes as an integral part of engineering practices.

These microbially driven events occur naturally within the biosphere as sometimes perceived to be "chemical and/or physical" functions within and upon the crust of the planet. These microbially events include:

- Corrosion initiation.

- Bioimpedence of hydraulic or gaseous flows (e.g., plugging, clogging).

- Bioaccumulation of chemicals (e.g., localized concentrations of heavy metals, hydrocarbons and/or radionuclides within a ground water system).

- Biodegradation (e.g., catabolism of potentially recalcitrant or harmful organic compounds).

- Biogenesis of gases (including methane, hydrogen, carbon dioxide, nitrogen which can lead to the fracturing of clays, displacement of water tables and differential movement of soil particles).

- Acidic leaching particularly of sulfide-rich materials (e.g., copper ore) which has been subjected to an oxidative regime in the presence of water. The acids may be inorganic, such as sulfuric acid, or organic, such as acetic acid.

- Water retention within biofilms primarily causing slime formation (which, in turn, can influence the rates of desiccation and/or freezing of soils).

- Concretion can also occur as a result of consortial microbial activity forming a complex framework of inorganic deposits often dominated by oxidized ferric salts and/or carbonates. These structures can become very large (e.g., rusticles) or can generate very tight forms of plugging.

There are a number of ways of summarizing the manner in which these events are observed in practice. These include the following major groupings:

1. MIF (microbially induced fouling)

MIF occurs where the microorganisms begin to generate a confluent fouling over the surfaces of the porous media particularly at the reduction-oxidation interface (redox front). As these biofilms expand and interconnect, so the transmissivity of water through the system becomes reduced (plugging).

2. MIA (microbially induced accumulation)

MIA will occur at the same time (as MIF) and will involve a bioaccumu-lation of various ions such as metallic elements in the form of dissolved or insoluble salts and organic complexes.

3. MGG (microbial generation of gases)

With the maturation and thickening of the biofilms, there is likely to be extension of the anaerobic growth zone. With this increased anaerobic activity, gas formation is much more likely. Gases may be composed of various ratios of carbon dioxide, methane, hydrogen and nitrogen and may be retained within the biofilm as gas fills vacuoles or collects in the dissolved state. Radical gas generation can cause radical biofilm; volume expansion can cause foam barriers to form which could quickly reduce hydraulic transmissivity through the biofouled porous structures.

4. MIC (microbially induced corrosion)

Corrosion poses a major threat to the integrity of engineered structures and systems (particularly those employing iron) through the induction of corrosive processes.   Where the biofilm have either stratified and/or incorporated deeper permanently anaerobic strata there develops a greater risk of corrosion. Such corrosive processes are often initiated with the generation of hydrogen sulfide which stimulates electrolytic corrosion) and/or organic acids are formed which trigger drops in pH causing the solubilization of metals.

5. MIR (microbially induced relocation)

MIR occurs where the biofilm growth and biological activities associated with these activities cause a physical shift in the redox front. Microbes will often "chase the redox front" and this has been used to manage microbial fouling events. Other shifting in the environmental conditions (e.g., nutrient depletion) can also cause a relocation of biofilms. Here, the biofilm begins to sheer and slough, carrying away some of the biomass and accumulates with that movement. Relocation may also be a managed function when the environmental conditions are artificially manipulated.

At this time, the focus of attention in science has been directed to the use of specific strains of microorganisms that have been selected or manipulated to undertake very specific bio-chemical functions under controlled conditions. While the biotechnological industry has achieved successes in producing useful products, it has failed to fully recognize the potential for manipulating the natural microflora to achieve a "real world" conclusion (such as the bioremediation of a nuisance or potentially toxic chemical). These events occur naturally around water wells and may influence the resultant water quality more than is generally recognized. Today, there is a growing interest in the processes of "natural attenuation" in which the ongoing processes in the natural world are managed to achieve the desired effect. Commonly, this approach is used to remediate some of the more recalcitrant pollutants such as oils and heavy metals.

MANAGEMENT OF MICROBIAL ACTIVITIES

The physical and chemical constraints are generally unique to each strain and consortium of microorganisms. As a result, generalizations can only be made only for major groups of bacteria with natural variations always occurring. Before these variables can be considered in depth, the form of microbial growth has to be considered. A popular conception of microorganisms is that: (1) these are cells usually dispersed in water, and that (2) many may be able to move (i.e., be motile). In actuality, microbial cells can commonly be found in three states. These are:

- **Planktonic** (dispersed in the aqueous phase),

- **Sessile** (attached within a biofilm to a solid usually immobile surface),

- **Sessile particulate** (incumbent within a common suspended particle shared with other cells).

In soils and waters, the vast majority of the microbial cells are present in the sessile phases rather than in the planktonic. Sessile microorganisms are normally found within a biofilm created by the cells excreting extra-cellular polymeric substances (ECPS). These polymers act in a number of ways to protect the incumbent cells. This action includes the retention of bonded water, accumulation of both nutrients and potential toxicants, and providing a structural integrity (which may include gas vacuoles) through the deposition of primarily extracellular inorganic structures and water attracting organic polymers. Upon the shearing from the biofilm, the particulate structures afford protection to the incumbent cells. These sheared suspended particulates migrate until reattachment and colonization of a pristine (econiche) surface.

To control these microbial activities, there is a need to exercise control over the rate of selected microbial activity within the defined environment. It becomes essential to build a management structure that will allow the control of all of the major factors of influence. These can primarily be subdivided into four major groups: (1) physical, (2) chemical, (3) biological, and (4) structural. Each group not only reacts significantly with the other major groups but includes a series of significant subgroups each of which can become dominant on particular occasions.

For each individual microbial strain or consortium of strains, there is a set of environmental conditions within the environment that relates to these four major factors. These allow particular definable econiches to become established. Such microbial associations are often transient in both the qualitative and quantitative aspects due largely to the dynamic and competitive nature of the incumbents. In chronological terms, there is a very limited understanding of the maturation rate of these natural consortia. For example, what is the length of time that an active microbial biomass can remain integrated within an econiche positioned within the totally occluded zone around plugged water wells? This has not been addressed. Nor has the question of how long would the period be before water wells could become biologically plugged.

Many consortia of fungi (molds) and algae can form very specific and unusual forms of growth, for example, surface dwelling lichens growing at rates as slow as 0.1 mm per year often in an environment (i.e., the surface of a rock) in which nutrients are scarce and water rarely available. Here, the microbial activity is at a level where maturation may be measured in decades rather than days, weeks or even months. This illustrates a condition in which microorganisms have adapted to an apparently non-colonizable hostile environment and yet formed an econiche where growth of a consortial nature can flourish for perhaps centuries.

Of the physical factors known to impact on microbial processes, the most recognized factors are temperature, pH, Redox and water potential. Functionally, microbial processes appear to minimally require the presence of temperature and pressure regimes that would allow water to occur in the liquid form. Low temperatures cause freezing in which ice crystals may form a matrix around any particulates containing microorganisms and the liquid (bound) water retaining polymeric substances. Water may then be expressed from the immobilized liquid particulates to the solid ice fraction leading to a shrinking of the particulate volume until a balance is achieved between the bound (liquid) water and the solid (frozen into the ice form) water. Such low temperatures inhibit biochemical activities causing a state of suspended animation to ensue. At higher temperatures, water can remain in a liquid phase under increasing pressures up to at least 374°C rendering it a theoretical possibility that microbial activities could occur in such extreme environments.

Frequently, microorganisms in suspended "sessile" particulate structures can adjust the density of these entities causing them to rise or fall within a water body. Elevation of a suspended sessile (resulting from a lowering of density) may be caused by the formation of entrapped gas bubbles or by the releasing of some of the denser bioaccumulates. Density may increase (particle sinks) where water is expressed from the particle causing the volume to shrink into the structure.

Normal surface dwelling microorganisms, however, generally function over a relatively limited temperature range of 10° to 40°C with the optimal activities occurring at a point commonly skewed from the mid-point of the operational range of growth. These optima are to some extent different for each strain of bacteria and therefore any temperature changes can cause shifts in the consortial structure of microorganisms.

Optimal and functional ranges of activity also exist when the microorganisms interact with the environmental pH regime. Most microbes function most effectively at neutral or slightly alkaline pH values with optima commonly occurring at between 7.4 and 8.6. Acidoduric organisms are able to tolerate but not necessarily grow in much lower pH values such as low as 1.5 to 2.0 pH units but are generally found in specialized habitats. Where biofilms have been generated within an environment, any pH shifts may be buffered by the polymeric matrices of the biofilms to allow the incumbent microorganisms to survive and function within an acceptable pH range being maintained by the consortial activities while surrounded by a hostile pH.

Redox has a major controlling impact on microorganisms through the oxidation reduction state of the environment. Generally, an oxidative state (+Eh values) will support aerobic microbial activities. On the other hand, a reductive state (-Eh) will encourage anaerobic functions which, in general, may produce a slower rate of biomass generation, a downward shift in pH where organic acids are produced, and a greater production potential for gas generation (e.g., methane, hydrogen sulfide, nitrogen, hydrogen, carbon dioxide). Aerobic microbial activities forming sessile growths are often noted to occur most extensively over the transitional redox fringe from -50 to +150 mV commonly called the "redox front."

Chemical factors influencing microbial growth can be simply differentiated into two major groups based upon whether the chemical can be stimulatory or inhibitory to the activities within the biomass. The stimulatory chemicals would in general perform functional nutritional roles over an optimal (most supportive) concentration range when the ratio of the inherent nutritional elements (e.g., carbon, nitrogen, phosphorus) are in an acceptable range to facilitate biosynthetic functions. These concentrations and ratios vary considerably among microbial strains and form one of the key selective factors causing shifts in activity levels within a biomass. In the management of a biological activity, it is essential to comprehend the range and inter-relationships of any applied nutrients in order to maximize the biological event that is desired.

Inhibitory chemicals can generate negative impacts on the activities of the biomass taking the form of general or selective toxicities. For example, heavy metals tend to have a generalized type of toxicity while an antibiotic generated by one member of a consortium may be selectively toxic to some transient or undesirable microbial vectors. It is surprising that the biofilm itself may tend to act as an accumulator for many of these potentially toxic elements which may become concentrated within the polymeric matrices of the glycocalyx (structure forming the biofilm). When concentrated at these sites, these toxicants can form a barrier to predation. At the same time, these accumulates are presumably divorced from direct contact with the incumbent microorganisms and therefore perform a protective function.

In recent scientific progress, some new ideas are being evaluated using biological vectors that can be used to directly control the activities of a targeted group of organisms. Traditionally, the concept of control focused on the deliberate inoculation of a predator (to feed upon) or pathogen (to infect) the targeted organisms with some form of disease. In the last two decades, there have been impressive advances in the manipulation of the genetic materials, particularly in microorganisms, which has generated new management concepts. These involve the inoculation of genetically modified organisms which have been demonstrated, in the laboratory setting, to have superior abilities (e.g., to degrade a specific nuisance chemical) compared to microbial isolates taken from the environment. The objective of these biotechnologies is that superior functions can be transplanted into the natural environment and achieve equally superior results to that observed in the laboratory. Such concepts still have to address the ability of the inoculated genetically modified organism to effectively compete within the individual econiches (some of which may be as small as 50 microns in diameter) with the incumbent microflora and still perform the desired role effectively. The drive has unfortunately been driving down to the cellular and molecular scale of interactions at the expense of the understanding of the individual, community and population factors. This has led to a devaluation of the roles of consortial activities within manageable environmental or pathological events.

Another major factor influencing any attempt at microbial manipulation within a natural environment is the impedances created by the physical structures between the manipulator and the targeted econiche. These structures also influence the environmental characteristics of that habitat. Manipulations may be relatively easy to achieve where the target activity is either at a visible and accessible solid or liquid surface/air interface or within surface water (preferably non-flowing). Subsurface interfaces involve a range of impedances not only to the manipulator but also to the incumbent organisms. In order to manipulate a subsurface environment, there are two potential strategies. The first is physical intercedence where a pathway is created to the targeted site by drilling, excavation, fracturing or solution- removal processes. Here, there will be some level of contamination of the target zone with materials and organisms displaced by the intercedence processes. As an alternate method, the second strategy involves a diffusive intercedence where the manipulation is performed remotely by diffusion or conductance, to achieve the treatment conditions towards the targeted zone. Such intercedent techniques may involve:

- the application of gases (e.g., air, methane),
- solutions (containing optimal configurations of the desired chemicals to create appropriate conditions through modifying the environment),
- suspensions of microorganisms in either vegetative, sporulated or suspended animation states (such as the ultramicrobacteria),
- thermal gradients (through the application of heat or refrigeration) or permeability barriers that will cause a redirection or impedance of hydraulic conductivity flows in such a way as to induce the desired effect.

Diffusive intercedence can offer simpler and less expensive protocols but may be more vulnerable to the effects that any indigenous microflora may have upon the process. These effects could range from the direct assimilation of the nutrients, modification to the physical movement of the materials being entranced, to the releases of metabolic products which may support the desired effect.

Engineering structures and processes in an environment free from microorganisms may be relatively easy to design and manage. On this planet, the microbial kingdom forms a diverse group of species that occupy a wide variety of environments. When engineering projects are undertaken, these indigenous organisms respond in ways that are often subtle and often go undetected or wrongly diagnosed. With an improvement in the ability to recognize and control microbial processes, there should develop not only ways to manage these microbial events, but also, to recognize the benefits as well as the risks of having such events occur.

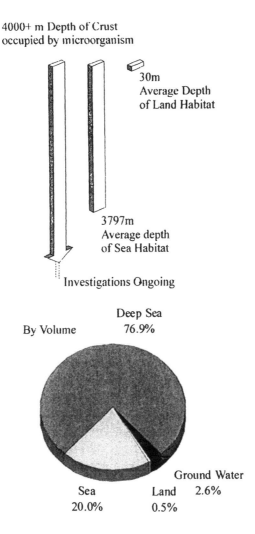

4000+ m Depth of Crust
occupied by microorganism

30m
Average Depth
of Land Habitat

3797m
Average depth
of Sea Habitat

Investigations Ongoing

By Volume

Deep Sea
76.9%

Sea
20.0%

Land
0.5%

Ground Water
2.6%

Figure One, Diagram depicting the distribution of the volume of living space on planet Earth. The concepts are based on "The Universe Below" by William J. Broad, (1997) published by Simon & Schuster, New York., pp. 44 - 47. This data was modified to include the volume of living space in the ground water.

BIOMASS IN THE CRUST

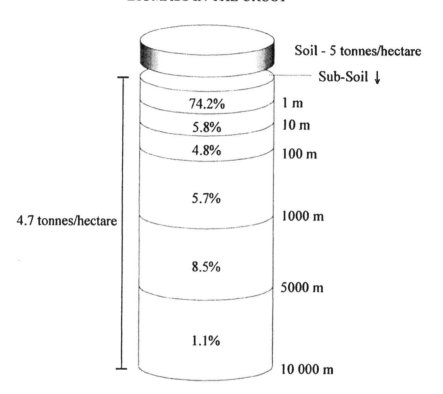

Soil - 5 tonnes/hectare

Sub-Soil ↓

74.2%   1 m

5.8%   10 m

4.8%   100 m

5.7%

4.7 tonnes/hectare   1000 m

8.5%

5000 m

1.1%

10 000 m

Figure Two, Theoretical projections of the relative distribution of the biomass on planet Earth. Little attention has been paid to the "living crop" that is the microbial biomass resting in, and upon Earth. Here, the diagram shows the relative mass of microorganisms in, and under, one hectare of prairie soil.

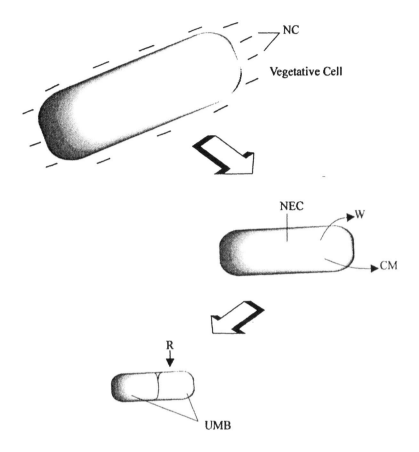

Figure Three, Life cycle of ultramicrobacteria (UMB). Most bacteria, when impacted with severe stresses, enter a survival phase. The objective is to achieve a neutral electrical charge (NEC) rather than a negative charge (NC) in order to prevent attachment to positively charged surfaces. Because of the reduction in size, UMB cells can cloister in fine porous media (e.g., clays, plutonic rock). To do this, water (W) and non-essential cellular materials (CM) are discarded. All of the metabolic mechanisms are then shut down and the last, and very symbolic, act before dormancy is reproduction (R) to generate two daughter cells. UMB are thought to be able to survive millions of years in this dormancy state.

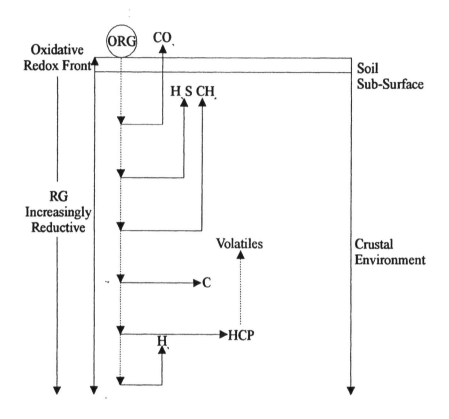

Figure Four, Diagram of the forms of stratification of the different biozones that have formed within the crust of the planet. Fundamentally, going down into the crust of the planet involves descending a redox gradient (RG) going continuously more reductive. Organics (ORG) biosynthesized   mostly at the surface permeate downwards to be degraded along various   micro-bially induced pathways leading to various forms of reduced end products from carbon dioxide (CO_2) to hydrogen sulfide (H_2S), methane (CH_4) and hydrogen (H_2). Other end products include the hydrocarbon polymers (HCP) and coals (C).

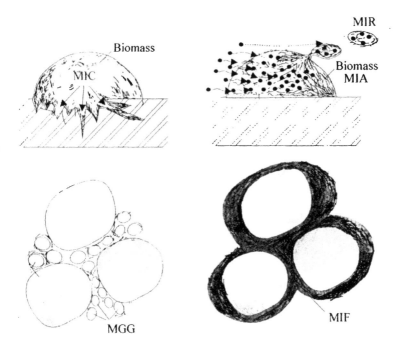

Figure Five, Diagram of the principal forms of microbial activity that can be associated with biofouling. They are illustrated as the microbially induced accumulation (MIA, upper right) where recalcitrant chemicals are accumulated within the slime, microbial gas generation (MGG, lower left) where gas is entrapped in foam pockets, microbially induced corrosion (MIC, upper left) where the microbes within the biofilms enter into corrosive processes, microbially induced relocation (MIR, upper right) in which the biofilms slough away primarily as suspended biocolloids and carry accumulated chemicals in the structures. Biofilms can so fill the voids in porous media that plugging is generated as a microbially induced fouling (MIF, lower right).

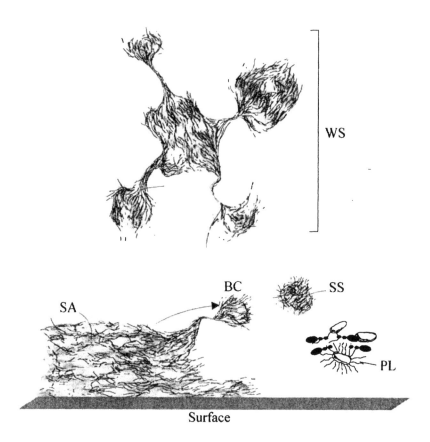

Figure Six, The principal forms of bacterial growth in ground waters include planktonic (PL), sessile suspended (SS) and sessile attached (SA). It is the sessile attached (biofilms) that form the bulk of the mass as slime attached to surfaces. When biofilms slough as biocolloidal (BC, suspended sessile) particles, they will move with the water flow. This can form the "well snow" (WS) that is often very evident in water columns in bore holes. Planktonic bacteria are "free living" in the water and are relatively rare compared to the other growth.

2

CONCEPTS

Marcus Aurelius Antoninus wrote in his Meditations published in the second century A.D. (Volume V, 23 and Volume IX, 36):

"For substance is like a river in a continual flow, and the activities of things are in constant change, and ... causes work in infinite varieties; and there is hardly anything which stands still."

"The rottenness of matter which is the foundation of everything! water, dust, bones, filth, or again, marble rocks, the callosities of the earth; and gold and silver, the sediments; and garments, only bits of hair; and purple dye, blood; and everything else is of the same kind. And that which is of the nature of breath, is also another thing of the same kind, changing from this to that."

He recognized many of the microbiological events that, at various rates, recycle the elements found within and upon the surface crust of this planet.

Much of the microbiological activities associated with these events occur within complex consortial communities formed within surface dwelling biofilms. Popularly referred to as "slimes," these communities often stratify, act as chemical accumulators of both nutritional and toxic chemicals and cause plugging and corrosion. These events have been witnessed over the course of history. Darwin, for example, writing in 1845 in "The Voyage of the Beagle" refers to a mine at Jajuel, Chile and reports on the tailing piles left after the gold had been separated:

"a great deal of chemical action then commences, salts of various kinds effloresce on the surface, and the mass becomes hard (after leaving for two years and washing)...it yields gold...repeated even six or seven times...there can be no doubt that the chemical action...liberates fresh gold."

Darwin was reporting a mine ore leaching process which was, for many years, thought to be a chemical process but is now recognized to be biological and to involve the formation of biofilms. Ionic forms of gold are now known to be microbially amended to the elemental state that may become deposited in the cell walls. Within such growths, there are incumbent strains of microorganisms that compete with each other for dominance in the maturing biofilm and cause secondary effects such as plugging, corrosion and bioaccumulation.

The terms, "biofilms" and "biofouling" can be defined as relating to the *cause* and *effect* respectively of any microbial physical and/or chemical intercedence into a natural or engineered process by a biologically generated entity. The structure of the biofilm is defined in terms of the incumbent qualitative and quantitative aspects with particular emphasis on the phases of maturation through to a completion of the life cycle (e.g., occlusion of a porous medium, total plugging).

Major environmental factors influencing the rate of growth of a biofilm are the C:N:P ratio, temperature, redox potential, and pH. These are generally considered to be the particularly important factors. Dispersion of biofilms would allow the consortial organisms to migrate to fresh as yet uncolonized econiches as either suspended particulates (due to shearing) or non-attachable ultramicrobacteria (generated as a response by many microorganisms to severe environmental stresses). The net effects of biofouling are usually considered in terms of causing various plugging, corrosion, bioaccumulation, biodegradation, gas generation activities, and also causing changes in the characteristics of water retention and movement in porous media such as those found in soils, ground- and surface- waters, and in engineered structures. Often biofilms are more commonly associated with corrosion and plugging events than with the hygiene risk factors.

Traditionally, the major concern in relationship to ground water microbiology was to recognize any potential health hazard which may be present. These hazards were generally viewed as significant where there was a distinct presence of coliform bacteria identifiable at the significance level of at least one viable entity per 100 ml. Coliform presence can be linked to the potential for direct (fecal coliform) or indirect (total coliform) contamination of the water by fecal materials. In practice, the most common concern relates to contamination with domestic wastewaters, manure leachage, liquid waste holding tanks and industrial discharges.

Today there is a growing realization that there is a wider range of microorganisms which can present a significant health risk and yet may not be associable with the presence of indicator organisms for fecal material (e.g., coliforms). In many cases, these organisms are nosocomial pathogens in that they are normal inhabitants of particular environments and yet are also able to cause clinical health problems in stressed hosts. The form of stress can vary from nutritional (e.g., starved), physiological (e.g., exhaustion or surgical after effects) or immunological (e.g., suppressed). Nosocomial pathogens are commonly present within natural environments and are often difficult to distinguish from the natural background microflora. This is particularly the case where the nosocomial pathogen forms a part of the consortium within a biofilm growing attached to the various surfaces in the econiche.

MICROBIAL DYNAMICS OF BIOFILMS

Biofilms are dynamic and there is an ongoing competition among the incumbent microbial strains that may result in three possible effects. First, the range of strains that may be present can become reduced or stratified (spatially separated). Qualitative reductions will occur where one or more of the strains become more efficient than the competing organisms and gain dominance. Young natural biofilms can often contain as many as 25 or 30 strains which will, during the process of maturation, become reduced in number to as few as five strains or less. Another response commonly seen in a maturing biofilm is the stratification of the biofilm in which the primary layering differentiates an exposed aerobic (oxygen present) stratum overlaying and protecting lower anaerobic (oxygen absent) layers. These lower surface interfacing layers often will biologically generate hydrogen sulfide and/or organic acids which may then initiate the processes of corrosion.

The second possible effect relates to the incumbency density of cells within the biofilm that may fluctuate with environmental conditions. This incumbency density (id, as viable units per ml of particulate material) may be measured by undertaking an enumeration of the viable units by a technique such as extinction dilution spreadplate estimation of the colony forming units per ml (cfu/ml). Concurrently, the volume of the biofilm can be estimated using the dispersed particulates from the biofilm and measuring the volume with a laser driven particle counter. The volume can be summarized as mg/L of total suspended solids (TSS). From the estimation of the number of viable units per ml (vu) and the volume of the suspended solid particulate material per ml (vp, computed as the TSS divided by 1,000), it becomes possible to project the incumbency density (id):

$$id = vu / vp$$

Here, the id would commonly be recorded as the incumbency density of colony forming units per microgram (equivalent to 10^{-6}g). This assumes that the particulate density was identical to that of water (i.e., 1.0). An example of the effect of stress on the id would be where a gasoline spill had occurred in a ground water system. Biofilms would be stressed, for example, by a significant bioaccumulation of gasoline (e.g., 16mg/ml biofilm). This impact has been recorded as causing id values of as low as 20 cfu/10^{-9} g. When nutrients were concurrently applied, the id values increased by as much as one thousand fold (i.e., 20,000 cfu/10^{-9}g).

Competition among the incumbent microbial strains in a maturing biofilm in response to environmental conditions can lead to a third effect which is related to the degree of imposed stress. Responses can occur at the individual cell level or to the biofilm itself. Where stress impacts upon an individual cell as a result of trauma, it can cause a response. These traumas can be created by a shift in one or more environmental factors to conditions that will no longer allow growth. Such controlling factors can include a shortage of nutrients or oxygen, a build up in toxic end products of metabolites or antibiotics, a physical shifting in the redox front and losses in structural integrity in the biofilm. In general, cells under extreme stress will void nonessential water and organic structures, shrink in volume with a tenfold reduction in cell diameter, generate an electrically neutral cell wall (to reduce ability to attach), undergo binary fission and enter a phase of suspended animation. These minute non-attachable cells are able to remain viable for very long periods of time as ultramicrobacteria (UMB). In this state, bacteria can move with any hydraulic flows as suspended animation particulates until the environmental conditions again become favorable. Once such facilitating conditions recur, the cells again expand to a typical vegetative state and colonize any suitable econiches. A newly developed water well can provide a suitable (inviting) econiche for the transient aerobic ultramicrobacteria passing through the ground water.

BIOFILM MATURATION

General concepts of bacterial growth may often envisage the cells as simply floating or "swimming" in water. In reality, bacteria tend to attach onto surfaces and then colonize the surface entirely to form a biofilm. The bacterial cells tend to be negatively charged and are attracted to positively charged surfaces. When this happens, the bacterial cells become anchored to the surface by extending organic polymers (long chained molecules) which make the primary attachment. Subsequently the cells will reproduce and colonize the surface using such mechanisms as "jumping," "tumbling" or they may simply clump to form a micro-colony on the surface.

Biofilms may respond to stress as a normal part of the maturation cycle. From studies undertaken using biofouled laboratory model water wells (1-L capacity), a sequence of events has been recorded following a basic pattern for the development of the iron-related bacterial biofilms. Initially, this biofilm will contain a randomized mixture of a wide variety of microorganisms. Over time, however, these organisms will either stratify into distinct parts of the biofilm, become dominant or be eliminated from the consortium forming the biofilm. A maturing biofilm is sometimes referred to as a glycocalyx.

Biofilms form a biological interface with the water passing over the surface. As the water passes over, chemicals are extracted and concentrated within the biofilm. These chemicals may be grouped into two major categories: nutrients and bioaccumulates. Nutrients are utilized by microbes for both growth and reproduction. Nutrient concentrations usually do not continue to build up indefinitely. Bioaccumulates however are there because they are not being used. By and large, therefore, they accumulate within the polymeric structures of the biofilm. Commonly, accumulates include non-degradable (recalcitrant) organics and various metallic ions such as those of iron, manganese, aluminium, copper and zinc. The role of these bioaccumulates would appear to be relatively "passive," but it is generally believed that these compounds reduce the risk of predation by scavenging organisms.

In this maturation cycle, a number of distinct events could be observed. These can be categorized into a number of sequential phases after the formation of a confluent biofilm:

1. rapid biofilm volume expansion into the interstitial spaces with parallel losses in flow;
2. biofilm resistance to flow next declines rapidly causing facilitated flows which can exceed those recorded under pristine (non-fouled) conditions;
3. biofilm volume compresses, facilitated flow continues;
4. biofilm expands with periodic sloughing causes increases in resistance to hydraulic flow (intercedent flow). There is a repeated cycling with a primary minor biofilm volume expansion followed by secondary increased resistance to flow ending in a tertiary stable period;
5. interconnection occurs between individual biofilms to now generate semi-permeable biological barriers within which free interstitial water and gases may become integrated into form an impermeable barrier (plug).

During the phase 4 increases in resistance to flow, it can be projected that the polymeric matrices forming the biofilm may now have extended into the freely flowing water to cause radical increases in resistance to hydraulic conductivity. At the same time as the resistance is increasing, some of the polymeric material along with the incumbent bacteria will be sheared from the biofilm by the hydraulic forces imposed. Such material now becomes suspended particulates and forms "survival vehicles" through which the incumbent organisms may now move to colonize fresh econiches. In studies using laser driven particle counting, these suspended sheared particles commonly have diameters in the range of 16 to 64 microns (in-house laboratory data). In some cases, the variability in particle size can be narrow (+/-4 microns) around the mean particle size.

BACTERIOLOGY OF PLUGGING

Biological impedance of hydraulic flows, whether in a water well, heat exchanger or cooling tower, involves a sequence of maturation within the biofilm prior to the plugging or other severe biofouling event. The types of microorganisms involved in such events vary considerably with the environmental conditions present. In water wells, the iron-related bacterial group is commonly associated with plugging (i.e., significant loss in specific capacity due to a biofouling event involving biofilm formation). It should be noted that an alternate term to "plugging" is "clogging." This latter term (clogging) is taken to mean any event dominated by chemical and/or physical factor which is causing a severe restriction or a total obstruction of flow from a water well.

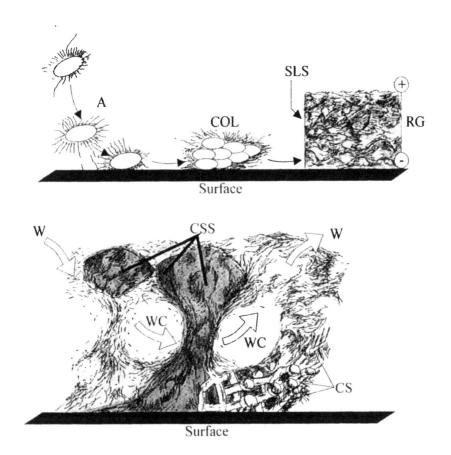

Figure Seven,   Sequences in biofilm maturation. A biofilm goes through a process of maturation from attachment (A), through to colonization of the surfaces with a thin biofilm (COL). This biofilm now grows into a series of simple layered structures (SLS) each operating on a separate part of the redox gradient (RG) through the biofilm. Finally, the biofilms mature into a complex set of slime (CSS) and crystalline structures (CS) that allow greater movement of water (W) through conduits (WC) that pass through the structure and occupy as much as 40% of the volume.

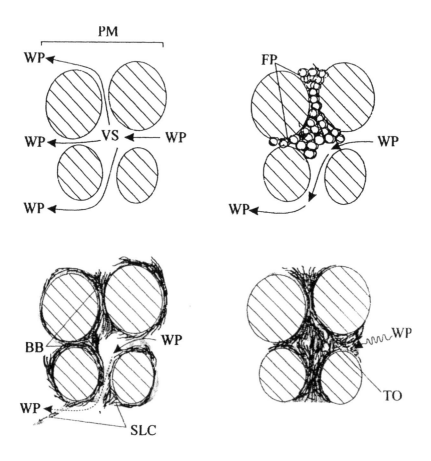

Figure Eight, Microbial plugging of the void spaces (VS) in a porous medium (PM) can take a number of forms. These range from "foam" plugging (FP) through to total occlusion (TO) of the water pathways (WP) through the medium to form a biological barrier (BB) and the stringing of slime-like columns (SLC) down through the water to impede flow (SC).

3

DIRECT EVIDENCE OF A
BACTERIAL EVENT

It has often been thought that a clear sparkling water sample is one that has to be free of bacterial contamination. This is not always the case since there can be as many as 400,000 bacteria in one milliliter of such a clear looking water. However, while the eyes may deceive on some occasions, other senses or time may be used to recognize a bacterial event through some obvious signals. Evidence that a bacterial problem may be occurring can take the form of a generating slime coating, floccular material or cloudiness which gradually intensifies in the water sample accompanied perhaps by changes to the taste, odor and color in the water. Each of these symptoms may relate to different problems being generated microbiologically.

Slimes are produced frequently where water may be flowing freely over a surface after passing through a closed system such as a pipe. Slimes are usually considered to be gelatinous in nature and often glisten in reflected light. The most common colors for such slimes range from white through grey to black, or through orange to red and dark brown. On occasions the red to brown dominated slimes will shift to and from a black slime in response to changes in environmental factors. For example, some slime growths will go brown in oxidative conditions and shift to black in reductive conditions. These slimes are almost a natural "litmus paper" for determining the ORP (i.e., oxidative, brown; reductive, black).

These glistening slimes tend to be of a nongranular texture in nature and do not have a gritty texture (rough sandpaper feel when rubbed between the fingers). One exception to this rule is that some of the brown slimes which will generate considerable granular deposits of iron and manganese oxides and hydroxides which are very gritty in texture. Here the surfaces of the slime may form hardening crusts. Such slimes actually bioconcentrate these iron oxides and hydroxides so that these growths may develop into forms of encrustations that become hard and brittle as they expand. Within the encrustation, or tubercle, microbial activity may continue and also initiate the processes leading to corrosion.

Regardless of the type of slime, all bear a common origin in that there was an initial growth of attached bacteria on a surface. The individual cells throwing out string-like molecules (called polymers) cause attachment.

These polymeric strings lock onto the target surface and anchor the cell to that surface. Once anchored, the cells grow and multiply to colonize the surface with a coating (commonly referred to as a biofilm) generated through the cells rolling, leaping and spreading during reproduction and colonization. At the same time, more and more polymers are produced which bind the growing numbers of cells down to the surface being colonized. In addition to providing support, these polymers also take up and bind large volumes of (bound) water into the structures developed within the biofilm. A slime-like mat is thus created within which bacteria can now develop complex associations (consortia) involving many different strains of bacteria.

VISIBLE MANIFESTATIONS

As the slime mat grows and matures, it goes through a complex life cycle involving rapid volume expansion, compression, a slow pulse-like secondary growth with periodic sloughing during which some of the slime may become suspended in and passes along with the flowing water. Slimes are therefore complex in nature and are ever changing in their makeup. Each major type of slime represents a different group of bacteria that, as a consortium, are dominating the biofilm. The following generalizations may be made for the slime types defined by color:

White or Clear Slime

These slimes have not taken up any significant amounts of metallic salts that would cause pigmentation to occur. Concurrently, the bacteria are not generating any pigments of their own which could impart color to the slime. Care should be taken in the determination of color generated by bacteria because some colors may only be apparent in the ultraviolet light waveband. When present in high concentrations, these pigments may give a lightish green, yellow or blue color. Application of an ultra-violet light will then show a strong color reaction of the same type. Such a phenomenon occurs when the fluorescent pseudomonad bacteria are very prevalent in the slime. These types of slimes may be very rich in heterotrophic (organic nutrient using) bacteria and can harbor a wide range of bacteria including coliforms where environmental conditions permit. Sulfur oxidizing bacteria may also form these types of slimes particularly in a hydrogen sulfide producing ground water.

Grey Slime

This type of slime forms an intermediate between a white (clear) slime and a black slime. Close inspection of the slime reveals that it will often consist of a white or clear slime within which there are intense black granules. These granules may range in size from smaller than a grain of sand to larger particles up to 7 or 8 mm in length. Such black deposits are most commonly metallic sulfides (usually dominated by iron and manganese) generated by interactions between the various metallic salts and hydrogen sulfide. This hydrogen sulfide may be generated by bacteria growing in the absence of oxygen (anaerobic) and reducing sulfur, sulfate (sulfate-reducing bacteria, SRB) or via protein degraders. Protecting these oxygen-hating SRBs is the clear slime in which oxygen-using (aerobic) bacteria grow in complex consortia.

Black Slime

When the slime is intensely black throughout, this means that the bacterial flora is dominated by SRBs. In all probability this type of slime will be found in an environment depleted of oxygen (reductive) but rich in the essential organic materials necessary to support these bacteria. Corrosivity is most commonly associated with the black slime due to initiation of electrolytic corrosion by the hydrogen sulfide generated by the microorganisms.

Orange, Red, and Brown Slimes

These slimes most commonly occur in environments that are relatively rich in dissolved oxygen (aerobic) and are exposed to transient amounts of iron and manganese in the water. A dominant group of bacteria in this type of slime are the iron-related bacteria (IRB). Here, the bacteria are unique in that they are able to take up iron and manganese in excessive quantities. The excess is deposited either around the cell, within the slime or special structures protecting or extending from the cells. Two such structures are the tube-like sheaths and the ribbon-like stalks. These are deposited as various oxides and/or hydroxides. Depending upon the precise combination of these oxides and hydroxides, the slime will generate an orange, red or brown slime as the bioaccumulated concentration increases. These iron and manganese rich deposits appear to perform number of roles from being protective (against predation and physical disruption) to forming a nutrient reserve. These slimes form at sites of intensive oxygen consumption that may frequently allow anaerobic bacteria to survive and grow within those parts of the slime where oxygen fails to penetrate. In consequence, these slimes may shift to a black phase when there is oxygen depletion in the environment.

In examining the bacteriology associated with these slimes, two approaches to the examination can be pursued. These are:

(1) determine the bacterial loading of the product water which has passed over the slime; and

(2) directly remove some of the slime from the original site for direct microscopic examination.

Examination of the product water suffers from a severe disadvantage in that bacteria will slough off from the biofilm (slime) into the water in an irregular manner. Such sloughing events tend to occur in a random pattern from the slime to cause a very variable population to be recorded in the water over a period of sampling. Higher populations are likely to be observed on occasions when the water flow is suddenly generated by pumping over a quiescent slime. This would be a different effect from a continuous stream of water constantly moving across the slime layer. However, it is often much more convenient to sample this product water rather than attempt to obtain samples directly from the slime which may be growing at a relatively inaccessible site. For example, it would be difficult to obtain a sample from a brown IRB-rich slime situated at the outer edge of the gravel pack around the well screen. Indeed, in some circumstances, even with the assistance of camera logging using closed circuit television (CCTV) or fibreoptics, it may still not be possible to even view the slime let alone obtain a sample.

In the practical world, the most common confirmation technique is to take a flowing or static water sample from as close as possible to the (suspected) site of the biofouling. This would appear to be the less reliable but it remains the only practical method to obtain such a sample.

In addition to the formation of slimes, ground water can also be subjected to various forms of growth attached to surfaces or freefloating in the water. These types of growth are categorized below:

Tuberculous Growths

These are a variety of slime growths in which the active bacterial growth occurs within a hardened salt and oxide rich encrustation. The outer shell takes on the form of a series of bulbous extensions stratified above and/or wound around each other. In color, these tubercles tend to be (externally) brown in color due to the high intrinsic concentrations of iron and manganese oxides and hydroxides. Frequently, fissures and cracks develop in the tuberculous shell so that bacterial slimes may become observable down these fissures. Corrosion is often associated with such tuberculation since that lack of oxygen causes anaerobic (reductive) conditions, a natural precursor to hydrogen sulfide development and electrolytic corrosion, can frequently occur deeper into the tubercle. IRB and SRB groups of bacteria are commonly associated with these tubercles.

One theory is that the tubercle begins as a gas bubble (such as methane) entrapped under a biofilm. This biofilm subsequently hardens to form the shell and, once the gas has been utilized, the biofilm enters and grows within the "shell" created by the gas and the hardening biofilm. The vacated gas vacuole is therefore finally colonized by at least some of the incumbents of the biofilm.

Cloudiness

Cloudiness is a loss in clarity in the water due to suspended particulates. Not only can slimes be seen growing in association with water saturated porous media but microorganisms may also be found growing directly in the water. This can cause cloudiness (an even haziness in the water), form flocculent growths (visibly distinguishable buoyant particles often with ill-defined edges) and/or cause discoloration of the water. Each of these three characteristics can be used as evidence of a different type of microbial event occurring in the water. General interpretations of these observations are listed below.

> **General cloudiness** - When the water is held up to a white light or a light is shone up through the water sample, the occurrence of a cloudy appearance may indicate that either there has been a heavy salt pre-cipitation or that there is an active dispersed microbial population in the water. In the latter case, the bacterial population that will cause this type of cloudiness may be in excess of one hundred thousand colony forming units (cfu) per ml.
> Direct microscopic examination of the water can be used to reveal the cause of this cloudiness. Salts would appear as large crystalline structures in the water with clear hard edges. Bacterial activity would tend to appear as small indefinite shapes often oscillating in the water randomly by Brownian movement or moving in specific directions if the bacterium is capable of sustaining directional movement (motile) or as larger irregularly shaped structures. In either event, bacteri-ological examination of the water should be pursued.

> **Flocculent growths** - Some bacteria are able to grow in the water as large diffusive objects that appear to have a shape and are able to control their intrinsic density close to that of the water ("Sea Snow" and "Well Snow" are good examples of this type of growth). This allows the bacteria to "float" in the water in large masses. On frequent occasions, these flocs may be colored from a light orange to a brown color due to the accumulation of iron and manganese salts. One group of bacteria found displaying this feature is the sheathed iron-related

bacteria. These bacteria exist for a part of their life cycle in a slime tube (a sheath) and, on other occasions, the cells emerge from the sheath to produce a copious slime formation within the water.

Coloration of Water

This can occur in water either with or without cloudiness. Water will, under some conditions, pick up solubilized organic and/or inorganic material that colors the water. In ground water, one of the most common sources of this color are the iron and manganese salts released after bioaccumulation within microbial growths around the well. Once the growth masses become saturated, the surplus salts may become solubilized or slough from the biomass and appear in the product water. Where these solubilized forms become present, there may not be a parallel increase in microbial numbers. Where the coloration event is primarily due to sloughing and is visible as a general cloudiness, an increase in bacterial numbers may be expected. In either event, the IRB bacterial group may be suspected to be involved at least to some extent. When there are suspended particulates in the water, it is possible to use this characteristic as a monitoring tool. This may be performed in a number of ways. These include the measurement of turbidity (for cloudiness) in the water, or enumeration and sizing of the particulates by laser driven particle counter (which will also yield the total suspended solids volume, TSS). Where there are a large number of pigmented particulates, a reflectometric evaluation of the dried filtered particulates can also be undertaken.

ODOROUS SIGNALS

Different bacteria have been well known to produce some odors that can be considered distinctive for particular groups of bacteria. These odors can aid in the initial determination of the types of bacteria that may be dominating in the water. To detect these odors, the water needs to be collected within an odor-free container (such as a sterile and clean glass-sampling jar). The water should occupy roughly 50% of the total volume of the container that should then be sealed. Vigorous shaking of the container will create an aerosol of water droplets that should harbour some of the odorous material. Clear the nasal passages with two or three deep breaths, loosen the seal (e.g., unscrew and lift the cap) on the water container and gently inhale. Any odoriferous chemicals are now more likely to be in the gaseous phase and hence be more detectable. Repeat to confirm the type of odor. If no odor can be detected, then warm the water to roughly 45°C and repeat the smell test. The higher the temperature, the greater the potential for some of the more volatile odors to be detected.

There are a number of different odors that can be linked to the activity of different groups of bacteria. This can be very useful in reaching an early decision as to which bacteria groups are likely to be causing the problem and should be further investigated. The major odors which can generated by microbial activities in water are listed below.

Rotten Egg smell

This odor is commonly generated by anaerobic bacteria functioning in oxygen free environments and reducing either sulfates or sulfur to hydrogen sulfide (SRB) or breaking down the sulfur containing amino acids in proteins to the same gas (by proteolysis). Water containing this bad smell may contain one or both of these groups of bacteria. Caution must be exercised when this gas is smelt since the nose can become quickly saturated with the hydrogen sulfide gas and no longer able to detect the presence of the gas. One conclusion that could be drawn from this type of odor event is that the water bearing this gas almost certainly originated from an oxygen free (reductive) zone where organic nutrients were present. Some water wells, when "rested" for a prolonged period, can generate string "rotten egg" odors since the well has become reductive and the SRB have become dominant in the stagnant water.

Fish smell

This is a very subtle odor that requires careful screening. The smell is similar to that commonly encountered around a fish retail outlet or processing plant. Many of the pseudomonad bacteria frequently generate these off- odors during periods of intensive growth. These pseudomonads are oxygen requiring (aerobic) bacteria that are able to utilise different and often specific organic nutrients. Where pollution of water occurs involving some specific compounds such as those associated with a gasoline release, these pseudomonads may become dominant. On some occasions, members of these pseudomonads can generate kerosene or oil-like odors.

Earthy smell

Some microorganisms belonging to the streptomycetes (mold-like bacteria) and the cyanobacteria (algal-like bacteria) produce a group of odors, which very much resembles the typical smells emanating from a healthy soil. These odors are grouped in a class called geosmins. When these odors are detected, there is a probability of aerobic growth (if originating from the streptomycetes) or it could be derived from surface water based algal growth (if cyanobacteria were involved).

Fecal (sewage) smell

Raw sewage and septic fecal wastes often generate a very typical odor. These odors can be generated by enteric bacteria that commonly occur in fecal material. These bacteria are more commonly called the coliforms. The presence of such odors in water is a very strong indication that fecal pollution has occurred in the water and coliform testing should be performed as soon as possible since an acute hygienic hazard may exist.

Fresh vegetable smell

Many of the green algae, diatoms and desmids will generate odors resembling different fresh or rotting vegetables such as lettuce and cucumbers. These odors indicate that there may have been some recent entry of algal-rich waters where a bloom was just forming. Sour odors of rotting vegetation indicate that the algal bloom was in an advanced stage of decay and high bacterial populations may be expected to be present in such waters.

Chemical smell

Odors such as those resembling gasoline and solvents are more likely to originate from a pollution event rather than from the ground water system where it could have been generated by microbial activity. Where these types of odors are observed, a detailed chemical analyses for BTEX (benzene, toluene, ethyl-benzene and xylene) and/or hydrocarbons should be considered as a very high priority to minimize health risk. When considered necessary, a gas mixture composition can be determined by a GC-MS analysis of headspace sample.

DIFFERENTIATION OF MICROBIAL FORMS IN BIOFOULING EVENTS

Iron bacteria has been a common term traditionally used to identify the biological component in well plugging but these organisms can also be defined as iron-related bacteria (IRB). These IRB participate in both the assimilation of iron and/or manganese into natural accumulates (iron oxidizing or precipitating bacteria), and the dissimilation of these elements away from natural accumulates (iron reducing bacteria) back into solution. Many IRB are able to perform both of these major functions. In the process of accumulation, the oxidized insoluble iron (ferric) and/or manganese (manganic) deposits are formed. This can occur within either the cell or the surrounding extracellular polymeric matrices (glycocalyx) in such a way that the product oxides and hydroxides render visible orange to red to brown colors to the conglomerated growths (e.g., slimes, tubercles).

These IRB bacteria can be subdivided into three major groups based upon the nature and site of the accumulated iron and manganese. These may be broadly grouped as:

1. Ribbon formers
2. Tube formers
3. Consortial Heterotrophic Incumbents (CHI)
   a) Aerobic heterotrophs
      i. Unsaturated zones
      ii. Saturated zones
   b) Anaerobic heterotrophs
      i.   Corrosivity generators
      ii. Gas generators
      iii. General anaerobes
   c) Health risk heterotrophs
      i.   Aggressive transitory pathogens
      ii. Covert opportunistic pathogens
      iii. Marker organisms of health risk
   d) Heterotrophic iron mobilizers
      i. Iron oxidizing bacteria
      ii. Iron reducing bacteria.

While groups 1 and 2 can be easily classified by obvious characteristics, group 3 includes many bacterial genera that function within more than one subgroup. Consequently, groups 1 and 2 will be described specifically while the CHI group will be characterized by classical systematics.

RIBBON FORMERS, *Gallionella* (Group 1)

In group 1, the ribbon formers, the iron and/or manganese oxides and hydroxides are accumulated within a spiralling ribbon of polymeric material which is excreted from the lateral (side) wall of each individual bacterial cell. Such ribbons can commonly exceed the length of the cell by ten to fifty-fold and the ribbon can include ten or more harmonic cycles. Where these twisted yellowish to brown ribbons (commonly referred to as stalks) are microscopically observed suspended in the product water from a water well, the dominant IRB in any associated biofouling is generally concluded to be *Gallionella*.

Little is known of the life cycle for this bacterium and it has yet to be grown in a pure culture. It is, however, clear that the bacteria producing these ribbon-like stalks are Gram-negative with rod- or vibrioid-shaped cells. While occasionally the cells can be seen to be still attached to the ribbon, more often than not, the ribbons do not exhibit an attached cell. Essentially these ribbon-like stalks are dead products from the growths of the cells and are not viable. This raises questions: why are most of the ribbons found in the water without the cells attached? And where can a bacterial cell grow within a heavily biofouled and turbulent environment and yet still produce such a ribbon?

One hypothesis that would address these questions is that these cells have grown attached to the surface structures of the biofilm from which the ribbon-like stalks (known in microbiology as "false" prosthecae) extend outwards into the zone of free hydraulic flow. The ribbon-like nature of the prosthecae would cause some of the transient flowing water to "spiral down" towards the attached cell. This flow would be in the same manner as an Archimedian screw-like action. This would carry both dissolved oxygen and nutrients into the proximity of the cell. This feature would give the cells of the only genus in group 1 (*Gallionella*) an advantage over the cells that are incumbent in the biofilm itself.

As the prostheca (stalk) lengthens through growth and the accumulation of polymers and oxidized iron, so the shear forces imposed by the flowing water would increase until eventually the ribbon would break off normally at the "root" close to the cell. Once sheared, the ribbon would become an inanimate suspended particle in the water. By this means, fragmented ribbons would enter the water phase where they could easily be recognized and identified microscopically. Once the prostheca has sheared, the cell may now excrete a replacement stalk and continue to benefit from the ecological advantages that this structure brings. The iron and/or manganese oxides and hydroxides would give a greater structural integrity to the prostheca and increase the survivability of the ribbon.

It is generally believed that *Gallionella* is able to gain at least some energy for ongoing maintenance and synthetic functions through the oxidation of these metallic ions from their reduced (ferrous) states. In general, this group is found in waters that have stressed nutrient loadings in the product water. Total organic carbon (TOC) concentrations in waters bearing a dominance of *Gallionella* is usually relatively low ranging often down to as little as <0.5 ppm and most common up to a maximum of 2.0 ppm TOC together with low concentrations of nitrogen and phosphorus.

TUBE FORMERS (Sheathed Bacteria)
Crenothrix, *Leptothrix* and *Sphaerotilus* (Group 2)

The second group of iron-related bacteria also produce special structures that are formed as tubes of extracellular (outside of the cells) material. These structured tubes are called sheaths and the bacteria form filaments of cells are able to move within the tube and even migrate out of the tube. Where this migration occurs, individual cells may become divorced from the filament and move away. This movement where it occurs, is caused by flagella which may be either tufted or act as a single polar flagellum. These organisms form part Section 22, Sheathed Bacteria in *Bergey's Manual*.

Iron and/or manganese oxides and hydroxides can become accumulated on and within the sheaths while the bacteria are able to function within the hollow central core of the tube-like sheath. Frequently when scanning electron microscopy is conducted upon biofouled material, abandoned sheaths are commonly seen embedded in the surfaces of the biofilm. These bacteria tend to occur in waters with a relatively low organic carbon loading (<2ppm total organic carbon).

When these sheathed bacteria grow, the sheaths may, for some genera, be attached to a surface by a holdfast. This allows the sheaths to retain the same position on a surface being subjected to hydraulic flows. Other sheathed bacteria either become enmeshed into the biofilm or form a particulate growth suspended in the water.

Cells are sometimes difficult to view through the sheath unless viewed by phase contrast microscopy. Wet mount microscopic examination using 95% ethanol can improve the transparency of the sheaths and allow the incumbent cells and/or filaments to be more easily viewed.

The three dominant genera of sheathed bacteria in ground waters may be differentiated using the following dichotomous list:

1. Sheaths encrusted with iron oxides over at least a part of the length. If yes, go to 2; if no, go to 6.
2. Individual cells clearly visible as cylinders or disks within the sheath. Tip of sheath clear while the base is coated with iron oxides. If yes, go to 3; if no, go to 4.
3. Genus *Crenothrix*.
4. Sheath heavily impregnated with iron and/or manganese oxides, sheaths not attached to surfaces by a holdfast. Cells in filamentous arrangement are difficult to view in the sheath? If yes, go to 5; if no, go to 8.
5. Genus *Leptothrix*.
6. Sheaths commonly attached to surfaces by a holdfast, normally transparent with little or no iron oxide encrustation. Cells, where visible, are in filamentous form and the individual cells are not easily recognizable? If yes, go to 7; if no go to 8.
7. Genus *Sphaerotilus*.
8. Is there a defined sheath which can be clearly seen resembling a tube in which individual or rows of cells can be seen with some branching of the sheath occurring? if yes, go to 9; if no, go to 10.
9. Possible sheathed bacterium of another genus.
10. May be a copious slime producing bacterium that casually resembles a sheathed bacterium.

Sheath forming bacteria are very common nuisance bacteria in water systems that have a relatively low nutrient loading. *Sphaerotilus* is an exception to this rule and is more commonly found in high organic environments. It is sometimes considered that, where there is a residual iron and or manganese in the postdiluvial water exceeding 0.2 and 0.01 mg/L respectively, there is a greater probability of the sheathed bacteria belonging to *Leptothrix* or *Crenothrix*. Where there is a lower concentration, the dominant organism may be *Sphaerotilus*. These bacteria will tend to undergo a flocculation when samples are taken from the well and subjected to very different environmental conditions.

Changes associated with a radical floc formation may cause an apparent visible "growth" to be generated that can cause system malfunctions within filters and treatment processes. Exposing the water to sudden shifts in the nutritional loading can encourage radical floc formation. Such floc formations can occur within hours of the water sample being taken. The floc formed commonly is of a white to reddish brown color. Upon microscopic examination of a wet mounted slide of the floc, it appears to be formed of a mixture of long very defined flexible tube-like structures that may have zones of darker brown encrustations around the tubes. In addition, there are poorly defined irregular masses of brown floc formed into conglomerates with the tubes.

These tube forming iron-related bacteria are aerobic Gram negative organisms and tend to be found at the redox fronts (such as in and around the well water column in the well). These organisms may also be found growing in the capillary interface between the saturated and vadose zones above an aquifer. Where a well screen has been set close to this interface then periodic releases of the sheath forming IRB may be expected to occur particularly if there is a significant vertical downward movement of ground water from that zone.

CONSORTIAL HETEROTROPHIC INCUMBENTS (Group 3, CHI)

Many microorganisms within ground water systems do not function independently in separate niches but cooperate with other species to form communities involving often many more than one species. These community structures are referred to as consortia. For energy, most microorganisms in ground water utilize the various organic fractions within the environment. These organic compounds may yield energy through various breakdown mechanisms to support the maintenance and growth of the incumbent microorganisms. Obtaining energy from such organic sources is referred to as heterotrophic function. The consortial heterotrophic incumbents (CHI) in ground water can be found growing within co-operative community structures generally within biofilms utilizing organic materials and other nutrients dissolved or suspended in the surrounding waters or within the glycocalyx structure itself.

It may appear surprising that such complex communities can form within environments where there would appear to be so little in the way of organic materials. The reason for this apparent anomaly is that the biofilm acts as very aggressive accumulators of any passing organic materials. Once retained within the structures of the biofilm, the various incumbent microorganisms can selectively use the entrapped organic material. There are a number of mechanisms involved in these degradative events. Indeed, the classification of microorganisms is, in part, based upon which groups of organics are or are not used. Some of the major characterizations are given below.

Proteolytic

Some microorganisms will degrade and use some or all proteins as a source of energy and amino acids.   Proteins are usually in very low concentrations in ground waters, and so the most likely source is other biological entities within or associated with the biofilm.. Cannibalism, or the direct assimilation of cellular material from another viable entity, is thought to be a relatively frequent occurrence within a consortial community. This can lead to changes in either the dominant species, reductions in the incumbency density and/or reductions in the variety of species present within the consortium.

Hydrocarbonoclastic

Microorganisms are sometimes able to degrade different fractions of hydrocarbons. These organic compounds are reduced organic entities predominantly composed of only carbon and hydrogen. Aerobic degradation (either molecular oxygen or nitrate appears to be required) does occur but at rates relatable to the complexity of the molecules (i.e., the more complex, the slower the degradation). Anaerobic degradation may also occur but very slowly for even the more complex cyclic and aromatic hydrocarbons.

Saccharolytic

Refers to the breakdown of various sugars. There are a wide variety of these sugars ranging from relatively simple forms such as glucose and sucrose to complex polymers such as starches. In general, these compounds are readily degraded by a wide variety of microorganisms aerobically. Glucose is one of the most universally utilised sugars. Where degradation is aerobic, the products are to a large extent either assimilated or released as carbon dioxide. Under anaerobic conditions however, these compounds are only partially degraded often with the releases of considerable quantities of organic acids. These acids are able to cause the pH in the environment to drop by as much as 4 pH units to create localized acidic regimes.

Cellulolytic

Cellulose is one of the most commonly synthesised organic materials on the surface of this planet. The molecular structure renders the cellulosic polymers difficult to degrade and degradation is generally performed by relatively specialized organisms. Aerobic degradation is generally faster than under anaerobic conditions but is still a relatively slow process extending into weeks, months and even years.

Lipolytic

The basic sub-units in lipids are glycerol and fatty acids falling into the general category of fats. Lipolytic microorganisms are able to selectively degrade specific lipids under either aerobic and/or anaerobic conditions.

Ligninolytic

Wood contains a dominant amount of cellulose but additionally includes up to 30% of lignin. These lignins provide some of the structural integrity to the plant. The degradation of lignins has been found to be predominantly an aerobic function. It is performed efficiently by many filamentous fungi, in particular, the white rot fungi (WRF). These organisms tend to function where there is a considerable amount of oxygen such as in soils and the vadose (unsaturated) and capillary zones in aquifers. The WRF are thought to include some species also able to degrade polychlorinated biphenyls (PCB) since there are some similarities in the molecular structure.

Structure of CHI Bacterial Consortia

Before discussing in detail the various genera within the group 3 CHI organisms, it is important to comprehend the factors that cause these microorganisms to grow as complex consortial structures within the biofilms. These organisms can become stratified within a biofilm. The aerobic phase tends to be dominated by pseudomonads belonging to Section 4, family 1 of *Bergey's Manual* while the anaerobic foci within the biofilm may become dominated by Sections 7 (sulfate-reducing bacteria), 6 (anaerobic Gram-negative rods) and 5, families 1 and 2 (the enteric and vibrioid bacteria). The precise dominance sequence is a reflection of the nutrient and oxygen loadings of the causal water being delivered to the specific econiches. Current researches reveal that the biofilm is a much more complex structure than the "layered cake" first envisaged. Hydraulic conduits and focal sites of consortial activity abound to make the structure much more sophisticated than originally envisaged.

Bacteriological Analysis of Water, Slimes and Encrustations

The amorphous nature of the sheared suspended particulates in water samples from such a biofouling renders microscopic examination commonly frustrating and can lead to misdiagnosis. The more traditional bacteriological examinations using selective culture media such as R2A and Winogradsky Regina culture media or simple biological activity and reaction tests (BART™, Droycon Bioconcepts Inc., Regina, Canada) can be used to aid in the identification of the causative organisms. It is intended to concentrate on simple procedures in this book that can either be performed in the field or in a simple microbiology laboratory with the minimum of equipment. In the latter case, it would be expected that the operator of the tests would be competently familiar with all of the appropriate basic methods employed in microbiology.

Water samples being taken of pumped supplies from a well may not necessarily give an accurate indication of the microbial events occurring at the sites of biofouling. The water could also contain the intrinsic planktonic populations along with whatever sessile organisms that may be present in such sheared particulates as are suspended in the water. These occurrences are frequently random, but a greater chance of observing these events can be achieved by entrapping the particulates in a moncell filtration system. Alternatively, applying a pristine surface (such as a glass slide) to water within the well water column onto which such organisms may now attach to form observable growths can also be used.

Application of Analyses

In practice, an understanding of the nature of a biofilm can be utilized in the management of particular intercedent events (i.e., through the generation of a biobarrier) such as in the control and bioremediation of a gasoline plume, removal of potential toxic heavy metals or aromatic hydrocarbons. These would be alternative events to the more appreciated roles performed when biofouling causes plugging and corrosion.

Environmental Factors

Many environmental factors will interact with the growth dynamics of a maturing biofilm. These range through physical, chemical and biological factors. Of the physical factors, the three pre-eminent constraints relate to temperature, pH and the redox potential where the site of activity is in a saturated medium. Temperature influences microbial growth in a number of ways ranging from the extremes where low temperatures cause an inhibition of the cellular metabolic processes, to high temperatures which may impact particularly on the protein constituents of the cell through thermal denaturation to cause the death of the cell. Between these two extremes lies a temperature range within which the cells can be metabolically active. Such a range will include a narrower band within which growth and reproduction can occur with maximal growth responses often being observed within relatively narrow temperature ranges (e.g., 10°C). For temperate region ground waters in shallower aquifers of up to 300 meters depth, the normal temperatures experienced without geothermal influences could range within a band from 1° to 25°C. This range is commonly associated with bacterial activities dominated by the lower mesotrophs (>15° to <45°C with optima at <35°C), facultative (eurythermal) psychrotrophs (grows both above and below 15°C, and strict (stenothermal) psychrotrophs (grows at <15°C only).

Temperature will therefore influence the makeup of an incumbent consortium within a biofilm. For example, water with a temperature of 10°C (+/-2°C) may be expected to include both strict and facultative psychrotrophs but no low mesotrophs except perhaps around the heat generating pump motors. This does not preclude the potential for mesotrophs surviving prolonged periods under these conditions. Shifting ground water temperatures may be expected to cause changes in the dominant microbial strains within the biofilm.

The polymeric structures in the biofilm will tend to impart a buffering activity that reduces the influence of any pH shifts in the water upon the activities of the incumbent organisms. In general, the maximum species divergence in incumbent flora should occur when the pH of the passing water is between 6.5 and 9.0. Where the pH falls periodically to below this pH range, the biofilms have some capacity to buffer these effects. Additionally, the pH within a biofilm may become stratified with higher neutral or slightly alkaline conditions occurring in the upper aerobic zones while the anaerobic zones may become more acidic and support, as a result, a narrower spectrum of microbial activities. In severely acidic conditions (e.g., pH <2.5) the spectrum may become so narrow that it includes only one or two different genera of microorganisms. For example, many of the sulfur oxidizing bacteria belonging to the *Thiobacillus* species are able to function at very low pH values during the oxidation of sulfur and reduced sulfur compounds leading to the exclusion of other species.

In ecological investigations at sites of magnified biofouling in saturated porous media, it has often been noted that these sites tend to be concentrated in the transitional zones between reductive and oxidative regimes.

Here, the redox values shift from a negative to a positive Eh value (e.g., from -50 to +150 Eh).

Sequential dominance of the various CHI bacteria has been observed occurring along these redox gradients. Such sequential events can be manipulated through a shifting and/or enlarging of the redox transitional zone. This can be achieved within an aquifer by recharging the edge of the reduction zone with oxygenated water to force the biofouling outwards (and away from the producing water well). This extension of the area occupied by the biofouling away from the well causes the bioaccumulation associated with the microbial activities to take place more distantly from the well. The postdiluvial water may consequently be of a higher quality since there would have been a slower passing of the influent water through the biozones which are now further away from the turbulent zones of influence created by the pumping action within the well.

Nutrient Factors

Chemical factors can also influence the spectrum of CHI activities associated with these biofouling events. These factors can be summarized into nutritionally supportive or inhibitory groupings. The nutritionally supportive group would include compounds able to be utilized by the microbial cells for catabolic (energy yielding) and synthetic functions essential to the continued survival, growth and dissemination of the species. Inhibitory groupings may be defined as any chemical which will, at a given concentration, directly interfere with any of the supportive functions of a given species in such a way as to at least minimally retard specific competitive functions, or maximally cause the death of that species group.

In the processes involved in natural competition between microorganisms, the supportive chemicals for one strain may, in fact, be inhibitory to another species. It may, therefore, be commonly expected that, where an inhibitory compound is introduced into an environment, there will be a radical shift in the component species within the impacted flora culminating in a tolerance of and, secondarily, a direct utilization of the compound by the surviving species. Additionally, the polymeric matrices of the biofilm will form a protective barrier through which such inhibitors would have to pass prior to directly impacting on the incumbent microbial cells. Such matrices can often form a repository for toxic bioaccumulates away from sites where there could be direct inhibition of the activities of the incumbent cells.

Nutritionally supportive chemicals are essential to the ongoing survival and growth of any given species. The nutrient elements essential to growth processes can be determined by an examination of the ratio of these elements in the microbial cell. A typical ratio for the elements C:O:

N:H:P:S:K:Na:Ca:Mg:Cl:Fe can be expressed gravimetrically as percentages of the dried weight (for *Escherichia coli*) as 50:20:14:8:3:1:1:1:0.5: 0.5:0.5:0.2 respectively.

Particularly critical to the growth of microorganisms is the C:N:P (idealized above at 50:14:3) ratio since these are the three macronutrients most frequently observed to be the critical controlling factors in microbial growth. If there was a 100% efficiency in the utilization of these elements, these idealized ratios in the nutrient feed could be expected to be optimized around N(=1) as the C:N:P ratio of 3.6:1:0.21 respectively. In reality, for the heterotrophic microorganisms, the organic carbon consumed far exceeds the maintenance requirement due to the heavy catabolic demands to create energy. The totally oxidized carbon is often venting as carbon dioxide or, when totally reduced, it can be vented as methane. The net effect of catabolism is to shift the C:N ratio downwards.

An additional diversion of carbon would be in the synthesis of ECPS formation outside of the cell. The optimal ratio for the efficient growth of heterotrophs has not been established for C:N ratio due to the natural variations that occur between the catabolic and synthetic functions. However, for the N:P ratio where the bulk of these elements are retained within the cells, it has generally been considered to be optimal within the range of 4:1 to 8:1. Excessive levels of phosphorus (i.e., ratio of <4:1) cause possible build ups in the reserves of stored polyphosphates or a greater probability of the flora shifting to those species able to undertake dinitrogen fixation (a mechanism for fixated nitrogen to correct the N:P ratio). Another concern relates to the use of different forms of phosphorus in various proprietary water treatment chemicals. While the products may perform their role effectively, unless 100% of the phosphorus has been removed after treatment, this phosphorus will then be utilized to support any reoccurring biofouling.

In generating a comprehension of the impact of organic carbon on the growth of sessile or planktonic microorganisms, concentrations of 1.0 ppm (mg/L) of organic carbon would appear at first to be insignificant (i.e., too low to support growth). However, where a saturated environment contained one million viable units of bacteria per ml and the dried weight of each cell was 2×10^{-13} g and a 50% gravimetric carbon composition, the combined net weight of carbon in each of these viable units would be 1×10^{-7} g. If these organisms were in an environment with a total organic carbon (TOC) of 1×10^{-6} g/ml of total volume, the ratio of cellular: free organic carbon would be 1:9. If one thousand bacteria were present per ml as viable units, then the cellular: free organic carbon ratio would shift to 1:9,000. It can therefore be projected that the critical dissolved organic carbon that could influence microbial activities in water should be considered to lie within the ug/L (ppb) range rather than the traditionally accepted mg/L (ppm) range. In developing a system to generate the bioaccumulation and/or bio-degradation of specific nuisance chemical compounds in the environment, it becomes important to set the C:N:P ratios in a manner beneficial to the required activity. If the target compound is organic, it may contribute to the carbon ratio as a result of degradation and/or assimilation. A typical C:N:P ratio to be established to maximize microbial activity would range from 100: to 500:1:0.25 depending upon the amount of carbon that may become incorporated into the ECPS, the rate of biological activity on the targeted compound and the availability of alternate carbon substrates.

CHI GENERIC GROUPINGS
Significant in Ground Waters

The understanding of the bacterial groups that play a part on occasions in ground water systems is complex. It is complicated by the fact that there are two major groupings:

(1) **Indigenous Bacteria** that occur naturally in ground water systems and there is a growing body of evidence that there are stratified within the earth's crust a series of distinctive types of microbial activity (intrinsic).

(2) **Opportunistic Bacteria** that naturally occur in another part of the biosphere but can, under suitable conditions, invade and compete within ground water systems (extrinsic). This latter group can include possible pathogens that may be subdivided into two groups on the basis of their relative pathogenicity.

(3) **Pathogenic Bacteria** include those bacterial groups which are able to routinely cause, where the numbers of invading cells are adequate, clinical symptoms of the disease in an infected host. Examples of pathogenic bacteria include those causing cholera, typhoid and dysentery, all of which can be water borne.

(4) **Nosocomial Pathogens** (opportunistic) is the name given to those bacteria which occur in the natural environment as a normal part of the ecosystem but which can, upon entering an immunologically or physically weakened host, cause clinical symptoms of infection. These nosocomial pathogens may not necessarily show consistent clinical symptoms of the disease and so have traditionally been more difficult to diagnose. Most attention has been paid to hospital induced nosocomial infections (where a patient has become infected with a nosocomial pathogen while resident in the hospital). Community induced nosocomial infections also occur but are much more

difficult to diagnose due to the dispersed nature of the infected hosts as compared to a hospital.

While there is a natural interest in the health (hygiene) risks that may be associated with ground waters, there are also risks to the "health" of the ground water system itself. These risks would relate to the influence that microbial growths and activities may have on the productivity of aquifers and the transmissivity of those waters within, between and into such aquifers. An aquifer in which microbial intrusions had in some way impaired productivity would be reasonably deemed to be a less "healthy" aquifer. Microorganisms may be subdivided into two major groups:

(1) Occlusive Microorganisms include those microbes that will grow within the porous structures of the aquifers and impede the transmissivity of the ground water through the system (and towards a producing water well). These organisms usually form in biofilms (slime) growing over surfaces and gradually forming a coherent growth that may totally block water movement (biological barrier or plug).

(2) **Nuisance Microorganisms** interfere with the quality (rather than the quantity) of the ground water passing through and/or being delivered from the aquifer. These interferences may take the form of noxious products of microbiological activity such as detectable tastes and odors, cloudiness and color generation. These effects may be associable with either the products of growth (chemicals or dispersed cells) or the biochemical modification of the chemistry of the water. This latter event may be due to such events as specific enzymatic activity, shifting reduction-oxidation states, and temporary microbiological entrapment of chemicals. There can be a phased release of the bioaccumulated materials.

Both occlusive and nuisance microorganisms tend to grow predominantly in consortia. Such growths involve the growth of a variety of microorganisms within a common biofilm or "slime." These forms of growth tend to focus at the interface between an oxidative (oxygen present) and reductive (oxygen absent) regime. The reduction-oxidative status can be measured or calculated and expressed as the redox potential (Eh value or the Oxidation-Reduction Potential, ORP) and the dominant microbial activity appears to occur at the redox front. One reason for this focus is that the oxygen (for aerobic growth) comes along a gradient from the oxidative state while the nutrients diffuse along the gradient from the reductive side of the gradient. In highly oxidative environments, there would tend to be shifts in the dominant microbial flora in favor of the molds particularly where the porous medium is not permanently saturated. In this event, taste and odor problems may become more pronounced with earthy and musty odors often dominating.

Under some conditions where radical contamination (pollution) of a ground water system with an organic chemical has occurred there is likely to be phased microbial response to this event. The phases could include entrapment into the biofilm, partial or complete degradation, suspension and mobilization of any non-degraded or partially degraded chemical product incumbent in any detaching biofilm and radical dispersion of any volatile or gaseous insoluble products. Three major events may, therefore, be associated with the following:

Bioentrapment
The act of a chemical or group of chemicals being taken up (and potentially bioaccumulating) in a biologically derived structure (e.g., biofilm, suspended sessile particulate).

Biodegradation
The destruction of at least a part of a molecular structure by a biologically derived chemical process (e.g., enzymatic action). The products of this degradation are normally one or more smaller molecules such as carbon dioxide, lactic and acetic acids, methane, benzene, hydrogen, and nitrogen.

Biomobilization
An event where a bioaccumulated compound is released back into the water flow through either a sloughing of the biofilm containing some of the accumulates or the direct release of the compound from the biofilm into the water.

These events would be in addition to the restriction of contaminant mobility by occlusive (bio-) barriers. In general terms, the microorganisms associated with such events are often considered to be **Biodegradative Microorganisms**. This would be true where there is pollution of a ground water system by a specific group of organic chemicals. Here, it can be expected that the range of microorganisms focusing in the biodegradation zone will contain fewer types in the consortium. These will be restricted to those strains able to "cooperatively" degrade the compounds with a maximum efficiency given the environmental conditions presented. The density of the incumbent degraders in the biofilm may reflect the level of aggressivity with which the targeted compounds are being degraded. In cases where some of the pollutant chemical is now becoming dispersed with shearing fractions of the biofilm, the level of degradation may be reflected in the density of cells recoverable from the suspended particulate (sheared biofilm) material.

It has to be remembered that when a water sample is being examined for microbial presence, the water sample will contain those components that either naturally occur in the water or have entered the water in sheared material from the biofilms attached to surfaces.

Absence of microorganisms from a particular water sample cannot be construed to mean that there would be no microorganisms present within the ground water system itself. Much of the observations on ground water microbiology do relate to the microorganisms recovered from product water samples and may not reflect the total range of organisms which may be actively associated with the ground water system itself. There are, however, a number of major groups of consortial heterotrophic microorganisms which have been associated with various major events in ground water systems. Each will be described briefly below but it should be noted that references to sections and families (bracketed) refer to the current classification of bacteria as described in Bergey's Manual (9th edition). Additionally, it should be recognized that there are two major groups of bacteria differentiated by a staining technique known as the gRAM (or Gram) stain. RAM refers to the ability of the stain to differentiate (R)eaction, (A)rrangement and (M)orphological characteristics of the bacteria. Particularly important in the gRAM reaction is the reaction differentiation which may be negative (gRAM negative, G-) or positive (gRAM positive, G+).

Pseudomonads, (Section 4 of *Bergey's Manual*)
These bacteria are all gRAM negative relatively primitive rods and cocci. All are aerobic and need oxygen (some can use nitrate as an alternative) in order to grow. Many are able to break down specific organic materials into inorganic substances (mineralization process) particularly under saturated oxidative conditions. These bacteria often grow in consortial biofilms in association with other microbial groups. One common example is the association of pseudomonads with sulfate-reducing bacteria in corrosive biofouling (MIC). Many pseudomonads are able to function very efficiently at low temperatures. Generally, these organisms are able to grow well at temperatures of less than 15°C are called psychrotrophs. In oxidative ground water having temperatures of between 2 and 12°C, the pseudomonads may be found to dominate particularly where there is a significant organic content (>0.5 ppm TOC) originating from a limited range of pollutants. In the case of liquid hydrocarbon plumes generating in ground water systems, the range of pseudomonads that are recovered can

become very restricted (i.e., 2 to 3 strains). Where a shallow monitoring well becomes biofouled under oxidative conditions, pseudomonads may well dominate the fouled zone and cause significant reductions in the recoverable pollutant withdrawn by sampling the well due to the preferential bioentrapment and biodegradation in the fouled zone around the monitoring well.

Some of the *Pseudomonas species* are pathogenic. Of these, *P. aeruginosa* is the most serious and can infect people with low resistance to urinary tract and lung infections, and can also invade burn areas. Two unusual families of pseudomonads may also commonly occur in ground water. One group is generally referred to as the methylotrophs. These are able to oxidize methane gas aerobically. Methane is commonly found in anaerobic soils, sediments and aquifers and is generated (Methanogenesis) where there is a reductive organic decomposition occurring. The second unusual family is formed by the pseudomonads that are able to grow in oxidative environments where the salt concentration reaches greater than 15%. These form the halotrophic pseudomonads.

Enteric Bacteria, (Section 5, family 1 of *Bergey's Manual*)

These bacteria are also gRAM negative rods but are able to grow under both oxidative and reductive conditions (facultative anaerobes). Most are fermentative degraders producing organic acidic products and often copious amounts of gas (usually CO_2 and H_2). The name "enteric" focuses on the fact that several of the genera inhabit the gastro-enteric tract of warm-blooded animals. They are also evacuated in significant numbers in fecal material. It is for this reason that the enteric bacteria have been used as indicator organisms for fecal contamination (and hygiene risk). As the major inhabitant of the human colon, *Escherichia coli* has been selected as a significant indicator of fecal contamination. *E.coli* can be pathogenic causing gastro-enteritis or urinary tract infections. However, several other enteric genera include major pathogens causing such diseases in humans as: typhoid and gastro-enteritis (*Salmonella*), dysentery (*Shigella*), pneumonia (*Klebsiella*) and the plague (*Yersinia*). Amongst the vibrioid bacteria in the related family 2 is another major gastro-enteric pathogen (*Vibrio cholerae*) the causative agent of cholera.

Common tests for hygiene risk in waters involves an evaluation for the presence of **Coliform bacteria**. These bacteria are defined as non-sporing gRAM negative bacilli which are able to ferment the sugar lactose with both acid and gas being produced in determinable amounts. It is interesting to note that the sugar lactose is not commonly found throughout the environment but appears to be synthesized most commonly in mammalian milk. Mammals when suckling during infancy, therefore, take in large quantities of lactose in the milk which biases the microflora in the gastro-enteric tract to those bacteria that can anaerobically ferment lactose (i.e., coliforms). These coliform bacteria also have to be able to resist the inhibitory effects of bile salts (excreted into the gastro-enteric tract) and be able to grow at warm-blooded temperatures (e.g., 35°C). Because of these restrictors, coliforms may be selectively grown in culture conditions where other contaminant bacteria would be suppressed. The restricting factors are the use of lactose as the major energy source, application of bile salts to selectively restrict competition to those bacteria normally found growing in the gastroenteric tract, and the use of an above environmental norm temperature for growing these coliforms (i.e., 35°C). Acid products are easily determined by color shifts in pH indicators and the gas may be entrapped for direct or indirect observation.

Confirmatory tests need to be performed if the presence of *Escherichia coli* is to be confirmed. The need to confirm that *E. coli* is present is important because other members of the coliform group could give posi-tives. Such microorganisms in a water system would, therefore, cause a positive coliform test. That test however remains presumptive until the presence of *E. coli* is confirmed. The most common genera causing these types of interferences are *Enterobacter* and *Klebsiella*.

Sulfur Bacteria, Sulfate-reducing Bacteria (Section 7 of *Bergey's Manual*)

Sulfur bacteria is an unusual name since it is used to describe two very distinct groups of bacteria. Commonly, the term is used to refer to the sulfate-reducing bacteria (SRB) which are associated with MIC. The term is, however, also used to describe the bacteria in that sulfur plays a major role in the activities of the bacteria (e.g., producing elemental sulfur, sulfuric acid). These will be addressed separately. SRB are serious nuisance organisms in water since they can cause severe taste and odor problems and initiate corrosion. These bacteria are called the sulfur bacteria because they reduce large quantities of sulfates to generate hydrogen sulfide (H_2S) gas as they grow. These bacteria are referred to as the sulfate-reducing bacteria (Section 7) and are often known by the initials SRB. The problems generated by H_2S are:

first, because it smells like "rotten eggs";

second, because it initiates corrosive processes; and

third, the gas can react with dissolved metals such as iron to generate black sulfide deposits.

For the plant operator each of these effects can become a serious nuisance. For example, the effect could be the sudden appearance of the smell of "rotten eggs" in water. Usually such events mean that somewhere upstream there has been a major aerobic biofouling that has removed the oxygen out of the water and allowed these bacteria to dominate. "Rotten egg" smells are not created by just the SRB group, but can also be sometimes generated by other bacteria such as many of the coliforms when suitable conditions occur. Hydrogen sulfide may be found in waters where oxygen is absent and there are sufficient amounts of dissolved organic materials present.

Forcing oxygen (as air) into the water can stress these SRB microorganisms since it is toxic to their activities. Commonly, the SRB protect themselves by co-inhabiting slimes and tubercles with other bacteria that are slime forming. These rotten egg smells will occur more commonly when a water system (or water well) is not used for a period of time. In these cases, various other bacteria co-inhabiting the biofilm thus protecting the SRB inside the consortial mass use up the oxygen in the water. Once these other co-inhabiting members of the consortium control the toxic oxygen, then the growth and activities of the SRB group can become rampant. Sometimes this is accompanied by the intense production of the "rotten egg" (hydrogen sulfide) smell.

The SRB group may establish itself attached to a solid surface within a biological slime or tubercle (a "bubble-like" structure composed of hardened iron-rich plates over-layering an active slime formation inside). Once established, these bacteria it will begin to generate the hydrogen sulfide gas. This gas can trigger a complex electrolytic corrosion process on some metallic surfaces. Such corrosion begins with pitting and terminates with perforation of the supporting structures (e.g., metal pipe wall) and system failure. At the same time, various slime forming bacteria co-inhabiting the site may magnify the problem by generating various organic acids which can also be corrosive. Once established, the corrosion is difficult to control since the slimes or tubercles "buffer" the effects of any chemical treatments such as chlorination, acidization or the use of cleaning agents. This means that higher dosages and longer exposure times need to be applied to try to control such corrosion events. Because the SRB organisms cause corrosion while growing on a surface, tests on the water itself may be negative. Positive tests will occur if some of these bacteria are present in the water sample while moving in an attempt to colonize another corrosion site.

Where there is a significant amount of iron, manganese or other metallic materials in the water, hydrogen sulfide can react with these compounds to form metallic sulfides. Many of these chemical compounds are black in color and can cause the slime to become black in appearance. When these blackened slimes break up, the water may contain threadlike strings of black slime-like material. On many occasions, these black growths are not accompanied by any "rotten egg" smells since the H_2S gas has been converted into the black sulfides. On some occasions, these slimes flow down walls, across surfaces and may even change in color through to browns, reds, yellows and greys as the oxygen in the air stimulates aerobic bacterial activity in the slime.

Identifying an SRB problem is easier to achieve by recognizing the "rotten egg" smell, finding tubercles and corrosion inside metallic equipment and/or the black slimes rather than performing bacteriological tests on the water that has flowed past the site. The reason for this is that the SRB bacteria grow in "protected" places often surrounded by other types of bacteria that may mask their presence. Microscopic examination is also made more difficult because of interferences caused by these growths. There are some cultural systems that can be used to confirm the presence of SRB but most rely upon these bacteria producing black sulfides that become visible. This may be seen as a deposit in a suitable liquid culture medium or as distinctive black growths in and/or around colonies containing SRB growing on agar (gel) media in the absence of any traces of oxygen. The lag time before these "blackenings" occur can indicate the aggressivity of the SRB group. A rapid (e.g., two-day delay) would indicate a much more aggressive population than in a case where there was a considerable delay (e.g., eight days).

Risks from an active SRB infestation are multiple in that there can be corrosion, severe taste, odor and colored water problems, losses or total failures in process efficiencies and increasing consumer complaints. Unfortunately, the ability of these bacteria to grow in places where they are protected by either copious "overburdens" of slime or within tubercles can make control very difficult. Treatments, which have been recommended, include various disinfection, acidization and cleaning practices. It is important when applying these various practices to ensure that:

(1) flush and clean as best as possible before starting the treatment program;
(2) apply the highest recommended dosage and the longest contact time for the selected treatment program in order to maximize the potential effectiveness;
(3) increase the dissolved oxygen in the water to suppress SRB activity; and
(4) consider the ongoing or routine application of disinfectants and/or penetrants to reduce the rate of recovery of the SRB from "shock" effect of the practice.

The routine monitoring for the redox (Eh) potential may form a useful monitoring technique. Where the Eh in the water dramatically declines (i.e., becomes more reductive), this may be taken as an "early warning signal" that conditions are now changing and becoming potentially more supportive for another SRB outbreak. There are also a number of simple cultural systems that can be used to monitor for the presence of SRB organisms in water. It may be convenient to routinely monitor the water (e.g., monthly) for the aggressivity of these bacteria.

Other Sulfur Bacteria

There are some other groups of "sulfur bacteria" which may cause problems in waters. They may be summarized as the sulfur oxidizing bacteria (Section 20), the colorless sulfur bacteria (Section 23), and the purple and green sulfur bacteria (Section 18). Of these groups, the sulfur oxidizing bacteria are well documented because of their association with the recoveries of metals through the leaching of ores and the problems with acidic mine tailings. These bacteria require oxygen to grow and convert various sulfides to acidic products such as sulfuric acid. The colorless sulfur bacteria do not usually produce acidic products but do convert hydrogen sulfide and other sulfides to sulfates. These are sometimes found in sulfur springs, water wells and distributions systems where sulfides are present but where there is also available free dissolved oxygen.

Oxygen is toxic to the purple or green sulfur bacteria. Like plants, these bacteria are able to photosynthesize but without the production of oxygen. Unlike plants, however, these bacteria reduce various sulfates to sulfides and elemental sulfur that becomes deposited in and/or around the cells. Common habitats for these bacteria include septic ponds where they occasionally dominate and turn the water red, and deeper lakes which have stratified (layered) and these bacteria form distinctive plates of floating growth within the strata.

The various sulfur bacteria can be summarized therefore as belonging to the following major groups and genera:

Sulfate-reducing Bacteria (SRB)
> *Desulfovibrio*
> *Desulfotomaculum*

Sulfur Reducing Bacteria (SRB but uses sulfur)
> *Desulfuromonas*

Sulfur Oxidizing Bacteria (producing acidic products)
> *Thiobacillus*

Colorless Sulfur Bacteria
> *Beggiatoa*

Thiothrix
Purple and Green Sulfur Bacteria
Chlorobium
Chromatium

Archaeobacteria (Section 25 of *Bergey's Manual*)

These are bacteria-like organisms that do not have the normal types of cell wall or genetic mechanisms found in the rest of the bacterial kingdom. Archaeobacteria tend to be found in some of the more extreme and/or unusual habitats such as swamps, acid springs and salt lakes. While there is a considerable diversity amongst this microbial group, there are, in reality, only three major groups known. These are the methanogenic archaeobacteria (able to generate methane), extremely thermophilic archaeobacteria (able to grow under conditions of high temperature), and extreme halophilic archaeobacteria (able to grow in concentrated salt solutions).

Methanogenic archaeobacteria (biogas generators)

Methanogenic archaeobacteria produce methane (natural gas, biogas) anaerobically by converting simple compounds such as CO_2, H_2, formate, acetate to either methane or a mixture of methane and CO_2. These bacteria thrive in a wide variety of anaerobic environments rich in organic materials. Well-documented examples are the rumen, gastroenteric tracts, anaerobic sludge digesters and sediments. Some methanogens even parasitize other microbes such as protozoa. The common occurrence of methane in ground water would indicate that the methanogens might also be active in anaerobic niches within the aquifer particularly where there is organic decomposition occurring. These microorganisms may well be the generators of the methane (CH_4) in landfill operations. Where these conditions are replicated in the laboratory setting, the gas can sometimes be observed to be biologically entrapped as separated bubbles with a biofilm forming the interface between the gas and aqueous phases. These bubbles then form into dispersed foam spread through the porous medium. Such occurrences may cause a reduction in the hydraulic conductivity through the affected zone thus creating a temporary biological barrier (foam plugging). Where methane is being bio-generated, it can be expected that the methanotrophic (methane utilizing) bacteria are likely to be particularly active at the aerobic (oxidative) interfaces. Common genera encountered among the methano-gens include *Methanobacterium*, *Methanothermus*, *Methanococcus*, *Methanomicrobium* and *Methanosarcina*.

Extreme halophilic archaeobacteria (Salt requiring bacteria)
These are defined as those bacteria which require at least 8.8% NaCl for growth, usually require 17 to 23% NaCl to achieve a maximal (optimal) growth. Higher salt concentrations up to saturation (32 to 36% NaCl) will gradually retard the growth of the extreme halophiles and some will still be able to grow albeit slowly at saturation. Two genera of aerobic bacteria (*Halo- bacterium* and *Halococcus*) are well known to be able to grow under these conditions where there is a sufficient organic content. These bacteria frequently produce bright red pigments that can color the water (seawater evaporation ponds often turn red where there are halobacterial activities).

These halobacteria are not the only microorganisms which can grow in salt rich environments. Algae (*Dunaliella*) and phototrophic purple sulfur bacteria (*Ectothiorhodospira*) can both thrive in surface saline water lakes with considerable primary production of organic material that then supports the halobacteria. In saline ground water (>8.8% NaCl) it can be expected that under aerobic (oxidative) conditions the halobacteria would be active in proportion to the available organic substrates. Under more alkaline conditions (pH between 9 and 11) where there is a low concentration of magnesium (Mg^{2+}), the dominant halobacteria may shift to two other genera (*Natronobacterium* and *Natronococcus*). While little research has yet been directed at the presence and activity levels of halophilic bacteria in extremely saline ground water, parallel studies on surface saline waters would indicate that microbial activity is likely to be present depending upon the availability of organic material.

Extreme thermophilic archaeobacteria
(bacteria which grow at extremely high temperatures)
These bacteria generally grow over a temperature range of 30C° with the minimum for growth being within the range from 55 to 85°C and maximum at between 87 and 110°C. Incredibly some of these bacteria can grow not only at these high temperatures but also under acid conditions (e.g., *Sulfolobus* and *Thermoplasma*). Most isolates have come from submarine and terrestrial volcanic sources but evidence would suggest that some thermophiles may be able to function in geo-thermally heated ground water systems. For example, *Staphylothermus* has been found to be widely distributed in thermally heated seawater around "black smokers" (marine depth, 2500 meters) and it is thought that this genus may also be widely distributed near hot hydrothermal vents.

From the present knowledge of the activities of these archaeobacteria, it is highly probable that many of the ground water systems thought to be "hostile" to microbial activities due to high temperatures, salinity or acidic conditions may, in fact, be able to support the activities of microorganisms such as the archaeobacteria.

Gram-Positive Cocci (Section 12 of *Bergey's Manual*)

There is a wide range of diverse types of gRAM positive coccal bacteria. "Coccal" indicates that the bacterium is spherical, ovoid or ellipsoid in shape. Genera that are composed of coccal bacteria usually bear the suffix -cocci. Aerobic cocci are dominated by the genus *Micrococcus* which occur in many natural habitats where oxygen is available. This range of habitats includes both surface- and ground water, and micrococci are sometimes incorporated into the consortia of bacteria involved in biofouling events (e.g., plugging of water well). Despite the frequency with which these bacteria are found, even on mammalian skin, only a few species can be pathogenic. Another (strictly) aerobic coccus of interest is *Deinococcus* which occurs widely in waters, ground meat products and feces. These coccal bacteria have an unusually high resistance to both desiccation and radiation effects. Where there is a saturated oxidative eutrophic regime being subjected to sub-lethal doses of gamma radiation, it may be expected that *Deinococcus* could become a major component in the consortial biofouling.

An alternative indicator bacterial group for hygiene risk due to fecal contamination is the streptococci referred to as the **enterococci** (group 3). These bacteria are normal residents of the gastroenteric tract of humans and most other animals and commonly appear in the feces. Sometimes the **enterococci** may be referred to as fecal streptococci including such species as *Streptococcus faecalis*. The ratio of **fecal streptococci** (FS) to **fecal coliforms** (FC) in water is sometimes used to determine the likelihood that the fecal material is of human origin. Where the water has been contaminated with human originated fecal material there is usually a preponderance of FC and the FC:FS ratio would favor the FC (e.g., 4:1 or 1:0.25). On the other hand, if the fecal contaminant had been of nonmammalian origin, the dominance would commonly shift to the FS group (e.g., FC: FS ratio of 1:4 or 0.25: 1). *S. faecalis* is another opportunistic (nosocomial) pathogen that can cause urinary tract infections and endocarditis. Normally, FS do not persist for long in ground water, but under brackish and saline ground water conditions, the enterococci are likely to survive since they are able to tolerate and grow in 6.5% sodium chloride under suitable environmental conditions.

Endospore Forming Gram Positive Rods (Section 13 of *Bergey's Manual*)

Some bacterial cells form a very heat resistant body within the cell where all essential components of the cell are concentrated. Once the body (spore) has matured, the cell material around it (hence the prefix endo-meaning inside of the body) is essentially dead and disperses. This leaves a small spore body that resists not only heat but also desiccation and other radical environmental shifts. These endospores are basically dormant viable entities. Some recent extrapolations suggest that the endospores of *Bacillus subtilis* may be capable of surviving between 4.5 and 45 million years in an interstellar molecular cloud while resisting the high vacuum, low temperatures and UV radiation present. This may be an interesting example of the potential for survival that the endospores have under adverse conditions. Similar events may be expected to occur in ground-water systems where the dormant endospores may travel over considerable distances through millennia of time within aquifer systems.

Endospore forming bacteria are divided into a number of genera. Three of these are of particular interest in relation to ground water environments. Of these, the most ubiquitous are the aerobic endosporogenous group called *Bacillus*. Many species of *Bacillus* occur naturally in oxidative ground water and are frequently components in biofilms that cause plugging. The only species pathogenic to humans is *Bacillus anthracis*, the causant organism for anthrax. This species is highly infectious to animals and humans, and is passed along through direct contact. Its endospores can remain viable in soils for decades but there is little evidence of *B. anthracis* endospores surviving in or causing a threat from ground water sources. While most species of *Bacillus* appear to be harmless to humans, there is a range of species that are pathogenic to insects (e.g., *B. thuringiensis*) and these often involve the generation of toxic agents lethal to the insects by these bacteria.

A second major genus in this group comprises anaerobic endosporogenous bacteria: the genus *Clostridium*. This genus includes a diverse range of specialized degraders able to efficiently break down cellulose, chitin, proteins and other organic material under anaerobic conditions. Oxygen is lethal to most species in this genus and so their habitat is frequently restricted to organically rich saturated niches. The presence of *Clostridium* species in ground water samples would indicate that the water was probably anoxic (contains no significant oxygen; <20 ppb) and had tracked through a zone of relatively intense organic decomposition.

Another source of organic material could be fecal contamination. The third major bacterial indicator species group for hygiene hazard is *C. welchii* due to its common occurrence as an inhabitant of animal gastroenteric tracts and fecal materials. A range of species of *Clostridium* produces powerful toxic agents that can affect man. These include *C. tetani* (tetanus) and *C. botulinum* (food poisoning, botulism). Neither of these species commonly use ground water as a vehicle of infection.

Some sulfate-reducing bacteria (SRB) possess endospores and hence are included in section 13 under the genus, *Desulfotomaculum* (de-sul"fo-to'mac"ool-um). Species of this SRB are frequently isolated as a component in consortia associated with corrosion.

Filamentous Actinomycetes, (Volume IV of *Bergey's Manual*) (Mycelial bacteria)

There is a large group of gRAM positive bacteria that form thread-like structures (mycelia). This is because the cells are in the form of filaments. These filaments may or may not branch. During successful growth, these filaments may form into a ramifying network called a **mycelium** (a feature of growth among another microbial group, the fungi). The nature of this growth is similar to the fungi in that the various filaments form an integrated network, usually by attaching to particles and forming networks between them. These commonly produce exospores (sporing bodies formed outside of the bacterial cells) which are not so resistant to harsh conditions as endospores nor as capable of prolonged survival.

In ground water systems, these filamentous actinomycetes are most commonly found at interfaces between saturated and unsaturated zones where oxygen is present together with adequate water. One focus site for such growths would be the capillary zone immediately above the saturated zone. Rates of growth would be controlled by the availability and type of organic materials present at the site either via the ground water or such recharges as may be occurring. While some actinomycetes are anaerobic many are aerobic. The largest group of actinomycetes which are associated with ground water belong to the genus *Streptomyces*. Species of this genus form a very aggressive component within a microbial community. Many species produce various antibiotics such as the tetracyclines and aminoglycosides (e.g., streptomycin) which give these bacteria a competitive edge. Some species of *Streptomyces* produce earthy odors during their growth. In fact, the musty odor of soil is produced by *Streptomyces* themselves. The compounds produced by these bacteria which cause this include the geosmin and isoborneol groups. Where an earthy musty odor is detected in a ground water sample, an infestation by species of *Streptomyces* can be suspected.

Fungi (molds)

Some microorganisms have a more complex cell structure, larger size in filamentous (mycelial) form, and often exhibit distinct asexual and sexual reproductive cycles (among other distinctive features). These make up the fungi. These microorganisms are mostly, but not entirely aerobic and play a very major role in the degradation of complex organic material into simple organic compounds and inorganic molecules (mineralization). Frequently, the fungi dominate biodegradation in unsaturated porous media and on organic surfaces under aerobic conditions. For example, in soils the fungi dominate the microbial biomass, which has been estimated to be 5 metric tons/hectare (2.2 tons/acre) in the top 20 cms. Fungi are involved in a wide variety of plant diseases (5,000 species of fungi are pathogenic to specific plants of economic importance) and also the deterioration of manufactured goods which contain organics and are exposed to the air and moisture (e.g., fabrics, leathers, paper and wood products). In ground water systems, the most likely site for fungal activity would be in an unsaturated zone where there are organic materials present. Examples of this would be the recharge zones from a surface sewage oxidation pond and in the capillary zone above a low-density gasoline plume polluting a ground water system. Working with fungi involves a different approach to scientific discipline, as compared to the other aspects of microbiology. The study of fungi is called **mycology** (Greek *mykes*, mushroom, and *logos*, discourse). The form of cellular growth and reproduction that occurs most easily differentiates the different divisions and genera within the fungi. Very commonly, it is the manner by which the exospores (asexual reproduction) and zygospores (a form of sexual reproduction) are produced and the environments within which mycelia are created is used in the classification of the fungi. Generally speaking, staining, microscopic and cultural techniques differentiate the fungi while the bacteria require a complex of biochemical methods along with staining and cultural techniques.

Algae (micro-plants)

Algae are some of the simpler members of the plant kingdom. They are all capable of photosynthesis using chlorophyll. Some algae are very large in bulk although they retain a simple form of tissue organization. These are the seaweeds (macro-algae) and are not found too often growing in wells! The simpler and smaller algae (micro-algae) are, however, sometimes recovered in water well samples. On occasion it is important to determine the origin and fouling potential relatable to these microorganisms.

In practice, the micro-algae can be relatively easily divided into a number of major groups in part by the color generated in the growth. These include:

(1) the blue green micro-algae also known as the cyanobacteria;
(2) the (grass) green micro-algae referred to as the chlorophytes;
(3) the brown micro-algae called diatoms and desmids; and
(4) the euglenoid micro-algae that are sometimes green or yellow and they may even be colorless as they shift between a "plant" form of existence and an "animal" form.

The cell wall varies considerably between these forms of micro-algae that would mean that different control strategies would have to be used depending upon the type of micro-alga recovered. Cyanobacteria are, as the name implies, bacterial and have a typical gRAM negative type of cell wall. Chlorophytes have a more typical cellulosic form of cell wall commonly found in higher plants. For the diatoms and desmids, the cell walls are often complex and highly structured (ornate) with a high silicate content. Euglenoids possess a cell wall more typical of protozoa (i.e., animals) and is more complex and dominated by proteins.

All of the micro-algae are capable of photosynthesis even under the very low levels of light that may be found in some wide bore shallow wells or in the soil. By way of a comparison, in soils the micro-algae appear to be able to grow phototrophically down to depths of 50 mm or more. The cyanobacteria tend to dominate in the soil horizon from 5 to 25 mm with the other micro-algae growing at greater depths. It is becoming apparent that many of the micro-algae are also capable of growing heterotrophically, that is competing with the mass of bacteria, fungi and other organisms for organic nutrients. This is particularly evident for the euglenoids and the chlorophytes. The occurrence of these micro-algae in well waters which has not been subjected to any level of light, therefore, becomes less surprising. Another mechanism by which micro-algae may enter a water well (particularly a shallow wide bore or lateral gallery well) is with the recharge water coming from soil or surface water reservoirs. From soils there is thought to be a continual "bleeding" not just of micro-algae, but of the microflora at large. These microorganisms (many possibly in the UMB form) move downwards with the recharge waters to enter and move into the ground water systems. Little is known of the distances over which these movements can be achieved.

The presence of micro-algae in a well water sample should raise a sequential series of questions (if the answer to any question is YES, then proceed to the next question):

1. Is there light penetrating into the water column of the well? Yes/No, if No go to 3.

2. The probability is that the algae are growing in the well. If there is a low total organic carbon, but relatively significant levels of inorganic nitrogen (e.g., nitrate, ammonium in low ppm range) and phosphate (in the high ppb range), then algal growth can be expected. The micro-algae are either competing heterotrophically for organic carbon with other microorganisms or they have "contaminated" the well water with surface recharge. Is the total organic carbon greater than 2.0 ppm? Yes/No, if No go to 4.

3. There may have been either a heavy algal population in the recharge water or the algae are competing heterotrophically for the organic carbon.

4. There is a likelihood that the algae present in the well have arrived via recharge water originating in soil or surface-water that supported algal growth.

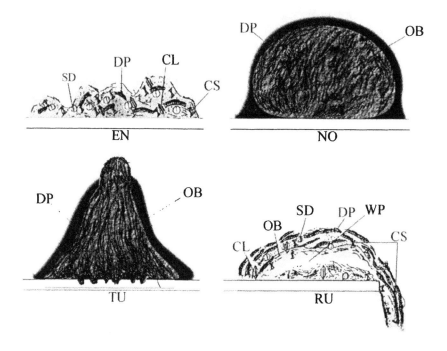

Figure Nine, Cross section diagrams of the structures commonly found in encrustations (EN, upper left), nodules (NO, upper right), tubercles (TU, lower left) and rusticles (RU, lower right). Common structures observed are occluded biofilms (OB), crystallized structures (CS), dense plate layers rich in ferric iron (DP),water pathways (WP) and entrapped materials which would primarily be sands (SD) and clays (CL).

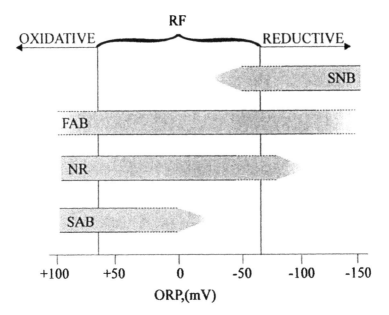

Figure Ten, Diagram of the impact that the oxidation reduction potential (otherwise known as ORP, and Eh) has on the biological activity around the redox front (RF) being the transitional zone between oxidative and reductive conditions. Strictly aerobic bacteria (SAB) cloister on the oxidative side unless they are capable of nitrate respiration (NR). Facultative anaerobic bacteria (FAB) generally grow more aggressively on the redox front while the strictly anaerobic bacteria (SNB) grow sequentially on the reductive side. It should be remembered that biofilms are capable of generating redox gradients within the strata and of allowing SNB to grow in what would appear to be unfavorable sites.

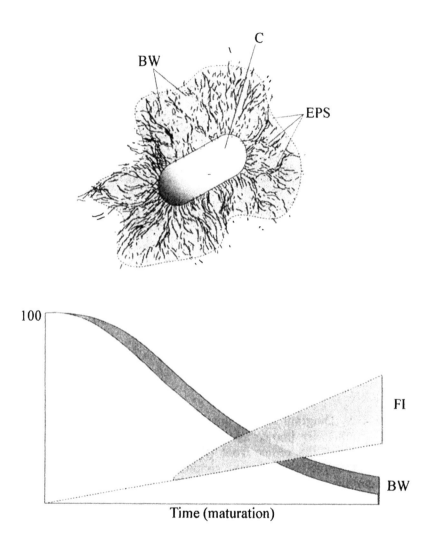

Figure Eleven, Much of the slime growth is actually caused by the binding of water (bound water, BW) in the extracellular polymeric substances (ECPS) around the cells (C, upper diagram). As the growth matures, it continues to accumulate metals such as ferric iron (FI), sands and clays to become denser. During the early growth, the ECPS and the bound water makes up as much as 99.9% of the volume but this falls over time (lower graph, y axis - % bound water, x axis - time).

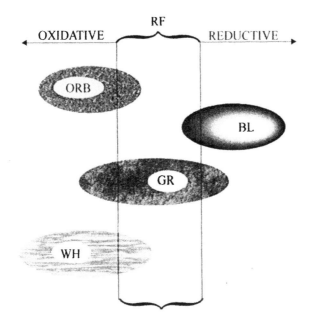

Figure Twelve, Chart depicting the common colors and textures of slimes. There are four major colors for slime/encrustations commonly observed. These are orange-red-brown (ORB), black (BL), grey (GR) and white (WH). ORB tends to occur at the redox front and through to the oxidative side. BL most commonly occurs on the reductive side of the front and is commonly associated with the activities of the SRB, GR and WH both tend to occur under oxidative conditions where there is a higher organic content and a low availability of iron. Both the WH and GR may at times become tinged to beige, yellow and pink hues and, very rarely, turn violet.

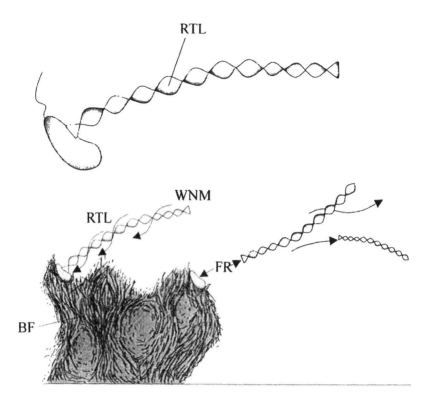

Figure Thirteen, The form and function of *Gallionella*. This iron-related bacterium is commonly recognized by its distinctive ribbon-like tail (RLT, upper). It grows in the outer edges of a biofilm (BF) and grows the RLT out into the water to encourage the water and nutrients to move towards the cell (WNM, lower). When the tail extends to become too long, it fractures (FR) and floats off with the water currents. Often these RLTs when seen microscopically in water samples leads to a laboratory diagnosis that the biofouling was caused by *Gallionella* when, in reality, it was a relatively minor contributor.

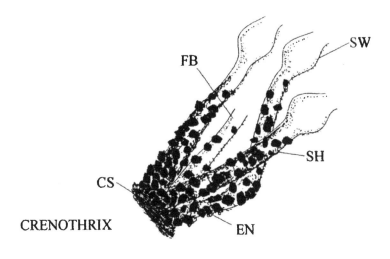

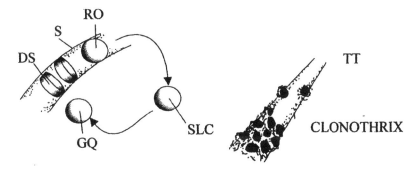

Figure Fourteen, Sheathed IRB belonging to the genus *Crenothrix* have thin sheaths (SH) that are attached often at a common site (CS) with evidence of false branching (FB). The sheaths are encrusted (EN) unevenly with iron oxides with thicker deposits being present closer to the attachment points. Cells inside the sheath are disc shaped (DS) but may become rounded (RO) further up the inside of the sheath. Cells can reproduce within the sheaths forming spore-like cells (SLC) which are released and frequently germinate quickly (GQ) on the outside of the sheath. Sheaths are swollen (SW) near the tips. If the tips are tapered (TT), the IRB belongs to a different genus, *Clonothrix*.

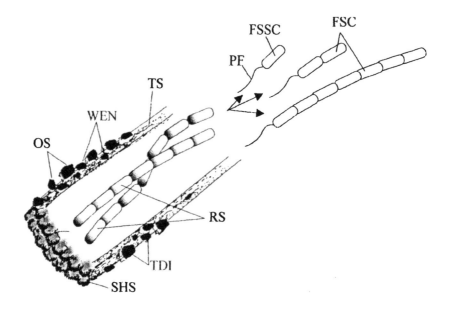

Figure Fifteen, Sheathed IRB belonging to the genus *Leptothrix* have thick sheaths (SH) that are rarely attached. Where this occurs it is by a sticky holdfast structure (SHS). The sheaths are encrusted both within the sheath (WEN) and outside of the sheath (OS) with thick deposits of iron oxides (TDI). Cells inside the sheath are rod shaped (RS) and can move up the inside of the sheath and leave the sheath as free-swimming single cells (FSSC) or in short chains of up to eight cells (FSC). They are motile by polar flagella (FG).

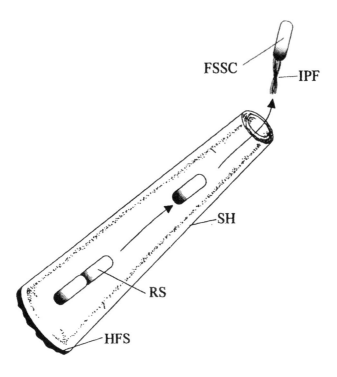

Figure Sixteen, Sheathed IRB belonging to the genus *Sphearotilus* have thin sheaths (SH) that are usually attached to surfaces using a holdfast structure (HFS). The sheaths are never encrusted with iron oxides. Cells inside the sheath are rod shaped (RS) and can move up the inside of the sheath and leave the sheath as free-swimming single cells (FSSC) with bundles of intertwined polar flagella (PIF).

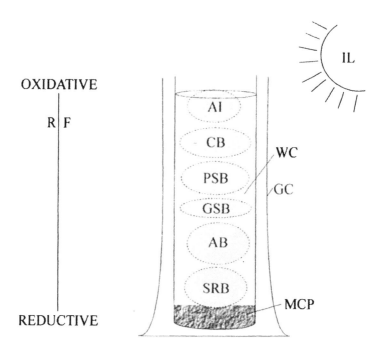

Figure Seventeen, Diagrammatic presentation of the classic Winogradski's column. In this illuminated (IL)vertical glass cylinder (GC), a water column (WC) is stratified vertically with the reductive side at the base and the oxidative at the air interface. To encourage stratification, a mixture of mud, gypsum (CaSO₄), and plant debris (MCP) is added before the water. Biological strata forms through growth with the anaerobes at the bottom (AB), aerobes at the top at the air interface (AI) while a cluster of facultative anaerobes form around the redox front (RF). Stratification of growth occurs over time in the following descending (oxidative to reductive) order: aerobic bacteria at the water:air interface (AI), cyanobacteria (CB), purple sulfur bacteria (PSB), green sulfur bacteria (GSB), anaerobic bacteria (AB) and sulfate-reducing bacteria (SRB).

ENVIRONMENT   HUMAN BODY

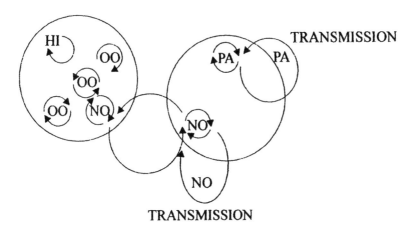

TRANSMISSION

Figure Eighteen, Illustration of the various forms of microorganisms that can occur in ground water. These include (from the health-risk viewpoint) the relatively harmless indigenous organisms (HI), organisms that are opportunistic (OO) and can adapt to different environments, nosocomial organisms (NO) that can cause infections in other organisms including humans, particularly when the immune systems are compromised, and the pathogenic (PA) organisms that will infect healthy targeted organisms. By far the vast majority are in the HI group but health risk assessments generally concentrate on the PA group with coliform bacteria being the common risk indicator group.

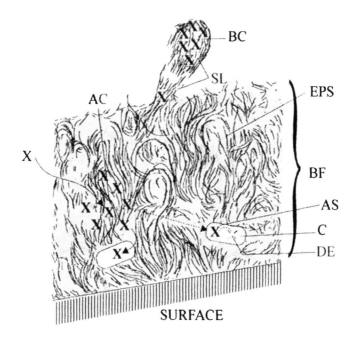

Figure Nineteen, Mechanisms by which chemicals (X) can enter a biofilm (BF) to simply accumulate (AC) within the ECPS, move through the ECPS to enter the cells (C) and become assimilated (AS) and degraded (DE). Where a biofilm breaks apart (lower) through sloughing (SL) the accumulates will be carried in the biocolloidal particles (BC) being carried through the water flow.

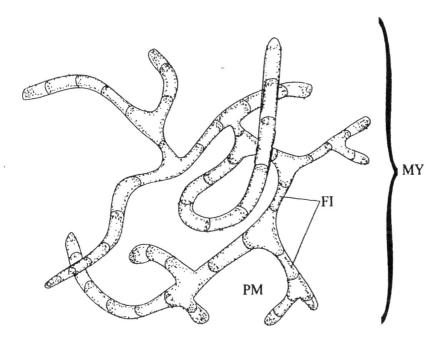

Figure Twenty, A very common group of bacteria found in porous media on the oxidative side of the redox front in semi-saturated and saturated waters are the mycelial bacteria (filamentous *Actinomycetes*). Growth is in the form of filaments (FI) which form a network (mycelium, MY) to cover the surfaces of the porous media (PM). The mycelium is unstable and may fracture to form small clusters of distorted rod-shaped cells.

4

MONITORING METHODOLOGIES FOR BIOFOULING EVENTS

INTRODUCTION

Perhaps the major challenge in attempting to monitor the occurrence of a microbially induced fouling (MIF) in a groundwater situation is the remoteness of the site from the sampling point. Rarely can satisfactory samples be taken at the actual site of the MIF because of the porous structures within which it is located. Downstream sampling (usually at the well head or beyond in a storage or distribution system) is therefore a convenient substitution. While convenient, such downstream samples may become subjected to anomalous events due to the addition of microbial entities functioning downstream from the site, and the loss of some of the fouling organisms due to such events as predation, competition or alternate site colonization. Therefore, such remote sampling may not accurately reflect the activity that is occurring in a focused manner at the MIF site.

Given the potential errors that are intrinsic in remote monitoring, it remains essential that the methodologies employed allow an appropriate assessment of the MIF. Such interpretations can occur over a range of domains from the simple presence/absence, through to approximate or accurate determinations of the numbers (quantitative) or types (qualitative) of microorganisms functioning both at the MIF site and along the conduit through which the water passes to the sampling site. In addition to an evaluation of the biomass in terms of numbers and variety, newer methodologies are focusing more and more on the level of activity being performed by these incumbent organisms. A very small but active (aggressive) microflora may have a greater impact on the dynamics of a particular MIF site than a large, but passive, microflora. These impacts may range through shifting hydraulic transmissivity (e.g., plugging and/or facilitative flows) to fluctuating bioaccumulation, biodegradation and sloughed releases of specific chemicals. Such complexities in the chemistry and hydrological characterization of product (sampled) water through this range of interactive events renders scientific interpretation that much more challenging where an MIF event is occurring. Indeed, the conjecture can be proposed that, where there is a randomized significant departure within an individual or a group of characteristics, an MIF event should be considered as a possible significant factor.

The development of an approach to determine the presence and significance of an MIF event can include a number of strategies. In summary, these approaches may be summarized into two pathways: (1) indirect, and (2) direct.

Indirect pathways of determination do not involve the direct determination of the biological entities which may be involved but, rather, examine shifts in the physical and/or chemical characteristics of the product water which could be attributable to any form of biological activity.

Direct determination of the involvement of biological entities within a groundwater system can involve many techniques. These include direct microscopy (possibly coupled with staining techniques), selective culturing and subsequent enumeration, chemical determination of specific compounds directly associated with microbial cellular activities, or an evaluation of the potential aggressivity of the entities when presented with standardized environmental conditions.

All of the methodologies being described are dependent, in part, for their validity on an acceptable system being employed for the sampling. Given the frequent remote location of the actual MIF focus site, it is very important that the sampling procedure takes into account the need to sample or "acquire" water from the suspected fouled site. There is, by the nature of the conditions prevalent, always going to cause a level of uncertainty that the sample obtained reflects accurately the microflora occurring at the site of concern. One net outcome of this uncertainty is that the absence of microorganisms from a given sample does not necessarily exclude these organisms from being involved in the biofouling event under investigation. It could simply mean that none of these entities happened to be present in the water at the time of sampling. Conversely, a positive indication for the presence of a given microorganism could be a reflection of a "contamination" event in which the identified organism entered the water from a site other than the fouled zone (e.g., downstream econiche point of use device). The degree of uncertainty can be reduced by more frequent and diligently controlled sampling program.

Sampling does not mean turning on the faucet or the sampling port and taking the first water that pours out of the port. A sequence of microbial fouling would have occurred, in all probability, in the discharge line, the valves and the pipes leading to the sampling site. These may be summarized as (from the moment of discharge going back into the water system):

- Within the cavity and orifices through which the water finally discharges, there is usually a buildup of microbes that form tight crystallized biofilms. A highly oxidative site in which the nutrients may come partly from sloughing materials coming downstream as the water is discharged, and any materials that might become attracted to the moistened biofilms growing within the discharge zone. Under severe circumstances, thread-like slimes may hang down from the port or the crystallized biofilm may extend to the outside of the port as a white or grey crust.
- The valve assemblies which close off the port become a zone where there is violent turbulent flow (valve open) or a stagnant reductive state (valve closed). Such a condition is likely to lead to the growths of anaerobic biofilms that become radically disrupted when the valves are opened.
- Behind the valves, is a "dead end" of water leading back to the main body of flowing water in the pipes. This zone has no flow except when the sampling port is open and so the biofilms tend to be looser and thread-like extending out into the stagnant water itself. The flowing water in the main line provides nutrients and oxygen to the "dead-ended" fouling. When the sampling valve is opened, much of these loose biofilms within the "dead end" pipe are flushed out.

There can, therefore, be a number of "flushes" of microbial growths when sampling through a sampling gate valve or tap. These flushes have the potential to give false information on the microbiology of the water in the main lines or contained body of water. The normal control of this is the use of a double protection. First, to reduce the discharge of living microbes from the open sampling port, boiling water or heat is applied to port using the standard recommended methods. This kills, or severely traumatizes, the microbes that had been growing in the crystallized biofilms and slimes downstream from the valves. Second, it is necessary to run the water for long enough to flush out any "dead-enders" that have been growing in the sampling line and on the valves. Normally, ten minutes is commonly considered enough time to complete this flushing action. Note the color of the water since this may be a good clue as to when the "dead-enders" have been flushed out. These microbes will accumulate iron and color the water a yellow to reddish brown color. Once they have been flushed, the color should decline to the background level of the main body of water to be sampled.

Sometimes, the operators of the water system become annoyed that water is being allowed to freely run for ten minutes prior to sampling and complain about the wastage. This wastage is nothing compared to the cost of the biofouling that can be occurring within the water being sampled and it is important to get reliable data on the water, not on the dead end line to the sampling port.

Water well sampling offers additional challenges because there are even more zones of biofouling arrayed within and around the water well itself. To gain knowledge on the biofouling in water wells, it is therefore necessary to undertake a sequence of sampling while the well is being operated and allowed to freely discharge for sampling. Common sequences involve shutting the well down for a preferred 24 hours (7 days would be the ideal) to traumatize the microbes causing any biofouling. Before the shut-down, one sample (control) is taken with a sequence of additional sampling when the well is started up again. The three possible scenario include:

Double-two

Water samples are taken before the well is shut down, and two minutes (2m) and two hours (2h) after the well is started up again. The 2m sample gives a good indication of the types of biofouling occurring within the well water column when compared to the "control" sample. Biofouling around and beyond the well screen may be observed by looking at the microbial characteristics of the 2h sample and comparing with the 2m and the control. An appreciation of the location and form of any biofouling can be obtained in a crude manner using the double-two strategy.

Triple-three

This is a more intensive sampling program involving sampling after three minutes (3m), three hours (3h) and three days (3d). While the disadvantage is clearly the length of the testing period, the advantages are that a better "picture" of the biofouling is obtained. When compared to the data for the control, the 3m would still reflect the biofouling occurring in the well's water column. The 3h sample would reflect the type of bacterial fouling occurring at, and behind, the well screen in the pack and formation. Background levels of biofouling would be assessable using the data from the 3d sample.

BAQC-sixteen program

For a high-value well (such as a municipal, irrigation or industrial well), a more detailed picture of the locations of the biofouling events in and around the well can be obtained using a sequence of sixteen samples including the control. Sampling sequence recommended for this program is (m-minutes, h-hours, and d-days): 2m, 5m, 10m, 20m, 30m, 1h, 2h, 3h, 4h, 6h, 9h, 12h, 1d, 2d, 3d, and 4d. Comparison of this data allows the position of the different biofouling zones in and around the well to be more precisely positioned along with the normal level of background bacterial activity that can be expected.

These sampling programs are designed to allow an increasing level of certainty to be obtained by structuring the sampling program to increase the potential for recovering organisms from any focused sites of the fouling. Of these three techniques, the double-two is the least sophisticated and the BAQC-sixteen is the most comprehensive. Where water samples have now been taken, the next concern is the nature of the bacteriological examinations that will then be conducted.

In the above premises, it is assumed that the major investigation may be via an investigation of the water samples obtained. Another investigative route is by downhole observation. Camera logging for the presence of biological growths within the water column or attached to the casing and well screens can be achieved using a submersible video-camera. Where these growths are attached, they become readily recognized as large, sometimes mucoid or plate-like structures that often extend out into the water column. If the growths are dispersed in the water, these may appear as visible floating (often "fuzzy") suspended particles or as a general cloudiness. These "floaters" often appear to be locked into a gel created by the colloidal nature of the water.

The color of the observed downhole growths may reflect whether there has been iron and/or manganese accumulation within the growths. Deeper orange-red to brown colors may reflect an iron dominance while blackening may indicate either a high manganese content or the presence of large concentrations of metallic sulfides. White slimes would indicate a low iron situation at that site.

The absence of any such growths when camera-logging a well does not automatically mean that there has been no biofouling of the water well and the surrounding aquifer systems. It could be that the site of the biofouling may be more distantly located from the well and, therefore, not directly observable. When such a remote site undergoes a sloughing event, some suspended particulates may be observed in the water. This is more likely to happen when the well has been taken out of production causing the natural microflora associated with production to go into trauma and start sloughing. (When the environment is radically changed by taking the well out of production, the biofouling bugs pack up their suitcases and move on!)

Some attention has been paid to the determination of microbial biofouling by the admission of devices on, or within, which attachment and growth would be encouraged to occur. Such systems may include glass, plastic or metal pristine sterile surfaces onto which any transient microbes may attach and possibly grow if conditions are favorable. Such growths may be examined for visually or by various microscopic techniques. Such down-hole recovery/incubation systems offer the advantage of replicating the conditions present in the well's water column. The disadvantages are, however, that the system has to be left downhole for the incubation period which may extend into weeks or even months if an adequate time is to be allowed for attachment and growth. Following this, the unit has to be recovered (at the appropriate time) and observed.

Another approach to overcome the inconvenience of downhole incubation is to divert some of the water being pumped from the well through a device filled with porous media relevant to the conditions associated with the well. For example, where a water well was installed in a sandy aquifer, then the porous medium would be a sterile sand medium. Water diverted through this porous medium would carry any transient microorganisms, some of which may attach or become entrapped in the porous medium. Two subsequent events would consist of physical (due to the entrapped particulates) and biological (due to the growth of biofilms within the porous medium) plugging. Both events would cause a loss in hydraulic transmissivity through the device and changes in the visible characteristics of the porous medium. The use of glass walls around the porous medium would allow a direct observation of these visible changes while changes (reductions) in the rate of discharge under standardized flow conditions could be used to record these effects. Where glass walls are used, the device should be kept out of direct light to prevent the covert nuisance growth of algae (as green slimes) which would be normally atypical for the well.

Given that a water sample has been taken and there is a need to determine the potential for a microbiological presence to be occurring, a range of strategies can be followed. These are summarized below and subsequently the various relevant methodologies will be dealt with in detail.

DIRECT INSPECTION

A clear water sample has traditionally been considered to be one that is free from major microbiological problems (particularly when it is also coliform negative). Where there is an initial cloudiness which clears steadily from the bottom upwards, this may be due to dissolved gases (under pressure) becoming released (i.e., the cloudiness) as bubbles rise and disperse. Where cloudiness gradually forms within the water sample after collection, this could be the result of microbial activity and growth in the sample itself. If the water sample came from an econiche where there had been an incomplete biodegradation of the organics because of such factors as anaerobic conditions, the taking of the sample would introduce aerobic conditions that would stimulate aerobic degradation. If the water is cloudy initially and it settles out leaving the water above clear, this may represent a denser type of particulate structure which may have originated from biofilms within the biofouled zone. Here, the density may be a reflection of the bioaccumulation of various organic and inorganic compounds together with the formation of "tighter" water holding structures within the growths. If there is an accumulation of inorganic compounds, this might influence the color and texture of the settled material. Ferric (iron) and manganic salts will, where these are bioaccumulated, cause the settled material to vary from a yellow to orange to red to brown to black coloration. In addition, the settled material may become more "flake-like" when shaken up.

Another direct inspection involves the detection of any odors in the water. These smells are sometimes very characteristic for different microbial events. Common odors microbiologically generated include "rotten eggs," "kerosene-like," "earthy-musty" and "fishy." All of these when detected can indicate that particular groups of microorganisms have been active within those waters. Perhaps the most serious odor is "rotten eggs" which is associated with hydrogen sulfide production, anaerobic conditions and corrosion of equipment and installations.

Cloudiness is, of itself, an indicator of possible microbiologic problems. Standard approaches involve the measurement of the degree of interference or reflectivity that occurs when a beam of light is passed through the water sample. Decreases in the light intensity passing through the water (absorbance) or increases in the light being reflected as it passes through the water (reflectance) can be used to estimate the density of particulates in the water.

Newer technologies utilize interferences that occur in a pattern of pulsed laser light passed through the water due to the presence of particulates. Laser particle sizing systems are not only able to count individual particles but also ascribe a size and, on some occasions, the shape. The typical range for such laser driven systems is from 0.4 microns (less than the size of a typical vegetative bacterial cell) to as large as 120 microns or more. The total volume for the particles can be calculated and given as ppm (v/v) of total suspended solids (TSS). Frequently, the sizes of the particles are placed into bins so that the average size of the particles and their distribution within the bins can be assessed. Often, when water has been treated, the aggregation of particles will shift to the smaller bins. This can be observed by the percentage of the total volume within each of the bins.

Often in cases where there is a biofouling event occurring, clustered particulate sizes may be recorded. These clusters may typically be over a relatively narrow range of particle sizes such as (microns): 8 to 12, 10 to 15, 12 to 16, 14 to 18, 16 to 24 and 18 to 26. Where there are ribbon-like particles (e.g., stalks of *Gallionella*) present in the water, these will often cause spurious clusters of particles in the larger sizes such as (microns): 31 to 34 and 59 to 64. The laser-driven particle sizing offers the potential to "view" any suspended material down to the size of a bacterial cell or smaller. A water sample showing a <0.009 ppm of TSS and recording no particles larger than 0.5 microns may be considered to be a water sample with a very low probability for the presence of microbial entities. It should be noted that very few waters are totally free of particulate material that will be recorded using the laser particle counter. Source water free of particulates is not easy to find. In Regina, Canada, the most suitable source is commercial bottled distilled water. Regular deionized and distilled water made in the laboratory has been routinely found to contain some particulate material. All water for use as diluent in LPC work is subjected to confirmatory evaluation to ensure the absence of any suspended particles.

INDIRECT INSPECTION

Other techniques of microbiological examination involve a more indirect approach that uses either a generalized or differential staining, determination of metabolic activities related to active microorganisms, or enumeration of cells able to grow on selective culture media under specific environmental conditions.

In the last century, there has been a movement through the use of staining techniques to microscopically confirm a microbial presence in water. This can be done through selective cultural techniques to sophisticated "bio-marker" systems which are able, under some circumstances, to detect the level of activity, viability or specific microorganisms of particular interest and concern (e.g., coliform bacteria in water samples).

Historically, the focus of attention in groundwater microbiology has been influenced by the development of medical microbiological monitoring technologies. In the latter part of the nineteenth century, the occurrence of iron bacterial infestations in water wells and distribution systems triggered a level of effort to determine the causant agents and the factors controlling these occurrences. In the late nineteenth century, an applied microbial ecologist, Sergei Winogradsky, developed selective media and determined the cultural conditions that would support the growth of these organisms. The elegant cell forms of some of the iron bacteria (e.g., the sheathed and stalked iron bacteria) led to these becoming a center of interest at that time. The ecology of groundwater systems and water wells did not develop significantly since the iron bacteria were considered to be nuisance bacteria of relatively little significance when compared to the very major significance of the coliform bacteria as hygiene risk indicator organisms.

Coliform detection methods were developed during the early part of this century and became well established so that now the presence of these bacteria is considered to be the prime indicator organisms for hygiene risk in waters. This mindset established the notion that the coliform test was an indicator for all nuisance bacterial activities. Consequently, the presence of coliform bacteria has often been taken to indicate that there is a bacterial problem while the absence of coliforms would indicate that there was not a microbial problem. Unfortunately, many of the bacteria associated with "nuisance" events in groundwater systems (e.g., plugging, corrosion, taste, odor and color problems) are not, usually, coliform bacteria. As a consequence of this, in practice, the absence of coliforms from water has been commonly construed to indicate that there were no bacteria present. This would then be interpreted to mean that the problems observed were of a "chemical" nature rather than biological.

Today, it is well documented that microorganisms are prime factors initiating corrosion, plugging, and the accumulation or degradation of specific chemicals. Such microbiological events have traditional been downplayed because of a well-meaning but, in this case, misdirected over-reliance on the coliform test as being an indicator of "nuisance" as well as "hygiene-risk" bacteria.

The net effect of this strategy of over-reliance on the coliform test has been that less attention has been paid to the ecological implications of microorganisms present and active in groundwater and groundwater-related engineered systems. A surge of research and developments was initiated in the mid-1970s and has progressed to date, but with a low level of funded effort. In the two decades, a surge of interest occurred generated, in part, by the growing recognition of the importance of managing the microbial components active within the hazardous waste control and biodegradation industry. Additionally, studies have been conducted on deeper subsurface microbiology. Present findings indicate that the crust of the earth is populated by microorganisms which have somewhat different characteristics to those found at, or close to, the surface of the planet. This "lays to rest" the second impediment to the recognition of microbial activity in ground waters which is that "groundwater is essentially sterile." Here, the term "sterile" has been confused with the term "hygienically safe" (i.e., free from detectable coliform bacteria).

The net result of the recent initiatives and the overall "rebirth" of microbial ecology is that the technologies for microbial monitoring and manipulation are in a state of growth. Traditional testing methods are rooted in developments that happened with the medical and food industries and were transferred as being applicable to the water industry.

The indirect technique most commonly used to enumerate bacterial numbers is the extinction dilution techniques followed by the spread-plating of the dilutions onto agar culture media in Petri dishes. These are now incubated at an appropriate temperature and in conditions that would favor the targeted microbial group(s). Such "favored" organisms are expected to metabolize, grow and form visible distinctive clumps of growth called colonies. Each colony is thought to have arisen from a single colony-forming unit (cfu). Often the assumption is made that each colony arose from a cfu that was, in fact, one microbial cell. The net end product of this form of analysis is that each cfu counted is reported as the number of cfu which could be back-calculated to have occurred in one milliliter or one hundred milliliters of water (i.e., cfu/ml or cfu/100 ml respectively). While there is comfort in obtaining a numerical value for the numbers of microorganisms of a specific type, there are some serious shortcomings in the application of this monitoring technique. These potential inadequacies relate to several aspects of the approach used and are discussed below.

VARIABILITY IN CELL NUMBERS IN A CFU

In waters, and particularly in ground waters, much of the microbial cell population may be incorporated into particulate (often in colloidal) structures. The diameter of such particles may vary considerably from as small as 4 to 8 microns, or as large as 80 to 120 microns. Small particles may contain no viable entities, a single cell or a few cells dispersed or clumped together within the particle. Large particles could contain far more microbial cells since there would be a larger volume of occupancy. In the extinction dilution technique, these particles become diluted within the diluent (e.g., Ringer's solution). This dilution may cause shifts in the size of the particulate structures due to disruption (particles break up into a number of smaller units) or aggregation (particles lock together to form fewer larger units). Additionally, particles may become temporarily attached to the walls of the apparatus (e.g., pipette) leading to the loss of those particles from the water phase permanently (if the particle does not detach) or temporarily (if the particle does detach later). In the latter case, the particle could appear in a much greater dilution of the series to cause an anomalous cfu growth. All of these events may cause the spread-plate data to be artificially low or high. While costly, the use of a fresh sterile pipette for each sequence in the dilution series would reduce the risk of this occurring.

Low cfu counts may be a reflection of the concentration of many of the viable entities within each particulate structure which may then produce only one colony incorporating the growth of one, some or all of the different types of microorganisms present in the particle. Additionally, low counts may also occur because some of the particulate structures have become attached to the surfaces of the diluting apparati and are no longer available to be recorded as cfu on the spread-plate surface.

NUTRIENT CONCENTRATIONS IN AGAR CULTURE MEDIA

The development of enumeration techniques was largely developed in the medical and food industries to determine the numbers of pathogenic and spoilage microorganisms respectively. These groups of organisms both infest environments that are generally very rich in nutrients, and consequently, the agar culture media were also high in nutrients. Commonly, the major sources of organic nutrients are added in the percentile concentration range of 0.1% to 5.0%. In groundwater systems, it is rare (except in the case of a severe organic pollution) to find environmental conditions in which the organics are in concentrations in excess of 0.001% (or 10 ppm). The exposure of such "deprived" microorganisms from such nutrient-poor environments to the nutrient-rich agar culture media can be traumatic or toxic. These organisms are termed **organo-sensitive** (i.e., not tolerant to radically higher concentrations of organics). Such negative responses cause the resultant enumerations to fall short of the true estimate of cfu (i.e., false negative). The trend in groundwater microbiology is to take into account these sensitivities and use agar culture media in which the nutrient concentrations have been reduced. Typically, the R2A or R3A agar medium used extensively in the enumeration of waters may contain only 10% or 1% (e.g., R2A/10 and R2A/100 respectively) of the nutrients normally recommended for use in this medium. Higher cfu recoveries can be achieved which reflect more accurately the number of viable microbial entities in the water sample.

ENVIRONMENTAL CONDITIONS FOR INCUBATION

While there is a well acknowledged selective effect through the type of agar culture medium used to enumerate microorganisms, there is another major consideration that relates to the conditions that the charged agar spread plates are incubated under. The microorganisms in a groundwater sample have almost certainly come from an environment in which the physical (e.g., temperature, pH, redox) and chemical conditions are relatively stable. Additionally, the intense consortial nature of growth for these incumbents would heighten the interdependence of each component strain on the other members. Clearly, there would be a radical shift in the environmental conditions on the surface of the spread-plate. Minimizing the shifts that would occur during incubation can lessen the shock this may create. Factors that are relatively easy to control include time, temperature, atmospheric composition and the composition of the culture medium.

Time is perhaps one of the most perplexing factors because there is a natural conflict between the need to get the data as quickly as possible and allowing enough time for the incumbent microflora to adapt and grow. This period of time for adaptation can vary considerably depending upon the nature of the organism, the form of consortial relationships and the animation status (i.e., active or passive) of the cells.

Animation status relates to the level of metabolic activity going on in the cells at the time of sampling. Cells that are actively growing and reproducing would be considered to be in a very active animation state while those that are totally passive would be considered to be in a suspended animation state. Generally, the less active (animated) the organism, the longer the time (adaptation phase) the organism will require before it begins to form a colony on a suitable agar culture medium if it is able to do so at all. Since it is unlikely that the microflora will be in a very active adaptable animated state in groundwater, there is bound to be a significant adaptation phase before growth. The first colonies to appear on an agarspread-plate are likely to be those that had the shortest adaptation time and these may not necessarily be dominant species within the sample. Other organisms would sequentially appear in order of increasing adaptation times. From the chronological perspective, the greater the delay (i.e., the length of the incubation period) before counting the colonies that had grown on the spread plates, the more complete would be the determination of the microflora.

Expediency often dictates a rapid "turn around" in the enumeration of microorganisms (i.e., "time is money"). This incorporates a "risk" that all of the incumbent organisms able to grow under the conditions of the test would not have had enough time to do so. In a radical condition where all of the microflora are in a low state of animation, a test may register 0 cfu/ml simply because the incubation period has been too short to allow these organisms to adapt, grow and develop into colonies which can then be enumerated. Prolonged incubation for spread plates is advised beyond the periods commonly practiced (e.g., 2, 5 or 7 days). Experiences in Regina would suggest that enumerations should extend to a minimum of 14 days with 21 or 28 days of incubation continuing to show a greater number of cfu present. Often the number of colonies rises in a step-like manner over periods of as long as six weeks (42 days).

Temperature is another major physical factor that can affect the range and rate at which the microflora adapt and grow on the agar culture medium. A common feature of groundwater systems is that the ambient temperature does not fluctuate to any great extent on a daily or seasonal basis when compared to surface environments (unlike surface waters). The net effect of this is that the incumbent microflora has adapted to these narrow ranges of temperature. Enumeration involves a number of stages that can be, to a varying extent, traumatic to the incumbent organisms. Typically, these stages would include temperature shifts during sampling, storage, preparation of the groundwater sample for testing and finally the temperature of incubation. Each of these stages could, by the nature of the temperature shift occurring, cause a traumatization of at least some component within the microflora.

For example, groundwater is sampled from a formation in a bio-zone operating at 8.3°C. This water mixes with groundwater from other formations during entry into the water well and subsequent pumping. This mixing causes the temperature to elevate to 11.6°C which is the temperature of water sample when taken. The ambient air temperature is 24.6°C and the water sample temperature rises to 14.1°C before the water is stored, packed in ice, in a cooler. Here, the temperature declines to 7.5°C over six hours. Once in the laboratory, the water sample is left standing for two hours at room temperature (22°C) before the extinction dilution is carried out. Water temperature rises to 17.4°C. The technician now performs the dilutions using sterile Ringer's solutions just removed from the refrigerator (4°C) which cools the water down to 8.3°C. After five minutes, the diluents are spread out over the selected agar culture medium which had just been removed from the refrigerator (4°C). The cell temperature falls rapidly to 4.2°C. Once the plates were moved to the incubator (30.3°C), the cell temperature rose rapidly to that of the incubator. After five days, the colonies were counted. This example shows that the microorganisms would have been subjected to a series of temperature shifts that would not have been experienced in the natural setting. Not surprisingly there would be a reduced potential for such organisms to grow under the conditions presented on the surface of the agar spread-plate.

In essence, the most appropriate manner to handle a sample would be to, where there is a relatively short storage and transit period, keep the sample at as close to its original site temperature as possible. In the case of example above, this would be 8.3°C. Secondarily, the temperature selected for the enumeration should be within the same general range as that of the original habitat. In the example shown above, 30°C was used to incubate the sample. This selected temperature is well outside of the immediate range of temperatures around which an organism naturally functioning at 8.3°C may always be expected to function.

As a guideline to the selection of suitable temperatures for the enumeration of the natural microflora functioning at a given site, the following structure can be employed (temperature range of the habitat, °C; incubation temperature range, °C low - high):

- 2 to 10°C; incubate at 8 - 25°C;
- 11 to 15°C; incubate at 12 - 25°C;
- 16 to 30°C; incubate at 25 -30°C;
- 31 to 40°C; incubate at 35 - 37°C;
- greater than 40°C; incubate within 5°C of the natural source temperature).

It is generally thought that the higher the selected incubation temperature, the faster the rate of growth. However, in groundwater situations, given the alien nature of the enumeration procedure, it would be preferable to allow a minimum of seven-day incubation (longer if possible).

Where there is an urgency to obtain data, early colony counts may be performed and the data referred to as **preliminary cfu/ml**. Such data should be "backed up" with **confirmatory cfu/ml** calculated from the number of colonies obtained after prolonged incubation where the colony numbers would, by repeated counting, be found to have stabilized. In general, a "plateau" in colony numbers may be reached in 21 to 42 days.

pH is a physical parameter that is well known to influence the types of microorganisms which may be grown on a spreadplate. Most microbial activity commonly occurs within the range from 6.5 to 8.75 pH units with optimal growth frequently evident at pH values from 8.2 to 8.4. Most culture media tend to have the pH adjusted to within this range with 7.2 to 7.4 pH units commonly being employed. Where the pH of ground waters falls outside (commonly more acidic) of the normal range for microbial activity, organisms within a biofilm often do have the ability to "buffer" the pH shift. This allows the pH of the biofilm itself to remain in a suitable range for the activities of the consortium of organisms present within the film. When groundwater pH values are acidic but not below 3.5, there is a possibility that a "buffered" biofouling may still be occurring in association with that system. Regular culture media may, therefore, be adequate to enumerate these organisms. For the more acidic ground waters (pH <3.5), there is likely to be a dominance of acidotrophic bacteria growing and specialized media will be required to enumerate these organisms.

Microorganisms are often very sensitive to the reduction oxidation (redox) potential of the environment in which they are growing. For many of the groundwater microorganisms, conditions are borderline between oxidative and reductive with oxygen in short supply (often ranging from

0.02 to 2.5 mgO$_2$/L) and respiration switching between oxygen and nitrate as the terminal electron acceptor or turning off if there is no substrate. In the typical aerobic spread-plate count, the surface of the agar is now exposed to atmospheric oxygen that saturates the surface agar strata. These conditions are more oxidative than the microorganisms may have been accustomed to growing in. As such there may be a period of adjustment prior to organisms growing in this very different environment. The atmosphere around the spread-plate can be modified prior to incubation to render a reductive condition through the removal of oxygen from the incubator. On occasions, the replacement gas may be a combination of carbon dioxide, methane, nitrogen and hydrogen in various ratios. These conditions now allow many of the anaerobic oxygen sensitive microorganisms to grow and form countable colonies. The selection of an "open" incubation will encourage aerobic growth and exclude the growth of any oxygen sensitive organisms (i.e., the non-aerotolerant anaerobes). Selection of a replacement reductive atmosphere will shift the growth forms to some fraction of the anaerobic population. The nature of these growths would be dependent upon the nature of the gases and culture medium used.

Traditionally, the agar base used in the "solid" medium has been accepted because of the convenience created by: (1) the distinctive colonial growths which evidence the visible presence of each cfu; and (2) the agar is not commonly degraded by microorganisms. This latter factor has supported the widespread adoption of agar for the evaluation of the microbial components in environmental samples. Only rarely will the agar be found to "liquefy" because of the activities of these "agarolytic" microorganisms. The convenience and historic use of agar-based culture media has led to the acceptance of data gathered by this route without perhaps an adequate consideration of the various limitations, some of which are discussed above. Frequently, the presence or absence of microbial activities is based on the application of a spread-plate evaluation rather than on a broader spectrum evaluation of microbial activities by direct and indirect techniques. The next sections of this chapter address the methodologies that can be employed to monitor the potential for microbial occurrences.

MICROBIAL EVALUATION, A Conceptual Approach

There are two fundamental methodologies for the evaluation of a microbial presence within a groundwater system, direct and indirect. Direct examination involves the determination of biomass presence at the molecular, cellular or conglomerate (particulate and biofilm) level. Indirect examination involves the determination of viable entity presence through subsequent cultural and/or metabolic activities of various types.

Traditionally, the concept has involved enumeration of the viable entities present within a given water sample expressed as either **viable units (vu)** or **colony forming units (cfu)**. The volumes commonly used for these determinations are either one milliliter **(/ml)** or one hundred milliliters **(/100ml)**. Where the concern is hygiene risk (potential presence of pathogenic microorganisms), the more sensitive scale (/100 ml) is commonly employed. For the "normal" microflora within the environment, there is a greater tolerance for these populations and these are normally expressed per milliliter (/ml). The term cfu is normally attributed to any enumeration involving the enumeration of colonial growths on an agar culture medium. This may be a generalized enumeration in which all of the visible (by eye or low magnification stereo microscopy) may be counted **(total colony count)**. Alternatively, a selective enumeration may be undertaken where only the **typical colonies** are included based on the specific criteria documented as appropriate to the selection procedure. Other colony forms that do not follow the criteria established for the typical colony count may be enumerated separately as **atypical colonies**.

VIABLE UNITS (vu)

These units do not involve the formation of distinctive colonies but rather the generation of some product of growth (e.g., cells to develop turbidity, organic acids to lower the pH). By sequentially diluting the water sample (usually in tenfold series or a statistically appropriate manner) in sterile culture medium, this will allow the required product to be generated where viable units were present and recognized. By noting which dilutions reacted because of the presence of active viable units, it becomes possible to calculate statistically the theoretical number of vu/ml or vu/100 ml. The most common application is in the determination of total coliform and fecal coliform bacteria by the **most probable number (MPN)** method.

AGGRESSIVITY

Another gauge for the determination, in a semi-quantitative manner, of the population size and activity level of a targeted microbial population is aggressivity. In this method, the water sample is placed in an environment (e.g., liquid selective culture medium) under incubation conditions that would encourage the growth and activity of the targeted microorganisms (e.g., iron-related bacteria). Here, the rate at which the onset of a determinable event (e.g., gas production, color generation, and clouding) occurs is considered to relate to the aggressivity of the organisms within the sample. For example, a very small population of microorganisms that are very active and rapidly generate a determinable event would be more aggressive than a large population of microorganisms that are in a passive state and do not respond to the test environment so quickly. It may be argued that the small population is much more aggressive potential than the passive larger population and, as such, present a greater fouling or hygienic risk.

SIMPLE FIELD TEST METHODS

When working in the field remote from laboratory support facilities, microbiologists have a range of simple techniques that have been developed which can give an indication of whether there are microbially driven events occurring (e.g., MIF, MIC). These techniques are by nature very simple and in some cases involve locally available material. Incubation is at room temperature as a matter of convenience.

Rodina Test

This test was described in 1965 and it functions through the ability of some IRB to generate a flocculent type of growth when the water is subjected to very aerobic conditions (Figure 23). The test itself consists of pouring water into a wide-necked flask or bottle and leaving the water to sit overnight. Do not fill the container to a depth of more than 10 cms. Here, the wide neck of the vessel generates a large surface area through which oxygen can diffuse into the water. This increasing oxygen presence together with a non-turbid environment causes some IRB to form into growths that resemble discolored cotton wool in appearance. Where this occurs, the presence of IRB can be suspected and confirmed by a direct microscopic examination of the growths.

Cholodny Test

In 1953 the Cholodny test was described where the IRB could be seen if these organisms attached to glass (Figure 23). The method recommends that some of the water and any sediment recovered be placed in a jar. A cork is then floated on the surface of the water with several cover glasses (slides or cover slips could be used as alternatives) attached on the underside. As the water clears through sedimentation, the glass attached to the cork becomes visible. The development of any "rust spots" or "cotton-like accumulates" either on the surface of the sediment or on the glass inserts is taken to indicate the presence of IRB. Where there were very large and aggressive populations of IRB, such events could occur in less than 24 hours of incubation. Where identification of the IRB was desirable, the glass inserts could be removed with the cork from the water, detached and air-dried. Upon staining using one of the recommended IRB procedures, the types of IRB present and active in the sediment-water can be identified.

Grainge and Lund Test

This test is identified with the need to develop a very simple test in which iron is added to the water sample in such a way that the activity of IRB could be easily identified (Figure 24). This technique was described in 1969 and recommended as a monitoring procedure for the effectiveness of control programs. A clean soft steel washer (preferably chemically cleaned) is placed in a conical flask (e.g., Erlenmeyer flask) along with an extruded plastic rod (e.g., stir stick) which is positioned vertically. The water sample is added sufficiently to cover the washer leaving the end of the plastic rod sticking out of the water. After being left for two days, the water line around the plastic rod is examined for any translucent string-like (filamentous) growths. Over time this growth develops a brown tinge (iron accumulation) and interpreted as positive indication of the presence of IRB.

GAQC Test

Otherwise known as the George Alford Quick and Crude test, this test involves the use of materials that are easily at hand to determine whether IRB are present or not (Figure 24). The method, developed in 1980, involves the admission of a non-galvanized iron washer to the water sample along with a carbon source (ethanol) to stimulate the activity of the IRB. To test a water sample, 150 ml is added to a clean glass and the washer dropped into the water to come to rest at the bottom. Two drops of Jack Daniels® whiskey are added to the water ("one drop for me and one drop for the bugs", George Alford, personal communication). Cover the glass loosely with aluminium foil (to reduce evaporation) and incubate on top of a refrigerator (warm), on a shelf or windowsill. After two days, a positive reaction is recognized by either a "fuzzy" growth around the washer or metallic "floaters" in the water. In both cases, IRB are actively either attaching directly to the source of iron (i.e., the washer) or accumulating the dissolved iron within suspended particulate masses (i.e., the floaters).

BART™ Test (Waverly Scenario)

Since 1988, the Town of Waverly in Tennessee has been using the BART™ testers to determine in a simple manner whether there was a recurring MIF problem developing in the water wells. This chapter is devoted to a comprehensive description of these various testers and Chapter Five gives examples of their application. The Waverly scenario represents the first application of the testers and so is included here as an example of a simple test method.

In the simple scenario developed at this time, tests were performed monthly on the water from the wells using the BART testers. Experience found, in this case, that it was the three units (SLYM-, IRB-, and SRB-BART) which could be used to monitor a potential biofouling problem (increased drawdown and unacceptable increases in manganese concentrations). The key factor was the time lag (days of delay) to the first observing of a reaction and the form of the reaction being generated. Where the system was functioning efficiently, the normal time lag was between 12 to 14 days. When the time lag fell below 12 days there was, through practical experience, an increased occurrence of biofouling observed. Normally, it would be the SLYM- and the IRB- that would drop first to 9 or 10 days. To check this, additional water samples were drawn on two consecutive days and re-tested. If these both showed the same foreshortened time lag, there would be an increased risk of biofouling and corrective action (e.g., shock chlorination or acidization) would be applied. Confirmatory tests were repeated after treatment. A successful treatment would be deemed one that had caused the time lag to return to the normal level (i.e., >12 days). It was found that where the SRB- [BART] also showed a shorter time lag, the treatment had to be more intensive to return the wells back to an acceptable time lag. On some occasions the increase in aggressivity of the SRB (i.e., shorter time lag) preceded rapid increases in the manganese concentrations in the water. This may have been due to a greater amount of sloughing from a more matured biofouling in which the more deeply seated SRB components were being released along with more of the accumulated manganese.

The Waverly scenario offers a very simple comparative way of monitoring the status of biofouling within a water system. It uses the premise that if the:

- time lag is getting longer - everything is getting better,
- time lag is constant - everything appears to be stable,
- time lag is getting shorter then the microorganisms are becoming more aggressive.

In the last case, duplicate samples on two consecutive days should be undertaken to confirm that there is a decline in time lag (i.e., the original test was not a "chance" event). Once confirmed by the duplicate tests, treatments should be undertaken to remediate the MIF occurrence.

It should be noted that the patented BART biodetector system was developed by Cullimore and Alford from the MOB (metallo-oxidizing bacteria) and the MPNB (metallo-precipitating non-oxidizing bacteria) test system described by Cullimore and McCann in *Aquatic Microbiology* (1978). At that time the test method had been designed for simple field use and differentiated six reaction types. Today, that test method is incorporated into the IRB-BART biodetector. It was found during that research phase that 95% of the ground waters tested in Saskatchewan, Canada were found to be positive for IRB. Microscopic examination found that, of the sheathed and stalked IRB, *Crenothrix, Leptothrix, Sphaerotilus* and *Gallionella* were frequently dominant types. Given the universal nature of the presence of IRB it becomes more critical to appreciate their relative aggressivity that can be measured by the time lag characteristic.

MICROSCOPIC INVESTIGATION FOR IRON-RELATED BACTERIA

Olanczuk-Neyman Method

This methodology was developed in Poland while working on IRB fouling of water intakes. A staining technique was employed which used membrane filtration to concentrate the suspended particulates prior to staining with a modification of the Rodina procedure.

1. Separately dilute 10, 1 and 0.1 ml of the water sample into 110, 119 and 199.9 ml of sterile Ringer's solution respectively to create final volumes of 120 ml. Gently mix by rotating each dilution in turn.
2. Filter each diluent through a separate 0.45-micron membrane filter and fixate the IRB to the filter by immersing each filter into a 1.0-% solution of formalin for a minimum of 20 minutes.
3. Following fixation, immerse the filter in a 2.0% solution of potassium ferricyanide for 20 minutes.
4. Rinse the filter off in 5.0% hydrochloric acid for 3 minutes.
5. Wash the acid off the filter gently using distilled water.
6. Cover the filter with a 5.0% solution of Erythrosin A in a 5.0% phenol based solution. (Note that a 2.0% solution of Safranin can be substituted for the Erythrosin A).
7. Gently rinse the membrane filter with distilled water to wash off the stain and air dry.
8. Examine microscopically.

Under low power magnification, the **bacteria cells will appear to be stained red** while the **iron deposits will stain blue**. Under oil immersion examination, the membrane filter will become transparent when saturated with the oil. This makes the observation of the IRB much easier. The Olanczuk-Neyman method calls for the numbers of IRB to be enumerated using 100 microscopic fields selected randomly and this is related back to the number of IRB/ml using the formula given below. It should be remembered that much of the material being viewed microscopically is dead including the stalks of *Gallionella*. Care must be taken to determine the location of viable cells.

To calculate the IRB/ml, it is possible to calculate this using the following formula:

Diameter of the microscopes field of view:	A microns
Average number of IRB on 0.1 ml dilution fields:	B
Average number of IRB on 1.0 ml dilution fields:	C
Average number of IRB on 10 ml dilution fields:	D
Number of fields counted at 0.1 ml dilution:	E
Number of fields counted at 1.0 ml dilution:	F
Number of fields counted at 10 ml dilution:	G

Note that counts are recorded where IRB cells were present; fields observed which did not display any IRB are also included.

Calculate the IRB/ml for each dilution field (asterisk means multiplication):

$$H = B * 10$$
$$I = C$$
$$J = D * 0.1$$

Total number of field counted:

$$K = E + F + G$$

Since the fields reflect only a factorial part of the total surface area, there needs to be a correction factor which would allow the total population of IRB entrapped on the filter to be calculated (and representing the total volume of water sample used). This factorial (L) may be calculated as the relationship between the area (AV) of the field of view (diameter A given) and the area (AM) of the membrane filter (diameter M given):

$$AV = 3.142 * (A/2 * A/2) \ 10^{-12} \ m^2$$
$$AM = 3.142 * (M/2 * M/2) \ 10^{-12} \ m^2$$
$$L = AM / AV \text{ factorial area increase}$$

Averaged Population of IRB (PIRB):

$$PIRB = (H * (E/K)) + (I * (F/K)) + (J * (G/K))$$

The **APIRB** is expressed as IRB/ml.

$$APIRB = PIRB * L$$

Meyers Stain (modified by Annette Pipe)

Meyers described a novel staining technique for IRB in 1958 which has been found in the experience of Annette Pipe to be superior to the other techniques presently available. Like the Ołanczuk-Neyman method, this technique stains **the bacterial cells red** while the **iron deposits are blue**.

This technique is primarily recommended for use with water samples which are showing some signs of a possible infestation (e.g., the water is discolored to a yellow, orange or brown hue, the water has a distinct colored deposit which may have an indistinct or fuzzy outline).

1. Separate the cellular material by centrifugation to the extent that a pellet is formed in the base of the centrifuge tube. The precise level of centrifugation will vary since a greater amount of g force (gravity) may need to be applied to "draw down" the cells in some waters.
2. Pipette off the water carefully so as not to disturb the pellet.
3. Using a sterilized clean bacteriological loop, withdraw some of the pellet and smear onto a bacteriological slide.
4. Air dry the slide and heat fix by passing the slide quickly three times through the lower part of a blue bunsen flame. Be careful not to overheat the slide. Only enough heat is required to make it warm to the touch causing the bacteria to become "fixed" to the glass.
5. Immerse the slide in absolute methanol for 15 minutes. While the slide is immersed, bring to boiling point a 1:1 mixture (v/v) of 2% potassium ferricyanide and 5% acetic acid both as aqueous solutions in distilled water.
6. Remove the slide from the methanol and immerse in the boiling mixture for 2 minutes.
7. Remove the slide and allow it to cool.
8. Gently wash the slide with distilled water.
9. Cover the slide with an aqueous 2% safranin solution and leave for 10 minutes.
10. Rinse off the stain with water, dry and examine microscopically.

This technique is not quantitative but does allow a variety of the IRB to be identified as present (or absent) and a relative relationship to be expressed between the species observed. The limitation of the technique is that it does not necessarily allow a clear observation of the CHI bacteria that may have accumulated iron within irregular amorphous structures. Additionally the technique is not suitable for waters with a low IRB population. Such waters are best stained using the Olanczuk-Neyman or the Leuschow and Mackenthum membrane filter methods.

Leuschow and Mackenthum Direct MF Technique
This technique was described in the Journal of American Water Works Association (1962, Vol. 54) for the investigation of waters that had a low number of IRB. Leuschow and Mackenthum in Wisconsin recovered IRB from 55% of the well waters tested with *Gallionella* and/or *Leptothrix* being the most common dominant IRB. In the reddish turbid waters, counts

by this method reached $>10^7$ IRB /ml. Unlike staining techniques, this technique involves the direct observation of the bacteria that have been entrapped on the membrane filter that has been dried and rendered transparent with immersion oil. It is restricted, as a technique, to only those bacteria which are occurring in large and/or distinctive structures and are pigmented (commonly by the entrapped iron oxides).

1. Filter 100 ml of the water sample through a 0.45-micron membrane filter.
2. Remove the filter and dry by exposing the filter to 100°C for long enough for the filter to appear to be "bone dry."
3. Saturate the dried membrane filter with immersion oil. Since the oil has the same refractive index as the filter, the disc filter will become transparent.
4. Place the filter on a bacteriological glass slide and examine under high power (x400 to x600 diameters).

Various common forms of stalked and sheathed IRB may be easily seen by this technique even when they are present in the water in relatively low numbers.

Negative Wet Mount Stain

One major problem with the staining techniques is that the "edge" of bacterial cells, particles, sheaths and stalks may be diffuse and difficult to observe. Additionally, where the bacteria are not pigmented or have not accumulated iron, the cells may be transparent to light and so not be recognizable. The negative stain uses a different concept in that the background is stained while the bacterial cells remain unstained. Thus, the cells stand out as being illuminated zones within a darkened (stained) background. This is achieved using **nigrosin** which is an acidic stain. Such stains do not penetrate and stain the bacterial cells since both the stain and the bacterial cells are negatively charged. These stains tend to produce a deposit around the cells, which forms into a dark background against which the bacteria appear as clear, unstained regions that relate to the shape of the cells. Where nigrosin is not available, India ink (25% aqueous solution) may be used.

This negative stain is more satisfactory for waters that are showing some visible signals of possible microbial presence (discoloration and clouding) and has the advantage that all types of bacteria may be observable by this technique. There are a number of mechanisms to obtain the sample for the negative stain. Approximately 0.2 ml of suspension is used in this staining technique. Very turbid waters or centrifuged pellets made from the water sample would probably have too many cells to be clearly definable.

A regular "clear" water sample may have too small a number of cells to be conveniently observed. In either event, some dilution (into a sterile Ringer's solution) or concentration (by passive settlement or centrifugation) may be needed to conveniently view a dispersed microbial mass. Trial and error may be required to find the appropriate dilution or concentration of the water sample.

1. Using a sterile inoculating loop, transfer two drops of the sample to the surface of the clean glass slide.
2. Add two drops of nigrosin and evenly mix the sample with the nigrosin.
3. Spread the mixture evenly over the surface of the slide using the narrow flat edge of a second slide held at an angle to the first slide. As the edge is moved along over the surface of the glass slide, so the suspension should spread out in its wake to form a narrow film.
4. Lower a cover slip of the central area of the film and gently lower one end first using forceps. Gradually reduce the angle of the cover slip until it is total in contact with the film. Take care that no air bubbles are entrapped under the slide.
5. Using high power and oil immersion assisted microscopy, examine the film through the cover slip.

Cells will appear to be either transparent (if there has been no excessive accumulation of iron) or orange to brown (where iron has been accumulated. The background will appear to be pale blue. Where there is a "tight" form of ECPS around the cell (e.g., capsule, sheath or a thick coat of ECPS), this will form the distinct boundary rather than the cell itself. With careful focusing and adjustment of the light, the cells can sometimes be seen within the extracellular structures.

A **dry mount** negative stain can be made as a "library" copy that can be kept and viewed at leisure (the wet mount preparations will dry out). Where this is desirable, the slide should be left to air dry after step 3 above.

Gram Stain

The Gram stain (gRAM) is a standard staining technique used for the preliminary identification of bacteria based upon **reaction (R),   arrangement (A)** and **morphology (M)**. These characteristics relate to the staining reaction that occurs (R), the form in which the cells are arranged (A) and the cell shape (M). Generally, this staining technique is used when working with pure cultures or natural samples where there is a high density of cells and relatively little interference factors (e.g., ECPS or inorganic deposits). It is considered to be a **differential** stain because there are two major reactions: (**pink-red: gRAM negative, G-; blue-violet: gRAM positive, G+**). Waters commonly have either too dilute a bacterial community or too many interference factors to allow the direct application of the gRAM stain to a water smear. This stain method, however, is commonly used in confirmatory procedures where microorganisms have been grown in liquid (**broth**) cultures or on solid (**agar**) media as colonies.

There are many variations for the gRAM staining technique. The one described below is recommended by Harley and Prescott in 1990.

1. Prepare a smear of the colony (using inoculating needle and dispersing the cells in a drop of water) or suspension of a broth culture over a glass slide. Heat fix (pass slide quickly through a Bunsen flame three times), cool and label slide.
2. Flood the slide for 30 seconds with crystal violet, and then rinse off the stain gently with water for 5 seconds.
3. Now cover the smeared part of the slide with Gram's iodine mordant and allow it to stand for 1 minute. Rinse off gently with water for 5 seconds.
4. Gradually (drop by drop) cover the smear with 95% ethanol. Note that violet streaks will begin to form as the crystal violet - iodine stain complexes bleed from the smear. Continue to apply ethanol until bleeding stops (if in less than 30 seconds). Rinse gently with water to remove the ethanol. Prolonged exposure to the ethanol may cause too much decolorization and even the G+ cells might then become affected to appear as G- cells. Apply gently the safranin stain until it covers the smear and leave for 1 minute. Rinse gently with water, blot dry with an absorbent (bibulous) paper.
5. Examine under oil immersion (x 1,000 magnification).

If there are too many cells to clearly see each individual cell, then repeat the procedure but use less cell inoculum and /or a more diluent. If too few cells, scan the stained smear microscopically to investigate whether there has been any clumping of the cells. If not, repeat the procedure but use a greater amount of colony or do not spread the water sample across the slide and allow longer for the smear to dry before being heat fixed. It should be noted that many of the IRB do not gRAM stain well because of the interference factors which prevent the stain and decolorizing ethanol to penetrate to the cells. Additionally yeast and molds, where these are present, routinely gRAM stain as G+.

SPREAD-PLATE ENUMERATION

Tsutomu Hattori in his book, *The Viable Count, Quantitative and Environmental Aspects,* states: "Microbial ecologists are often faced with the serious problem of large discrepancies between microscopic and plate counts, especially for cells isolated from natural environments. Using a sample from either soil or aquatic environments, the former count is usually 10 to 1,000 times larger than the latter count."

The most common technique used and generally recognized as being valid is the extinction dilution/spread-plate enumeration (as cfu/ml or /100 ml) using a variety of agar media. Over the past two decades, there has been a growing debate as to which of the vast array of culture media are the most suitable. This debate has centered on the most appropriate culture medium for the enumeration of aerobic heterotrophs, hygiene risk indicator bacteria (e.g., coliforms, enterococci, *Clostridium welchii* and species of *Klebsiella, Acinetobacter* and *Aeromonas*, iron bacteria, biodegraders, sulfate-reducing bacteria and anaerobic bacteria). Considerations also range through the use of the MF technique (where there are low populations in the cfu/100 ml range) and the correct methodology for extinction dilution (where there are high populations in the cfu/ml range).

Methodologies have tended to ignore the serious concerns expressed above with respect to the discrepancies observed between microscopic and spread-plate counts. These concerns would relate to the potential radical underestimation of the microbial cell populations in a water sample using the spread-plate techniques. However, a heavy reliance has been placed upon the relative value of the population numbers recorded in a sequence of samples and the cfu/ml generated has been considered by many users to closely reflect the number of cells.

In conducting a spread-plate enumeration from a water sample, there are a number of stages that are involved after the selection of the appropriate agar culture medium:

1. Vortexing to disperse particles (optional);
2. Serial dilution of the water;
3. Dispensing of diluents onto the agar surfaces;
4. Drying of the agar surface (optional);
5. Incubation under the correct temperature and atmosphere (reductive or oxidative);
6. Counting the colonies which are visible;
7. Calculating the cfu/ml or cfu/100ml.

Each of these stages will be addressed in turn.

Selection of the Appropriate Agar Medium

Agar-agar itself is a galactoside obtained from certain marine red algae (seaweeds). Most microorganisms are incapable of degrading agar-agar. It is used at concentrations of between 1.0 and 1.5% (w/v) depending upon the formulation of the medium and the quality of the agar-agar used. Lower quality agar-agar contains up to 0.5% phosphorus (as P w/w) while higher qualities contain less than 0.2% P. This translates into a potential supplementation of the agar medium with up to 75 and 30 mg P/L respectively. This would form a significant additional nutritional base for microbial growth and activities. Indeed, water agar utilizes the nutrients inherently present in the agar-agar and the water used to make the medium. On occasions where the microflora has a significant (if not dominant) organo-sensitive component, the water agar may give the highest counts of cfu/ml when compared to other recommended media.

Agar-agar was originally selected because of two very favorable factors. First, the agar-agar was not degraded by many types of microorganisms and, therefore, relatively stable (when compared to gelatin). Second, the agar-agar would remain as a "solid" gel at temperatures up to boiling point (i.e., >90°C) which would allow incubation in the thermophilic range, and yet would not set from the molten state until 46°C. This latter feature allows microbial suspensions to be mixed into the setting agar-agar for greater dispersion occurs in the colonies that sub-sequently form. This is known as the **pour plate technique**. The method is not widely used for environmental samples because the heat shock caused by adding the organisms to cooling molten agar-agar can be traumatic.

In the cultural techniques using agar-agar, the water and supplemented chemicals are dispersed. For a colony to form, the organisms have to "mine" both the water and these chemicals from the agar-agar base. If any organism is unable to extract adequate water and nutrients from the gel base, colonial growth would not occur. This may be a major factor restricting the range of microorganisms able to grow on agar culture media.

In regions with high relative humidities where the agar medium is not subject to radical evaporation rates and drying out, 15 ml is a common volume to dispense a molten agar medium into a standard Petri dish. However, in dry climates such as is commonly experienced in the American prairies, a greater volume of agar (20 to 25 ml) may have to be dispensed to compensate for the greater losses. Bagging the poured plates in a plastic bag may not function well since water can condense on the inside walls of the bag and drip back into the dish causing contamination and erroneous results. Inoculated agar plates are incubated upside down (agar facing downwards) so that water cannot condense on the upper surfaces and drip back down onto the growing colonies.

Each agar culture medium is developed to enumerate a particular spectrum of the microflora in the water and no single medium is capable of stimulating the active growth of all of the microorganisms that could conceivable be present. There are therefore a broad spectrum and selective forms of agar media that serve very different purposes. The broad-spectrum media generate a set of nutritional conditions in which a wide variety of microorganisms may be able to flourish (e.g., the heterotrophic bacteria). Selective culture media implement the use of inhibitory chemicals (e.g., antibiotics), marker dyes and restrictive nutritional regimes that encourage and support the growth of a narrow spectrum of microorganisms. The various agar media are listed below.

AA agar

A very selective medium for the gRAM negative strictly aerobic heterotrophic bacteria (Section 4 of *Bergey's Manual*) and, in particular, the genus *Pseudomonas*. Colonies are commonly discrete and may be pigmented, or if the pigments are water-soluble, these may diffuse into the underlying agar medium.

Brain Heart infusion agar/4 (BHI/4)

This quarter-strength Brain Heart infusion agar has been found to support excellent colonial growths of a broader spectrum of bacteria than the R2A agar medium. Colonies tend to grow more rapidly and to a larger size. It is a very useful medium to obtain a generalized enumeration of aerobic bacteria. If placed under anaerobic conditions (e.g., through the use of an anaerobic jar and gas pack), colony counts for anaerobic bacteria can be obtained.

Czapek-Dox agar (CD)

This is a broad spectrum medium for culturing many of the **fungi (molds)**. This medium is particularly suitable for the culture and enumeration of *Aspergillus, Penicillium* and related fungi. Heavy fungal populations may be present in groundwater that has interfaces with unsaturated zones (e.g., via recharge) and some level of organic input (e.g., from sewage lagoons or oxidation ponds). Typical fungal colonies appear to have a rough almost cotton-like appearance that spreads out over the surface of the agar. Visibly distinctive bodies within the mass of growth (called a mycelium) may form the sporing bodies.

LES Endo agar

This medium has been developed for the enumeration of **total coliforms (TC)**. It is a complex medium containing sodium desoxycholate (a bile salt) which will inhibit many of the bacteria that are not commonly found in the intestinal tract. Lactose is the major carbohydrate (sugar) which is fermented by a relatively narrow spectrum of bacteria, which includes the coliforms. Typical coliforms have a colony with a golden-green metallic sheen that develops within 24 hours. The sheen may be centrally located, around the periphery or cover the whole colony. This medium is often used as a basal agar for the incubation of membrane filters. It is generally considered that a valid colony count will be one where there are between 20 and 80 typical colonies per filter and no aberrant atypical overgrowths.

Lipovitellin-salt-mannitol agar

This is a medium used to determine the presence of *Staphylococcus aureus*. This bacterium is commonly a concern in confined recreational waters where bathing is practiced. When present, the colonies formed are surrounded by a yellow zone with opacity. The opaqueness relates to activity associable with the presence of coagulase positive staphylococci.

M-FCIC agar

This is a medium that is specific for the genus *Klebsiella*. Typical colonies of *Klebsiella* produce a blue to bluish grey colony. It is used to determine the presence of this genus in waters by membrane filtration. Atypical colonies are beige to brown colors. *Serratia marcescens* will grow but will be distinctive through producing a pink colony.

M-HPC agar

This is an agar medium that has been developed for use with the **membrane filtration evaluation for heterotrophic bacteria.** It is an unusual medium in that it contains glycerol and gelatin as two major sources of nutrients.

Mn agar

This medium incorporates both the ferrous and manganous forms of iron along with citrate to encourage the growth of **iron-related bacteria**.

M-PA agar

This is a medium used in conjunction with membrane filtration to detect the presence and numbers of *Pseudomonas aeruginosa*. Typical colonies are flat in profile with brownish to greenish-black centers which fade out towards the edge of the colony.

Pfizer selective enterococcus (PSE) agar

This is a medium that allows the differential culture of the enterococci (**Streptococcus faecalis** group). These bacteria form brownish-black colonies with brown halos.

Plate count agar (peptone glucose yeast agar)

This has been a major medium used to enumerate **heterotrophic bacteria** by the spread-plate method.

Potato Dextrose agar (PD)

This medium allows a broad-spectrum enumeration of both **yeast** and **fungi (molds)**. Yeast will tend to form large (3 to 6 mm in diameter) colonies that have a domed profile and may be pigmented a pink color. Some yeast emit fruity odors (esters) which are very distinctive. A broad range of fungi will also grow well on this medium.

R2A agar

This is a more complex medium that has been found to support a broad range of **aquatic heterotrophic bacteria**. It is, however, somewhat more difficult to prepare. This medium is being used more widely by spread-plate to determine the total bacterial count and is particularly supportive of many of the **pseudomonad bacteria** that frequently cause problems in controlled recreational water systems (e.g., whirlpools and swimming pools). Some laboratories use dilutions of this medium and claim to obtain greater recoveries (i.e., higher colony counts). Common dilutions are R2A/10 and R2A/100. R2A will also support the growth of some actinomycetes and, in particular, the *Streptomyces* (these are a common source of the earthy musty odors associated sometimes with taste and odor problems in water).

Starch-casein agar

This is an agar medium developed for the selective enumeration of the **mycelial bacteria** (*Actinomycetes*). These bacteria resemble molds producing cotton-like or chalky growths like fungi except that here the growths are tightly bonded to the agar and the texture is often leathery. The fluffy appearance is due to frequent production of raised spore bodies.

Triple sugar iron agar(TSA)

This medium was developed for the examination of food products for **enteric bacteria** (Section 5, Family 1 in *Bergey's Manual*) but is applicable to ground waters. Some ECPS forming bacteria will produce copious mucoid formations that may fill the void space between the agar and the lid of the Petri dish. It is a useful medium to determine whether enteric bacteria (turns yellow due to lactose fermentation, red if proteolysis occurs, black if hydrogen sulfide is produced) or slime forming bacteria (excessive mucoid colonies formed) are present and dominant.

Tryptone glucose extract agar

This is a general medium used for the culture of **heterotrophic bacteria**. A wide spectrum of heterotrophic bacteria will grow on this medium which is a minor variation of the standard plate count medium. Here, the yeast extract has been replaced by a slightly higher concentration of beef extract.

Wong's medium

It is a very selective medium for the culture and enumeration of bacterial species belonging to the genus *Klebsiella*. Colonies are pink to red in color and may be somewhat mucoid.

WR agar

This medium is based on the Winogradsky (Regina-formulation) for culturing **iron-related bacteria** (CHI group). This medium generates brown colonies for the typical heterotrophic iron-related bacteria due to the uptake of iron from the agar medium. Under circumstances where the groundwater is from regions with a high level of organic pollution (e.g., gasoline or solvent plume), the biodegraders may grow on the WR agar but produce an atypical colony type (commonly white or beige).

Choice of Culture Medium

This is dependent on the intensity of the examination to be undertaken and the perceived potential microbial activities that could be occurring in the groundwater. The selection of a medium for the determination of anaerobic bacteria remains difficult because many of the strictly anaerobic bacteria are both fastidious (require complex media) and sensitive to oxygen (require refined anaerobic dilution and cultural methods). These are difficult to culture in the routine microbiology laboratory unless it has been equipped to handle anaerobes. Anaerobic incubation of agar spread plates prepared "normally" under aerobic conditions will frequently allow the growth of facultative anaerobes and the aerotolerant anaerobes only. A fastidious medium such as BHI/4 will support many of these bacterial types. Typically, in an intensive evaluation of an aerobic fouling of groundwater, an acceptable range of media would include the following agar media:

BHI/4	-	general aerobic bacteria
WR	-	heterotrophic IRB
AA	-	pseudomonads
R2A/10	-	pre-screened with 15 min exposure to 500 ppm of chlorine as sodium hypochlorite for the *Streptomycetes*
TSA		radical slime forming bacteria and enterics
PD	-	fungi and yeast
LES	-	coliforms

Dispersion of Particulates (vortexing)

When a water sample is taken, it would be very unusual, but not impossible, for all of the microorganisms to be dispersed in the planktonic phase (separate cells freely suspended in the water). Commonly, a significant proportion of the cells will be entrapped within particulate struc-tures, many of which would be colloidal. This would be particularly true where the water has passed through a zone of biofouling and contained a high number of sessile suspended particulates. To maximize the enumera-tion of potential cfu/ml, it becomes necessary to disperse the particles so that, theoretically each particle would contain a single viable entity which would then each form one colony (cfu). This is clearly a theoretical "ideal."

In practice, the water sample can be subjected to vortical agitation using a vibrating horizontal plate that rotates. Such devices are commonly employed in microbiology laboratories to mix dilutions and suspensions evenly. There is no ideal time to "vortex" a water sample, but common practice employs between 10 and 60 seconds depending upon the clarity of the sample (clear to turbid respectively).

Serial (extinction) Dilution of the Water Sample

The objective of a serial dilution is to dilute out the water in a sterile solution (commonly Ringer's isotonic solution) so that there are sufficient viable entities to generate a statistically satisfactory number. For agar spread-plate enumeration, the usual range is between 30 and 300 colonies if the colony count is to be considered statistically valid. For membrane filters, the acceptable range is between 20 and 80 colonies.

Commonly, the dilutions are made in a tenfold series to (and often beyond) extinction. This extinction occurs when there are no colonies formed on the spread-plate due to the dilution being so great that there are no viable entities in the volume of dilutant being streaked onto an agar culture by the streak plate technique. That means there is a zero probability of even a single colony growing enumeration.

The technique for conducting a serial dilution and spread-plate are well described in the literature and will not be described in detail here. However, each diluent may be subjected to a streak plate enumeration. Diluent that are utilized are generally streaked from 0.1 ml of the diluent although some laboratories prefer to use 0.5 or 1.0 ml. In the interests of economy, it would appear to be important to only streak out those dilutions that are likely to have statistically acceptable counts. There is little point in trying to count spread plates which have too many colonies (i.e., greater than 300, **too numerous to count, T.N.T.C.**) or too few colonies (i.e., less than 30, **too few to count, T.F.T.C.**). Through the practice of the art, the appropriate range of dilutions can be selected which would maximize the probability of obtaining acceptable colony counts.

Selection of the dilution range may be initially decided using the given below. Dilutions are referred to in the tenfold dilution sequence by the symbol 10^{-x} (ten to the minus x) where x refers to the incremental dilution. For example, 10^{-3} dilution would mean three dilution increments of 10 (i.e., 10 x 10 x 10 or 1,000th dilution). In this system, x refers to the number of zeros following the one (1) by which the dilution was expedited. In general, the source and clarity of the water sample is instrumental in deciding the number of dilutions that will be performed. In simple terms, there can be three basic groups:

1. a clear water without any evidence of excessive biological activity being present;
2. a moderately clear water which may have passed through a zone of biofouling or shows evidence of biological activity (e.g., color, odor, cloudiness);
3. cloudy or colored water with evidence of fouling (e.g., septic odor, intense color, sediment).

Each of these types of water would require different dilutions to be "plated out" by the spread-plate technique if satisfactory colony counts are likely to be obtained.

Water sample group A may be expected to normally have a low bacterial population and the priority for spread plates can be the original water, 10^{-1}, and 10^{-3}. Sample group B priority for spread plating would be 10^{-1}, 10^{-2} and 10^{-4}. Group C water samples would naturally be expected to have much higher populations of microorganisms, and greater dilutions that should be subjected to spread-plate analysis are 10^{-1}, 10^{-3} and 10^{-5}. It should be noted that in some cases where is there is a larger amount of sediment and rust, very few microorganisms may be recoverable even though it would appear that the water should contain high populations. If there is a septic odor in the water and sewage pollution is suspected for a group C water sample, T.N.T.C. may occur even on the 10^{-5} dilution. In such events, a 10^{-7} should also be included.

Where the inoculum selected for the spread-plate is only 0.1 ml, then the selected dilutions should be reduced by one tenfold increment. For example, a recommended 10^{-3} would be replaced by a 10^{-2} dilution where 0.1 ml was used. Plating out duplications of each dilution will increase the confidence of the data obtained. However, it is generally regarded that at least five replicates of each dilution should be plated out to achieve an acceptable level of statistical validity.

Dispensing of Diluent onto Agar Surface (spread-plate)

Three common volumes of diluent may be elected to form the inoculum for spreading out over the agar surface. These are 1.0, 0.5 and 0.1 ml. The latter two volumes require a correction to the standard formula for computing the population as cfu/ml from the colony counts obtained. For the latter two volumes the correction factor would be times two (x2) and times ten (x10) respectively.

It is important in spreading the diluent over the agar to maximize the coverage of the agar surface with the diluent. This is so that the microbial cells may become dispersed to a maximum extent for each viable entity to have the highest possible opportunity to grow large enough to form a visible colony. The most common technique is to use an L-shaped glass spreader that is sterilized in a flame just before being applied to the droplet of freshly dispensed diluent on the agar surface. Next, the short side of the glass L is gently moved through the droplet and on over the surface of the agar in swirling motions. By this action, the microbial cells are dispersed over the agar surface in a relatively even manner.

Drying the Agar Surface (optional)

Where 1.0 or 0.5 ml of diluent has been used in the preparation of the spread-plate, a significant amount of liquid water may remain on the surface of the agar. Such free water can cause problems during the subsequent incubation through colony growths "running together" to become uncountable. There is also a greater risk of contamination. It is important to ensure that where free water is still evident on the agar surface, it is dried off before the plates are incubated. This can be achieved by tipping the Petri dish lid at an angle to the base so that there is a better movement of air over the surface of the agar to increase evaporation. The lid should, however, still cover the agar surface so that the agar is not directly exposed to any particles from the air. Unless conditions are very humid, drying should be complete within two hours. In these conditions, the smaller inoculum should be used to eliminate the need for drying.

Selection of Incubation Conditions

The next critical phase in the analytical procedure is providing an environment conducive to the formation of colonies of the desired types of organisms where these are present. Three major parameters involved in the selection are temperature, atmosphere and time of incubation before counting the colonies. These selections are critical to the appropriate examination of the water for the aggressive (active) microflora. For example, if a water sample were to be taken from a pristine groundwater source that had a stable temperature of 3.7°C, it would not be appropriate to incubate the spread-plates for two days at 35°C. The temperature would be literally toxic to many of the microflora and the few that could flourish at that temperature would not have time to adapt. The result would be, in most cases, a false negative.

Consideration has to be given to the conditions from which the water sample was obtained and the specific objective of the exercise. If the objective is to determine the potential for microbial biofouling or bio-degradation, the incubation conditions should "mimic" as closely as is conveniently possible the original environment. On the other hand, if the objective is to determine the potential likelihood of hygiene risk through fecal contamination, it may be appropriate to concentrate on environmental conditions known to be favorable to the contaminating organisms (i.e., 35°C with a minimum incubation period of 24 hours).

The former mandate for the indigenous microflora would naturally put a considerable strain on an analytic environmental microbiology laboratory because there would need to be a wide variety of incubation temperatures available. It is, therefore, essential to categorize the water into groups wherein each group is incubated at a different temperature.

For spread-plate analysis, the temperature selections automatically link the length of time to the incubation period. It is generally accepted that the higher the temperature over the range from 4 to 37°C, then the shorter the incubation period needs to be. However, the length of the incubation period before all of the capable cfu entities have actually grown to form a visible colony will vary. This variability will primarily be controlled by the length of time that the entity takes to adapt to the cultural conditions presented and assumes that the "territory to be occupied" does not become dominated (invaded) in the meantime by another colony or colonies. The ideal would be to continue the incubation period until the number of colonies counted is stable (i.e., all cfu entities able to grow have grown to form visible and recordable colonies). In reality, this is rarely practicable except as a research exercise.

Two incubation time mandates can be projected for each group separated by the source water temperature. The first incubation time relates to the "natural" fouling or degrading flora and time for the hygiene risk organisms is discussed later. Recommended incubation conditions are:

- Water group, 0 to 15°C. Incubate at 12°C for psychrotrophs for 21 days (count weekly), where the temperature is above 10°C additional or alternate incubation at 25°C for 21 days (count weekly) to determine the number of facultative psychrotrophs.
- Water group, 16 to 27°C. Incubate at 25°C for 14 days (count also at 7 days).
- Water group, 28 to 35°C. Incubate at 35°C for 7 days (count also at 2 days).

For the hygiene risk organisms (e.g., coliforms and nosocomial pathogens), the general guidelines specify incubation for 24 hours (extendable on occasions to 48 hours) at 35°C. This is based on the premise that these organisms are able to adapt rapidly and grow under these optimized conditions. In reality, the microorganisms may be under some level of trauma that may require either a more supportive culture medium or prolonged incubation. Such prolonged incubation does not usually extend to beyond seven days. Coliform colonies that could have appeared only after such extended incubation would, first of all, have to be confirmed by identification to be *Escherichia coli*. Once this had been established, the water would be considered to contain traumatized coliforms and a hygiene risk may now be established.

Counting Colonies on a Spread-plate

When a viable entity settles onto the agar surface, it may be a single planktonic cell, a particulate mass containing only one type of microbe, or a mass containing a consortium of several types of microorganism. Each of these has the potential to produce a single colony. For the latter event, the consortial members would compete for dominance in which the subsequent (mixed) colony formed would contain only the surviving strains. This may not reflect the original composition of the consortium in the water sample and underestimation and "false negatives" would abound. In the two former events, only a single strain of organism was involved and so these colonies are likely to be "pure."

Clearly, there would not be time or facility to completely determine which colonies are pure and which are mixed. In the colony count procedure, it is customary to count all of the colonies which are **typical** (i.e., fit the standard characteristics ascribed to the particular microorganisms expected to grow on the medium). **Atypical** colonies are those that do not fit the standard descriptors. Such atypical colonies are generally viewed as being anomalous events reflecting abnormal circumstances and are discounted.

In some cases, all colonies are considered typical (i.e., a total plate count). Enumeration of colonies in a total plate count includes all visible colonies. A number of digital image analysis systems are now in common use for such counting procedures. Manual counting is still very commonly practiced. This method generally places the incubated agar plate to be counted on a back illuminated dark ground screen (such as the Quebec counter). The background is dark and the colonies are illuminated from the side. Plates are placed in a rack with the base towards the observer. As each colony is counted, a small mark (e.g., dot with a permanent marker) is made on the base of the dish where the counted colony was observed. The technician records on a hand counter, electronic digitizer or remembers the count to record the total number at the end of the count. Where there are more than a hundred colonies on the plate, the base can be divided into four quadrants and only one quadrant of colonies counted. The number of colonies counted is then multiplied by four to give the total colony count for the plate.

Some selective media do differentiate out colony types as being typical of a particular group of organisms. Where these media are used, the colonies should be counted separately by typical group as specified in the media and standard methods documentation. Other colonies may be recorded as atypical. On occasions, these atypical colony forms may be so numerous that they cover the agar surface and mask any typical colonial growths that may have been occurring. This condition is referred to as an **overgrowth**. Greater dilutions can sometimes be used to resolve such overgrowth events but, in cases where the atypical cfu exceeds the typical cfu forms, extended dilution and spread plating does not resolve the issue.

Computing the cfu/ml or cfu/100ml

Data has now been gathered on the numbers of colonies recorded forming on the agar spread-plate at the various dilutions (as total plate, typical or atypical counts). The simplest interpretation is where there has been no colonial growth. This would normally be expressed as 0 cfu/ml or 0 cfu/100ml whichever would be appropriate. Unfortunately, such a result would suggest that none of microbial types under investigation were present whereas, in reality, this result indicates that these microbial types were **not detected (N/D)**. A number of factors (e.g., trauma, competition, inappropriate sampling, dilution and cultural conditions) could have prevented colony expression (and thus created a "false negative" situation). Care must therefore be practiced in the interpretation of the results and consideration should be given to the probability that the resulting data would be of **comparative** value rather than of **absolute** value. It is interesting to note that in the medical industry where microbial presence is likely to be associated with specific microbes (i.e., the pathogen), enumerations tend to be given an absolute value (i.e., if the pathogen is present in these numbers or, in absence, these microorganisms did not cause the disease). In groundwater there is a complex community structure which reduces the value of the data to a comparative level. Exceptions would include anomalous conditions such as where a major aerobic biodegradation was occurring. Here, the organisms associated with the degradation events would dominate and a more precise enumeration may subsequently become achievable.

There are two primary mechanisms by which the water sample may be enumerated using the agar spread-plate technique. One method primarily examines expressions of colonial growth from dilutions of the water sample (streaked spread-plate). The second method examines various volumes of the water sample itself to determine whether microorganisms are present through their entrapment and subsequent growth on a filter (membrane filtration technique). High population numbers (cfu/ml) are easiest recorded using the streaked spread-plate while low population numbers (cfu/100ml) are more conveniently enumerated using the membrane filtration technique. Each methodology involves a different manner of calculating the population from the numbers of colonies recorded. These will be discussed below.

Streaked spread-plate (calculation formula)

A series of colony counts (CC_d) are obtained for each of the dilutions (d). Where there has been an inadequate dilution of the water, the number of colonies may be **too numerous to count (T.N.T.C.)**. This is generally considered to be where more than 300 colonies can be enumerated by direct visual examination. If the dilution is too high, then the number of colonies would become **too few to count (T.F.T.C.)**. This may be considered to have happened where there are fewer than 30 countable (recognizable) colonies on a given plate. There remains the possibility that two tenfold dilutions may yield 301 colonies at one dilution and only 29 colonies at the next incremental dilution rendering it technically impossible to ascribe a count. Therefore, there is some flexibility built into the interpretation of the data from agar spread-plate counts. The ideal methodology would be to use a replicate analysis that should minimally include five replicates for each dilution and ideally incorporate fifteen. Clearly, this is not practicable for economic reasons.

Midpoint cfu/ml formulation

One formulation for obtaining the cfu/ml involves the selection of the dilution streak spread-plate that displays between 100 to 150 colonies after incubation. This is at the midpoint of the acceptable range for calculation of the cfu/ml. Where all other colony counts are outside the range (i.e., 30 to 300), the population may be calculated using the following formula:

$$\text{Population, cfu/ml} = CC_d \times Y$$

where CC_d is the colony count at dilution (d) and Y is the multiplier factor based upon 1 followed by a series of zeros equivalent to the tenfold dilution factor (d).

For example, where the dilution factor (d) is 10^{-4} then factor Y would be 1 followed by four zeros (0000) to make the number 10,000. Where the CC_d was 125, the population would be 125 x 10,000 or 1,250,000 cfu/ml.

Coupled cfu/ml formulation

Where there is more than one CC_d falling within the range of 30 to 300 colonies on the streaked agar spread-plate, the calculation has to be modified to allow an equal weighting to each of the dilutions which have fallen in the range. Calculate the cfu/ml_d for each dilution d based on the equation:

$$\text{Dilution Population (cfu/ml}_d) = CC_d \ \times \ Y_d$$

Once the cfu/ml_d have been calculated for each relevant colony count then the final cfu/ml may be calculated as the mean value of the sum of the cfu/ml_d values obtained where Z is the number of cfu/ml_d values used.:

$$\text{Projected Population (cfu/ml)} = \ (\text{Sum of Z cfu/ml}_d \text{ values}) \ / \ Z$$

Projected cfu/ml Formulation

This is used where there are no colony counts on the streaked agar spread plates but there are counts higher (i.e., >300) and the lower (i.e., <30). These counts are outside the normally acceptable range. In these events, the cfu/ml may be projected on the basis of the cfu/ml_d data obtained closest to the acceptable range (i.e., 15 to 29 and 301 to 500). Here, the cfu/ml would be calculated in the same manner as the coupled formulation given immediately above.

Fluctuating cfu/ml formulation

Where there is a considerable amount of fouled material (e.g., sloughed biofilm, and soil particles), it is quite probable that the CC_d values may fluctuate radically along the dilution scale. Spurious high colony counts in the greater dilutions may be the result of the break up of particulate (biocolloidal) structures which then generate a larger number of colony forming units. Secondary releases of particles (and cells) which had become attached to walls of the diluting equipment and had later released in much higher dilutions may also cause this effect. The net effect is that the CC_d values decline more slowly through the acceptable range and then, at greater dilutions, suddenly exhibit higher CC_d values. Streaked spread-plate enumerations from soils can sometimes be very prone to these "quirks." It has been known for dilutions as high as 10^{-10} to 10^{-15} to display abnormally high colony counts for the reasons given above. When this event occurs, the cfu/ml should be calculated using the midpoint cfu/ml formulation taking the CC_d data for the first dilution in the series that falls to within the acceptable range. It should also be noted that where the data is radically fluctuating this could indicate a major biofouling event or a significant presence of (biocolloidal) particles.

Spread-plate Enumeration of Bacterial Loadings (cfu/ml)

Projected populations of specific groups of bacteria can be determined by culturing various dilutions of the water sample on different agar media. Dilutions commonly recommended for this function are 10^{-1}, 10^{-2} and 10^{-4} using sterile Ringer's solution. Volume spread per plate may be 0.1 ml (high humidity) or 1.0 ml (low humidity). Incubation temperatures have to be related to the original temperature of the water. It is perhaps most convenient to utilize room temperature (20 to 24°C) where the water temperature ranges from 12 to 30°C. Incubation times should be long enough to allow the bacterial colonies to develop, this occurs commonly as a stepwise series of increases. It is recommended that the agar plates be enumerated after 14 days. However, if there is urgency in obtaining the data, then 7 days incubation period can be substituted recognizing that a lower number of colonies may be counted. If there is time, prolonged incubation for 4 to 6 weeks may reveal a higher colony count. Data will be presented as colony forming units per ml and the specific group of bacteria being enumerated will be dependent upon the type of agar culture medium being used. The following are a list of agar culture media that can be used to determine specific groups of bacteria:

HB	R2A agar	- aerobic	- Heterotrophic bacteria
IRB	WR agar	- aerobic	- Iron-related bacteria
GAB	BHI/4 agar*	- aerobic	- Gross aerobic bacteria
SRB	Postgate's B	- anaerobic	- Sulfate-reducing bacteria
PSE	AA agar	- aerobic	- Pseudomonads
STR	R2A/10 agar*	- aerobic	- Streptomycetes
MLC	Czapek-Dox agar	- aerobic	- Molds
MLP	PDA agar*	- aerobic	- Molds
CLF	M Endo LES agar	- aerobic@	- Coliforms

The agar culture media marked with an asterisk (*) possess some unique features. The BHI/4 is a Brain Heart infusion agar at 25% of the nutrient loading. R2A/10 is the standard R2A medium but with 10% of the normal nutrient loading; in addition, the water sample is pre-treated with 100ppm sodium hypochlorite solution. The chlorine is subsequently neutralized with sodium thiosulfate to reduce the bacterial vegetative cell count and allow the exospores of *Streptomycetes species* to survive and grow. The third medium highlighted with the (*) is PDA agar, potato dextrose agar, which supports the growth of a wide variety of molds. The (@) symbol refers to the M Endo LES agar which should be incubated at 35°C for 24 hours only to enumerate coliform bacteria.

Spread-plate Bacterial Population Relationships (SPBR)

As water is pumped from a larger and larger zone of influence, so there are changes in the microbial composition of the water as a result of the "fringe bio-zone" organisms arriving at the well. One manner in which these shifts may be observed is to examine the interrelationships of the various bacterial populations. This may be achieved by comparing the populations as SPBR factorials. Here, the SPBR is determined by the following formula:

$$\text{SPBR X factor} = (\text{Population}^2) / (\text{Population}^1)$$

Where the ratio would be population^1: population^2 and expressed as 1: X, X being the computed factor given as SPBR in the above equation.

For example, where the IRB:GAB ratio is to be determined based on respective populations (cfu/ml) of 5,000 and 15,000 the SPBR X factor would be computed as 3.0 and the IRB:GAB ratio would be 1:3.0. While all possible combinations could be compared in this manner, the dominant comparisons currently envisaged are listed below.

IRB:GAB

The largest numbers of IRB are commonly found in the bio-zones closely associated with the water well installation itself. The ratios in these zones will vary from 1:1 to 1:10 but further away from the well itself, the ratio may shift to 1:>100. Where water samples are taken from a producing water well, the IRB:GAB ratio may remain relatively constant and reflect, to some extent, the dominance of IRB in the biofouling event (i.e., 1:1 IRB are dominant; 1:100 IRB are suppressed).

GAB:CLF

In cases where the coliforms (CLF) are absent from the water sample, this ratio could not be computed. However, where a CLF population is recovered, there exists a hygiene risk as well as a biofouling risk. The higher the SPBR X factor (as it approaches 1.0, i.e., parity with the gross aerobic bacteria). the greater the level of concern. This approach to parity would mean that there has either been: (1) intrusion of fecal-rich waste water, or (2) substantial growth of some coliform-like bacteria within at least one of the bio-zones around the well. Concern should be expressed where CLF are recovered and a standard coliform test should be undertaken to confirm the event. Identification of the dominant coliforms should reveal *Escherichia coli* if event (1) has occurred, or *Enterobacter* and/or *Klebsiella species* in the case of event (2).

GAB:PSE

Where there is an aerobic degradation of a specific group of organic pollutants (e.g., gasoline, organic solvents), pseudomonads (PSE) may tend to dominate over other bacterial types. This event will cause the GAB:PSE to bias from a GAB dominance (i.e., 1:<1) to a PSE dominance (i.e., 1:>1). It should be noted that where there are extremely high pseudomonad populations associated with the degradation of specific organics, there may be very large numbers of atypical colonies (not brown) that are usually white or cream colored growing on the WR agar spread plates.

GAB:MLC

Molds may dominate in groundwater that includes a fraction of water from the capillary and/or unsaturated zones of recharge. This is particularly likely where these zones have become heavily contaminated with organic wastes (e.g., recharges from oxidation ponds around a pulp mill). Such waters can carry very high populations of mold (fungal) exo-spores that germinate and grow very readily on the Czapek-Dox (MLC) or Potato dextrose agar (MLP). In a saturated groundwater system, the GAB:MLC ratio may not be computable due to the absence of these exospores. However, when such contaminations do occur, the GAB: MLC ratio can switch dramatically from 1:<0.001 to 1:>1. Because of the abundance of exospores, it is not unusual to see the SPBR shift to 1:>1,000 in events where there have been a massive mold intrusions. The same conclusions can be drawn from the GAB:MLP ratio.

The composition of the bacterial flora within a bacterial interface around water wells will vary depending upon a number of environmental factors. These can be summarized by the major potential events:

- Aerobic degradation often involving a narrow range of pollutant chemicals and commonly dominated by *Pseudomonas* species (PSE),
- Mixed aerobic-anaerobic degradation often involving a broad range of pollutants and a wider spectrum of bacterial types (GAB),
- High population of molds (MLC) would indicate that the water may have passed through an unsaturated zone where there was a considerable amount of degradation of organics occurring dominated by molds,
- Where the coliforms (CLF) are found to be a significant part of the flora, then there is a serious potential for fecal material to be present and an enhanced hygiene risk,
- If iron-related bacteria (IRB) are found to be a major (>10%) part of the microflora, there is a heightened risk of plugging and/or corrosion.

Other ratios may also be computed where differences in the water characteristics from the different bio-zones are observed. Unfortunately, the SRB group of bacteria tend to live within tight consortia with other bacteria (often the pseudomonads) and the SPBR is difficult to apply to this group of bacteria.

Membrane Filter (calculation formula)

Generally, the membrane filter functions where there are expected to be low numbers of the targeted microorganisms which would therefore have to be enumerated in the cfu/100 ml range. For membrane filters, the surface area to be enumerated is smaller and the number of colonies for an acceptable colony count number (CC_v) is between 20 and 80 colonies. It must be remembered that volumes (v in ml) are passed through the membrane filter and that the population is enumerated on the basis of the filtered volume of the original sample. That sample can be diluted into sterile isotonic solutions in order to obtain counts where there is a higher density of microorganisms.

There is a major interference factor that can distort the value of data obtained by this means. That is the event where the filter not only entraps the microorganisms within its porous structures (normally 0.45 or 0.22 microns), but also particulate inanimate and viable (bio-colloidal) structures that become trapped on and within the porous medium. Some of the bio-colloidal particles will "squeeze" through the pores to pass on into the filtrate. Other particles will entrap on the surface of the filter. This would have a number of impacts on the subsequent colonies that grow. These include:

1. Shifting in the composition of available nutrients to include those entrapped in the colloids and inanimate particles,
2. Protecting any incumbent viable entities from any inhibitors present in the selective culture medium,
3. Bias created by the colloidal "burden" on the surface of the filter. This may change the forms of colonial expression that are likely to occur. One net effect of these interferences is that microorganisms (e.g., pseudomonad bacteria) not commonly expected to grow may flourish in a selective medium inhibiting their growth may actually generate **overgrowths.** This renders the counting of typical colonies more difficult if not impossible.

Enumeration of the CC_v follows a similar mandate to the CC_d for the streaked agar spread-plate. Calculation of the population uses a baseline value of 100 ml for the reference volume on most occasions. The same form of computation as the spread-plate can be used with the two exceptions:

- Acceptable CC_v range is now 20 to 80; and
- Calculations often do not involve dilutions and relates back to 100 ml.

An example of the computation of a calculation of the cfu/100 ml is given below:

$$MF_c = 100 / v$$

Where MF_c is the correction factor and the CC_v is the colony count falling within the range of 20 to 80 for the volume (v) which would be given as ml. From these data, the populations can be computed using the equation:

$$Population (cfu/100ml) = CC_v \times MF_c$$

Where the CC_v falls between 20 and 80 colonies and the MF_c is the appropriate correction factor.

Because the range of acceptable colony counts is so narrow (i.e., less than a tenfold scale), there is lower likelihood of obtaining two colony counts from within the range. There is a greater possibility, however, that no colony count will fall within the acceptable range. In such an event, the counts from 81 to 160 should be selected.

While the numbers produced by either the spread-plate or the MF techniques would appear to be valid and can extend to three decimals of accuracy, the user should remember that there are a number of limitations. These counts reflect only a part of the total microflora and, unless subjected to rigorous replication, can be inaccurate. When used in a comparative manner, these tests can present a reasonable level of certainty that a particular microbially driven event is occurring.

Coliform Testing
Coliform bacteria include a number of bacterial genera within the Section 5 of *Bergey's Manual of Systematic Microbiology*. These are grouped into the categories:

Fecal coliform   -   *Escherichia coli*
Coliform group -   *Klebsiella pneumoniae*
                            Enterobacter aerogenes

These bacteria are defined as being facultatively anaerobic, Gram negative, non-sporing, rod-shaped bacteria that ferment lactose with gas production within 48 hours at 35°C and are resistant to bile salts. Note that the reason for extending the incubation period to 48 hours where the tests have been negative at 24 hours is to cover those coliforms where there is a delay for whatever reason before gas is produced. These bacteria make up 10% of the microflora in the large intestine. *Escherichia coli* is universally present while *Klebsiella* and *Enterobacter* species are commonly present (in 40 to 80% of the incidences recorded). However, the coliforms include a range of species that may thrive in various soil and water environments (e.g., sediments). These can generate false positives. Where the fecal coliform test is operated at a higher temperature (44.5°C), there is a restriction of fermentative activity to only the fecal coliforms (*Escherichia coli*). The widespread acceptance of the coliform presence as an indicator of hygiene risk (fecal contamination) is flawed by the fact that these bacteria lose their viability in water at a relatively fast rate and so false negatives may be generated. There are two principal quantitative test methods for the coliforms:

Multiple Tube Method

Typically, a selective medium such as lactose/lauryl tryptose broth can be used. This technique involves a multiplicity of various dilutions of the water sample in fermentation tubes in a logical sequence which allows a mathematical projection of the most probable number (MPN) of coliforms to be calculated in the water sample. This can be done where only some of the tubes generate a positive reaction (i.e., gas in a Durham's tube). At the beginning of the twentieth century, Durham developed an inverted glass test tube in the liquid culture medium as a means to entrap and observe the production of gas by microorganisms. It was found that bacteria could be partially identified by the sugars they could break down (i.e., ferment to acid with or without gas). Coliforms were found unique in their common ability to ferment lactose to both acid and gas. Following a presumptive test such as the multiple tube method, there is an optional second phase to confirm the presence of *E. coli* by fermentation in brilliant green bile lactose broth with subsequent (more complete testing) on Levine's EMB or endo agar. This test (MPN) is expensive to perform due to the many tubes that have to be employed involving a high level of technician management.

Membrane Filter (MF)

The MF technique is conceptually simpler than the MPN methods since it involves the entrapment of any bacterial cells and suspended particles on the upper surface of a fine porous filter membrane (0.45 micron mean pore size). Theoretically, any volume of water could be passed through the filter and the incumbent cells trapped in and upon the porous material. When the filter is transferred to an appropriate culture medium and incubated, the number of colonies formed can be back calculated to the population size in the original water sample. The procedures in the MF test method do, on occasion, cause stress in the filtered cells which leads to delays before colony formation occurs. A variety of "holding" media has been suggested to reduce the stress on the coliforms and improve the potential for complete recovery of all coliform bacteria in a water sample. It is more economical than the MPN test, but is thought to be subject to a higher risk of false negatives where coliform bacteria are in stress.

Presence/Absence Coliform Test

To compensate for the economic costs and relative lack of precision in the MPN and MF methodologies, simpler tests have been developed based on the presence/absence (P/A) of coliforms in the water sample. The medium used is less selective for the coliforms and does provide an environment enriching for other genera of bacteria also sometimes associated with the gastroenteric tract. These genera include (% range of incidence):

Enterobacter species	40 - 80%
Klebsiella species	40 - 80%
Proteus	5 - 55%
Staphylococcus	30 - 50%
Pseudomonas	3 - 11%
Fecal Streptococci	100%
Clostridium	5 - 35%

This P/A test uses a broth culture comprised from lactose broth, lauryl tryptose broth with bromocresol purple as a pH indicator. It is usually performed using 100ml of water sample that dilutes the broth medium to the correct strength. The generation of yellow color indicates that fermentation of the lactose has occurred which is considered to be a positive acidic reaction indicating that bacteria are present which may be linked to fecal contamination. Where foaming is detected, gas production may be considered to have occurred (usually this occurs between 24 and 48 hours into the incubation period at 35°C). "This simple test is gaining acceptance around the world and is a standard procedure in some countries." (See Selected Bibliography, Prescott, Harley and Klein, *Microbiology*, W.C. Brown, DuBuque, IA, 1990).

There are a range of coliform tests available using the MF, MPN and the MUG concepts. The latest coliform tests to be approved are based on the MUG system (e.g., Colilert®). These tests are very well accepted and use a number of targeted molecular metabolic pathways that are selectively used by *Escherichia coli* or coliform bacteria.   These pathways are described in summary form below.

ONPG Test for Coliforms

In this test, the coliforms possess an enzyme Beta-galactosidase to metabolize ONPG causing a yellow color to be generated by the release of o-nitrophenyl. There are a few non-coliform bacteria that do possess this enzyme, but it is suppressed by the selective nature of the coliform media that are employed.

MUG Test for Escherichia coli

In many ways this test parallels the ONPG test with several differences in the metabolic mechanisms used. The enzyme here is Beta-glucuronidase that reacts with 4-methyl-umbellliferyl and Beta-d-glucoronide to produce the product 4-methyl-umbelliferone that fluoresces in U.V. light to give off a blue color. This methodology has been developed both at the P/A and at the quantifiable level and has been widely adopted as a standard coliform test around the world.

Enumeration by Sequential Dilution Techniques

Over the last fifty years, microbiologists have tended to rely mainly upon the use of agar culture media in the enumeration and primary identification of microbial agents. There have, however, been periodic initiatives to enumerate microbial activity in incremental dilution series using liquid culture media as both the diluent and culture medium. Two approaches to the determination of microbial numbers and/or activities are commonly practiced. These are the **tenfold serial dilution technique** and the statistically validated **most probable number (MPN)**. The former method involves the logical dilution of the water sample to beyond the realm where recoverable viable entities could be present (and recordable). In the latter technique, various volumes of water are investigated to determine which aliquots still contain a demonstrable microbial presence. Replicates of the volumes selected are used to generate a statistical determination of the microbial population.

The additional use of a diluent as the culture medium for growing the targeted microorganisms replicates perhaps more closely the natural environment in which these organisms usually flourish than the surface of an agar culture medium. It is naturally assumed that the density of the cells declines in a direct relationship to the degree of the dilution applied. For example, a water sample containing 3,300 cfu/ml would be expected upon a hundred-fold dilution to contain 33 cfu/ml. A further hundred-fold dilution of the dilution would theoretically contain 0.33 cfu/ml or it could be expected that there would be a one in three (1 out of 3) probability that any given ml of the diluent would contain a viable entity.

Application of the tenfold serial dilution technique is most likely to generate data that would indicate a differential point where one dilution (less dilute) would be positive while the next dilution (more dilute) would be negative. It can be extrapolated that the population must be greater than the critical population (i.e., at least one viable entity in the total volume of the diluent) than the dilution factor applied to the original water sample since growth had occurred. At the same time, the population must be smaller than would allow growth to occur in the next higher tenfold dilution. Generally, the population expressed by this technique would be given as greater than (>) the highest dilution factor showing growth. For example, in a tenfold series down to 10^{-6}, growth was recorded down to 10^{-3} dilution but not beyond, it can be extrapolated that the population is minimally 10^3 viable units per ml or 1,000 v.u. /ml. This would be expressed as >1,000 v.u. /ml. At the same time, the population would have to be considered as being less than 10,000 v.u. /ml otherwise the next highest dilution would also have exhibited growth. The tenfold serial dilution may therefore be considered to operate more at the semi-quantitative level of accuracy.

Most probable number (MPN) technology is more applicable to low populations of microorganisms that may be associable with a hygiene risk concern. Most commonly, the MPN is used for the evaluation of coliforms in water. Here, these bacteria may be present in numbers between 1 and 180 per 100ml. Typical water sample volumes to be employed are:

- Scenario A, 5 tubes each containing 10 ml of water plus the medium;
- Scenario B, 10 tubes each containing 10 ml of water plus the medium;
- Scenario C, 5 tubes of each of 10 ml, 1 ml and 0.1 ml of water plus the medium are used.

In all three cases the number of positives is used to determine the MPN index (likely number of coliform organisms per 100 ml) and the 95% confidence lower and upper limits. Nineteen times out of twenty, the number of coliforms present in the water sample will fall within these upper and lower confidence limits. Positives are taken to be those tests that show fermentation (visible gas formation) within the incubation period (usually 24 hours) at 35°C.

The MPN technology does have application to investigations where there is expected to be:

1. A low number of targeted organisms in the water;
2. A large potential population of other microorganisms which could interfere with enumeration procedures;
3. A selective medium or specific product display which will signal activity generated solely by the targeted organisms;
4. Incubation conditions which will suppress the non-target flora.

Chapter 5 comprehensively deals with the use of the Biological Activity Reaction Tests (BART™). These testers were designed to allow a simple method to determine the aggressivity and composition of the various microbial consortia that can occur in natural, biofouled and bioremediation situations. Experience has found that the BART testers are suitable for in-field and laboratory monitoring of the various communities of bacteria that can impact on water and soils. While Chapter 5 describes the operation and interpretation of the testers, Chapter 6 gives comprehensive examples of their use in surface, ground, waste and oceanic waters along with soils, remediation sites, and examples from the food and oil/gas industries. The major strength of these testers is that they function at the consortial level and are able to determine the major microbial communities functioning and their collective level of aggressivity. Treatments to manage these communities can be assessed by the shifting in the time lag to the detection of positive reactions (i.e., the sooner, the more aggressive, the greater the population, the more active that fraction of the microbial community). This shifting in aggressivity creates a time lag difference (TL_d) and this can be used to determine the effectiveness of a given management strategy to control the microbially driven events.

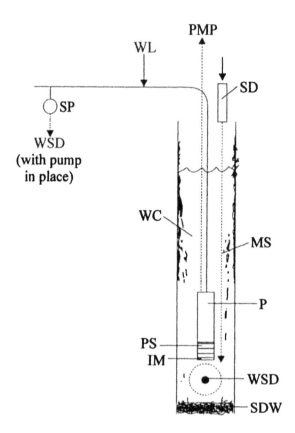

Figure Twenty-One, Depiction of some of the challenges faced trying to take a water sample from a specific site within a biofouled well. In this example, the water sample desired (WSD) is from below the pump (P) in the water column (WC). To obtain the sample, the options are to pull the pump (PMP) and lower a sampling device (SD) to that site, or pump the well. If the well is pumped, then the WSD would possibly become impacted by organisms infesting the sampling port (SP), the water lines (WL), the pump and especially the impellers (IM) and pump screen (PS), particulates stirred up from any sediment SDW) in the well, and materials sloughing (MS) from the walls of the well.

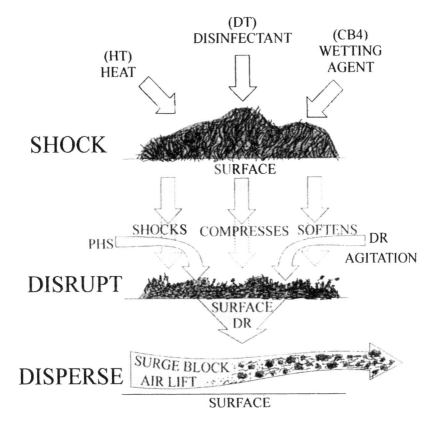

Figure Twenty-Two, Treatment stages for the patented blended chemical heat treatment (BCHT^{TM}) process. There are essentially three stages in which the plugged material is removed. They are, in sequence, to shock the biofilms (upper) through a combination of heat (HT), disinfectants (DT) and a penetrant (CB4 is the recommended form). This causes the biofilm to destabilize. The second stage (center) involves the disruption (DR) additional application of a pH shunt (PHS) either up or down by at least 3 pH units. These additional stresses cause biofilms to begin to split apart and detach from the surfaces. Physical forces are used as the principal agent in the final stage (lower) to disperse the shattered biofilms and associated "crud" from the well.

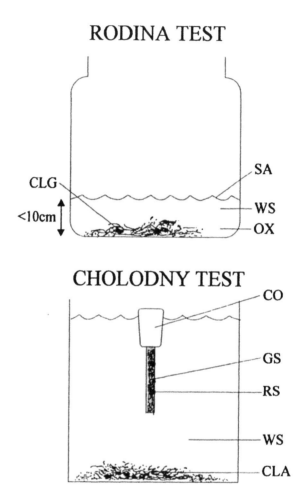

RODINA TEST

CLG
<10cm
SA
WS
OX

CHOLODNY TEST

CO
GS
RS
WS
CLA

Figure Twenty-Three, Two very simple field tests are the Rodina (upper) and the Cholodny Test (lower). The Rodina test simply uses a large surface area (SA) so the water sample (WS) becomes very oxidative (OX). When this happens, the IRB generate "cotton-like" growths (CLG). The Cholodny test is more sophisticated and is based on floating a glass slide (GS) on the underside of a cork (CO). "Rust spots" (RS) commonly form on the slide along with cotton-like accumulates (CLA) where IRB are present in the water sample.

GRAINGE & LUND TEST

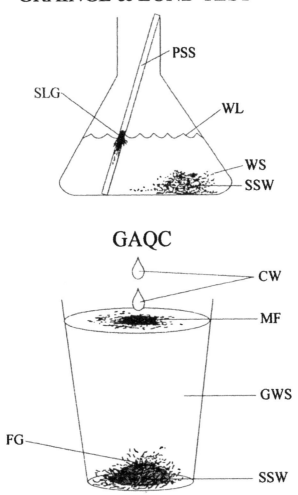

GAQC

Figure Twenty-Four, Two other simple tests for IRB are the Grainge and Lund (GL, upper) and the George Alford Quick and Crude (GAQC, lower). The GL test uses a plastic stir stick (PSS) leaning into the water sample. A soft steel washer (SSW) is added to water and string-like growth (SLG) around the PSS at the water line (WL) is considered positive. GAQC also uses a SSW dropped this time into a glass of the water sample (GWS). Detection of the IRB is where metallic floaters (MF) or fuzzy growth (FG) occurs around the SSW. To speed up this process, two drops of corn whiskey (CW) are added to the water sample as "nutrients."

GRAM NEGATIVE   GRAM POSITIVE

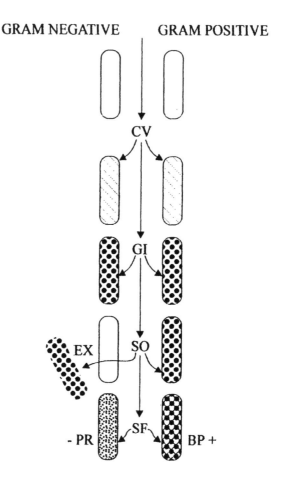

Figure Twenty-Five, Of all of the staining techniques used in bacteriology, the Gram stain has been the most widely adopted. This is because a small majority of the bacteria stain Gram negative (pink-red, -PR) while a large minority stain Gram positive (blue-purple, BP+). The differential sequence includes staining sequentially with two stains, crystal violet (CV) and then Gram's Iodine (GI). The next stage is critical where a solvent (SO) such as ethanol is added. The solvent will extract (EX) the stain from the cells in Gram negative (GN) but not in Gram positives (GP). To see the unstained (US) GP cells, a counter-stain such as safranin (SF) that will stain the GP cells pink-red.

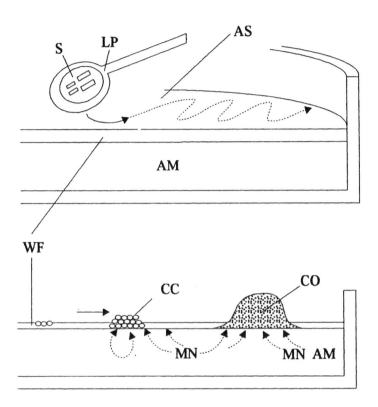

Figure Twenty-Six, Illustration of the steps involved in the enumeration of bacterial numbers using an agar spread plate technique. A dilution of the sample (S) is streaked over the agar surface (AS) using a 0.4mm loop (LP). Bacteria in the sample are smeared (upper) over the agar surface and cluster in the water film (WF) where they can, if able, "mine" nutrients (MN) from the agar medium (AM) and gradually grow to form clusters of cells (CC) which grow large enough to become visible as a colony (CO). Counting the number of colonies grown after an incubation period gives an indication of the number of colony forming units (CFU) in the original sample.

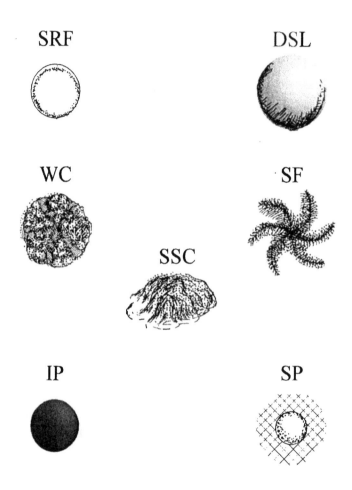

Figure Twenty-Seven, Diagram of the various forms of bacterial colonies that can form on some agar spread-plates. Generally, grown bacterial range from 1 to 10mm in diameter. Common shapes include: simple round flat (SRF), very domed slime-like (DSL), wrinkled circular (WC), swirling fronds with an irregular edge (SF), swarming slime-like colonies (SSC), pigmented colonies where the colony is colored by the insoluble pigment (IP) or the surrounding agar is colored by the soluble pigment (SP).

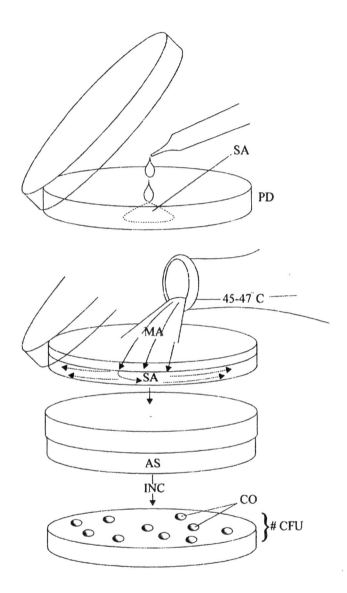

Figure Twenty-Eight, Method commonly employed in the poured agar plate technique. In this technique, a 1 ml sample (SA) is added directly to the petri dish (PD) in which the test is to be performed. Molten agar (MA) medium is poured over the sample at 45 to 47°C. This mixes the sample into the agar as it sets (AS) at 43 to 45°C. Bacterial colonies (CO) form in the agar during incubation (IN) and can be counted as colony forming units (CFU).

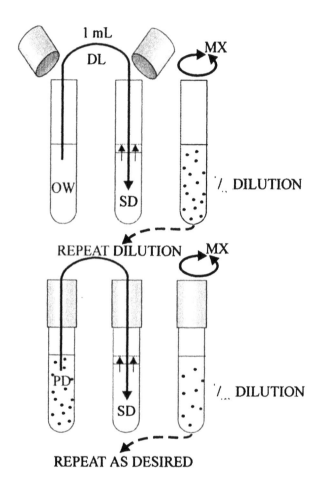

Figure Twenty-Nine, Schematic presentation of the steps required to conduct an extinction dilution of a sample. The original water sample (OW) is diluted (DL) by one order of magnitude (tenfold) commonly by mixing (MX) 1ml of the previous dilution (PD) into 9ml of a suitable diluent (SD) such as Ringer's solution or phosphate buffer. The potential number of colony forming units (CFU) are reduced by 90% with each stage of the dilution. Optimal data is generated where the CFU numbers (i.e., countable colonies) falls between 30 and 300 colonies on an agar plate.

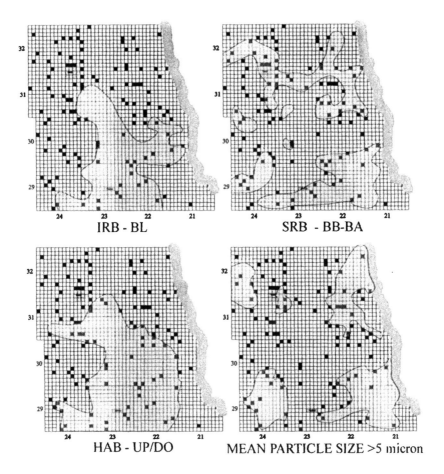

Figure Thirty, Distribution map for bacterial consortia infesting various water wells in Kneehill M.D. (see also Figure Sixty-seven). The three of the four shaded charts reveal the wells with very aggressive populations of iron-related bacteria (IRB, generating reaction BL, upper left), sulfate-reducing bacteria (SRB, generating BB - BA, upper right), and heterotrophic aerobic bacteria (HAB, generating UP, lower left). Mean particle sizes in the water samples were recorded by laser particle counting and mean diameters of >5 microns are recorded (lower right). The map reveals the potential to generate maps of the occurrences of biofouling using a biogeographical information system. Such an approach would substantially aid in the determination of the causes of plugging within the area surveyed.

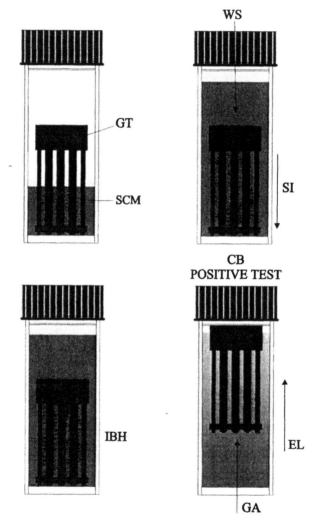

Figure Thirty-One, Diagrams depicting the principals employed in the patented gas thimble (GT) coliform test. The method uses a mixture of the water sample (WS) into a selective coliform medium (SCM, upper left). The GT is sufficiently dense that it sinks (SI, upper right). It is incubated at blood heat (IBH, lower left). If coliform bacteria (CB) are present, then they ferment the lactose in the SCM with the production of gases (GA). The gases cause the GT to float up (elevates, EL, lower right). This elevation defines the fact that the test is positive for coliforms. The time lag inversely relates to the population size (i.e., the larger the population, the shorter the time lag).

5

ABOUT THE *BART*^{TM}

The environment contains a myriad of different bacteria that are all capable of causing problems. These problems can range from slimes, plugging, discoloration and cloudiness to corrosion and infections. Such a wide variety of bacteria are not easy to detect and identify using a single test and yet their impact can make the water unsafe, unacceptable or unavailable due to losses in flow through plugging or equipment failure due to corrosion. The Biological Activity Reaction Test (BART™) is a water testing system for nuisance bacteria and can involve several different tests. These tests detect the activity (aggressivity) of these nuisance bacteria by the time lag (TL, measured in the number of days from the start of the test to when a reaction is observed). The longer the TL before the observation of activity, the less aggressive the bacteria are in that particular sample.

There are seven different tests that are recognizable by colored cap coding and the initial letters preceding the word BART. These include selective tests for:

Iron-related Bacteria	IRB-BART	Red Cap
Sulfate-reducing Bacteria	SRB-BART	Black Cap
Heterotrophic Aerobic Bacteria	HAB-BART	Blue Cap
Slime-forming Bacteria	SLYM-BART	Green Cap
Denitrifying Bacteria	DN-BART	Grey Cap
Nitrifying Bacteria	N-BART	White Cap
Fluorescing Pseudomonads	FLOR-BART	Yellow Cap

Each of these bacterial groups cause different problems and often a combination of these tests should be used to determine which bacteria are present and causing problems. In the event that further information beyond presence/absence is needed, information on these reactions can be accessed using the internet: **http://www.dbi.sk.ca**. To read all of the reactions, lift the inner test vial carefully out of the outer BART test vial and view through the inner vial against an indirect light.

METHODOLOGIES

A common list of the methodologies and applications would be:

IRB-BART   A test becomes positive when foam is produced and/or a brown color develops as a ring or dirty solution. The TL (time lag) to that event is the delay. A negative result has no brown color developing, no foaming or clouding. This test is commonly used to detect plugging, corrosion, cloudiness and color. The bacteria that may be detected by this test include iron oxidizing and reducing bacteria, the sheathed iron bacteria, *Gallionella*, pseudomonad and enteric bacteria.

SRB-BART   A very simple test to perform in which a positive test occurs when there is a blackening either in the base cone of the inner test vial (80% of the time) or around the ball (20% of the time). The culture medium is specific for the sulfate-reducing bacteria, such as *Desulfovibrio* and *Desulfotomaculum*. This is a more specific test and specifically relates to corrosion problems, taste & odor problems ("rotten egg" odors), and blackened waters. Slimes rich in SRB tend to also be black in color. A negative indication occurs when there is an absence of blackening in the base cone of the inner test vial or around the ball.

HAB-BART   There is a very real need to determine the amount of heterotrophic aerobic bacterial activities in some wastewater, particularly those that are aerobic. Here, biodegradation may be a primary concern, such as on a hazardous waste site. This test relies upon the ability of the hetero-trophic aerobic bacteria to reduce a methylene blue dye. To add the methylene blue to the sample, the test vial once charged is simply placed upside down for 30 seconds to allow the blue color to develop. A positive reaction is detected by the blue color becoming bleached (due to the activity of methylene blue reductase). Bleaching may begin at the base of the test vial or just below the ball. Note that a residual blue ring is likely to remain around the ball, but this does not mean heterotrophs are absent. A negative indication occurs when there is an absence of the blue color becoming bleached. This test is used to detect slimes, plugging, taste and odor, cloudiness and can also detect the amount of aerobic heterotrophic activity on hazardous waste sites.

SLYM-BART   Some bacteria can produce copious amounts of slime that can contribute to plugging, loss in efficiency of heat exchangers, clouding, taste and odor problems. This is one of the most sensitive BART tests. A positive reaction involves a cloudy reaction in the inner test vial often with thick gel-like rings around the ball. A negative test remains clear.

DN-BART Nitrates in water are a serious health concern particularly for infants. There is one group of bacteria called the denitrifying bacteria and many of these are able to reduce the nitrate to nitrogen gas. In this test, this gas forms a foam of bubbles around the ball, usually within three days. The presence of this foam by the end of day two is taken to be an indication of an aggressive population of denitrifying bacteria. Absence of foam, regardless of any clouding of the water, indicates that the test is negative for the detection of denitrifying bacteria. This test is applicable to any waters where there is likely to be potential septic or organic contamination. The presence of denitrifiers would indicate a potential health risk due to either septic wastes or nitrates in the water.

N-BART Ammonium is a common product of the breakdown of organic nitrogen under anaerobic (oxygen-free) conditions. The nitrifying bacteria oxidize the ammonium to nitrate in soils and waters. This nitrate now becomes a hazard if it builds up in the water. This test is a different form of BART test in that it is fixed length and involves the detection of nitrite as an indicator of the presence of nitrifying bacteria. Nitrate is not detected because the amounts can be transient where there is a considerable amount of denitrification. A reaction cap is used to allow the detection of the nitrite after a five-day incubation period. A positive reaction occurs when a coloring is produced in the reaction cap and a negative indication occurs when no coloring is produced in the reaction cap.

FLOR-BART A major group of aerobic heterotrophs are the pseudomonads. These bacteria are very well adapted to breaking down some chemicals such as jet fuel and solvents but also can infest recreational waters and cause conditions ranging from skin, eye, ear, and nose infections to pneumonia-like infections. The infectious pseudomonads produce an ultra-violet fluorescence that is usually a pale blue color. Presence for this test means that either a greenish-yellow or a pale blue glow is generated by the careful application of an ultraviolet light just below the ball. The degraders tend to generate the greenish-yellow glow while the health risk group generates the pale blue glow. A negative indication occurs when the sample remains clear.

THE SIX W'S OF THE BART TESTERS

There are numerous ways in which microbes can become a nuisance in water. Often these events are ignored, considered inevitable or put down to simple physical and chemical effects. Ignorance may be bliss, but it is expensive. These microbes can cause corrosion, plugging, failing water quality and the shortening in the lifespan of the installation. In today's world, disposability is being replaced with sustainability and ignorance replaced with knowledge. There has been a considerable lack of attention paid to the nuisance events caused by bacteria other than those associated with hygiene risks. In ground waters, it has been a common practice until a decade ago to consider the environment to be essentially sterile and so microbial events were not considered important. In surface waters, larger and more obvious organisms tended to receive more attention than the slimes and clouds in the water. Today, it is becoming recognized that microbes are present in all waters and that they have a nuisance impact that needs to be managed if sustainability is to be achieved. This document addresses the advantages of becoming more aware of the microbes and their activity in water. It should be remembered that there is no such place as a totally sterile water environment and that, if the microbes are active, there will be affects on the environment.

A. What Are BART Testers?

BART stands for the patented Biological Activity Reaction Test. As the name implies, the test detects biological activity by looking for activities and reactions. Activities relate to growth events such as the formation of clouds, slimes, and gels. Reactions relate to the manners in which the microbes interact within the BART test. These reactions may take the form of color changes, development of gassing, and precipitation. The unique nature of the BART test which makes it very different, and possibly superior, to the agar techniques is that the water used in the test all comes from the sample and contains the microbes still within their natural environment. The water in the agar methods comes with the agar but it is tightly bound. This means that the microbes have to be taken from the water, placed into contact with the agar surfaces, and expected to "mine" the bound water for growth from the agar. Many microbes in the environment are not able to do this easily and so may be missed using agar cultural techniques (i.e., no grow, no show, no count and so not important).

The BART uses a unique system for encouraging the microbes to grow in the test. First, there is normally no dilution of the sample.

Second, the sample becomes adjusted to a variety of different habitats by the nature of the BART. Third, the microbes that can be active and/or react with the selective conditions created within the BART test can be considered to belong to a specific group of bacteria (e.g., iron-related bacteria). These selective conditions are created using two devices. The first is a floating ball, floating intercedent device (FID), that restricts the entry of oxygen into the sample below. The second device is the use of a crystallized deposit of selective nutrients which sits in the bottom of the tube and encourages the activities and reactions by a specific group of microbes. In the first device, the oxygen enters around the floating ball to allow oxygen requiring (aerobic) microbes to grow. They will use all of the oxygen diffusing down so that the sample further down becomes devoid of oxygen. This volume underneath becomes suitable for the growth of microbes that do not require oxygen (anaerobic). The single BART provides environments which are aerobic (oxidative) and anaerobic (reductive). This is a reduction-oxidation gradient with a transitional zone (redox front) in the middle.

At the end of the nineteenth century, Sergei Winogradsky originally reported this type of phenomenon forming in waters kept in tall glass cylinders. The beauty of this device is that different microbes prefer to function at different sites on the redox gradient and so can be seen being active and reacting within that zone. Very often, the first sign of this is the development of a cloud of growth that may be fuzzy and diffuse or a fat plate floating in the watery medium.

The key to determining the presence of different groups of microbes is the crystallized selective medium attached to the floor of the BART device. This medium will begin to dissolve slowly when the sample is added. As the medium dissolves, a series of chemical diffusion fronts become established and move slowly up the BART tube. This slow upwards progression which can take as long as two days, and gives the microbes in the sample time to adapt to the increasing concentration of nutrients and, if suitable, begin to become active. Even the very sensitive microbes that would normally fail to grow on any agar media are better able to adapt and grow within a BART test if the crystallized medium is suitable for their growth. The location of the growth gives an early indication of the type of microbes involved. Activity in the base of the BART test would tend to suggest anaerobic organisms while activity at the top around the ball is more likely to be aerobic. Often the activity may center along the diffusion front for the dissolving crystallized medium. When this happens, the microbes are likely to be able to grow under aerobic and anaerobic conditions (facultative anaerobes).

Given that the BART test presents a whole range of environments for the microbes to grow, the key becomes the form of the crystallized culture medium that is in the BART. It is this factor that causes different communities of microbes to become active and, hence, be detected. The eight BARTs employ different culture media to make the test selective. These are listed below (Table One) defining the microbial group first (acronym is given in brackets) followed by the form of the selective culture medium used.

<u>Table One</u>

**Principal Microbial Groups Determinable
Using the BART Biodetector**

Microbial Community		Selective Culture Medium*
Iron-related Bacteria	**IRB**	Winogradsky's medium
Sulfate-reducing Bacteria	**SRB**	Postgate's medium
Slime-forming Bacteria	**SLYM**	Glucose Peptone medium
Heterotrophic Aerobic Bacteria	**HAB**	Sugar Peptone medium
Algae	**ALGE**	Bold's medium
Fluorescent Pseudomonads	**FLOR**	Peptone base medium
Denitrifying Bacteria	**DN**	Nitrate Peptone medium
Nitrifying Bacteria	**N**	Ammonium salts medium

*media have been modified to maximize the potential for recovery of the microbes using the BARTsystem.

Considerable attention has been paid to drying these media in a manner so that they do not lose their selective function and are not able to rehydrate until the sample is added. Each medium is dried in a different manner to ensure a stabilized form. To prevent rehydration, the test vials are immediately packed in foil pouches which act as effective moisture barriers. A three-year shelf life can be achieved when stored in a cool, dry place. Refrigeration is not necessary.

Reading the Results Using BART Testers
There are two important forms in which information can be obtained. These relate to the type of consortial (community) microbial activity that may be occurring and the determination of the *population*. The consortial microbial activity can be determined from the *reaction patterns* observed. Essentially, the reactions displayed can be used to build up a "picture" of the community (consortium) of microorganisms in the sample being tested.

Populations are determined by the length of the time lag (TL) with the proposition that the longer the TL to the detection of a reaction, the smaller the aggressive population of the microbial consortium being determined. This TL is normally measured in days to the first detection of a reaction. Since the BART tests each detect a different consortial population, the TL for one consortium does not directly relate to the TL for a different test type.

Interpreting the Test Data

It is relatively easy to interpret a negative test because the bacterial groups do not generate any signals of activity in the [BART] test. A positive detection means that: (1) a minimum number of bacteria must have been present to cause the observed activity and reaction, and (2) that the TL to that observation can be used to project the population size. The following tests are normally used at the presence/absence level:

N-BART   minimal population detected: 1,000 cfu/ml
FLOR-BART   minimal population detected: 100 cfu/ml

It should be noted that lower detection limits can be achieved by extending the TL for the N-BART before examining the contents for the presence of nitrite, a transitional by-product of nitrification.

The remaining tests can either be used as a presence/absence or as a semi-quantitative test by determining the TL at which the first positive activity/reaction occurred. The possible log. populations in cfu/ml are listed (Table Two) for the other five tests.

Table Two

The Relationship Between Time Lag (days of delay) to the First Reaction to the Population (log. cfu/ml)

Time Lag (days)	IRB	SRB	HAB	SLYM	DN
0.5	6.6	6.6	6.8	6.8	6.8
1.0	6.0	6.0	6.6	6.6	6.6
1.5	5.8	5.8	5.8	5.8	5.6
2.0	5.0	5.0	5.6	5.6	5.0
3.0	4.0	4.6	3.0	4.6	3.0
4.0	3.6	4.0	2.0	3.0	2.0
5.0	3.0	3.6		2.6	
7.0	2.0	2.0		1.0	
8.0	<2.0	2.0		1.0	

Note: To use this table, determine the time lag (TL) before the first reaction/activity was sighted. Notice that in this table, the tests were monitored twice a day. That may not be feasible in many cases so that, where readings are further apart, the test results may tend to err in favor of longer TL and hence somewhat lower predicted populations. These log. population can be converted to arithmetic population using Table Three below.

Recognizing Positive Reaction Patterns

Reaction patterns are the visible evidence that some activity is occurring in the BART tester as a result of the interaction between the active microorganisms in the sample with the redox gradient and the chemicals diffusing upwards from the base of the inner BART test vial. As a result, reactions range from clouded zones and gas bubbles to changes in the color of a part or the whole length of the sample. Each reaction pattern will be addressed by BART type. (Note that the traditional numbered reactions are shown in brackets and some reactions have been discounted as unreliable).

Table Three

The Relationship between the Log. and the Arithmetic Populations in cfu/ml

log. pop. arithm. pop	arithm. pop	log. pop.	log. pop.
6.75	6,000,000	4.0	10,000
6.5	3,000,000	3.5	3,000
6.25	1,750,000	3.0	1,000
6.0	1,000,000	2.5	300
5.75	600,000	2.0	100
5.5	300,000	1.5	30
5.0	100,000	1.0	10
4.5	30,000	0.5	3

B. Why Use BART Testers?

The BART testers have two major advantages:

1. You do not need a laboratory to set up the test to determine whether there are aggressive bacterial problems in the sample being tested. They are easy to read since the signals (reactions) generated are observable.

2. BART testers provide a greater variety of environments within which the bacteria of concern can grow. This is a very major advantage over the traditional agar techniques in common use in the microbiology laboratories today. This feature makes the BART testers far more sensitive and reactive to aggressive bacterial populations.

These are two main advantages in using the BART testers since they offer convenience, simplicity, sensitivity and durability. Convenience means that the techniques employed to set up the test are easy to follow. The BART testers are supplied in individual, moisture proof, foil pouches that prevent the tester from degenerating as a result of premature rehydration. Within the foil pack is the double tube tester. The outer tube acts as:

- Protection from damage to the inner test vial in which the test is actually performed.
- Security measure to reduce the risk of any odors and accidental leakage generated from the inner test vial escaping from the tester.
- Providing a bigger base for the tester so that it is more stable and less likely to be knocked over.
- Convenient determination of any reactions without having to directly handle the inner test vial after the test has been started.

Meanwhile, the inner tube offers all of the patented advantages of providing a very broad spectrum of environments in a watery environment where the different parts of an aggressive bacterial consortium can elect to grow. These environments can be described in broad terms as changing primarily with the descent down beyond the ball:

- There is a water film covering the top 20% of the surface area of the ball above the surface level of the sample. Biofilms can grow up into this very oxidative zone to be seen as a slime-like coating. Slime-forming bacteria and molds are two common groups of microbes that can grow at this location.

- The sample above the equatorial region of the floating ball. Here, the liquid medium remains saturated while oxygen diffuses downwards from the atmosphere above the floating ball. As a result, it is common for the aerobic slime-forming bacteria to grow into various types of slime-rich ring-like structures. Additionally, these biofilms that form slime rings entrap any gas bubbles being generated deeper down in the active inner test vial. These gas bubbles bounce up around the lower side of the ball and get caught up in the mass of biofilm growth to cause a foam that collects around the ball. Molds (fungi) will also grow in a commonly fuzzy manner down into this zone. This site is very oxidative and the chemicals may be moved into colored oxidized states (e.g., reds, browns and yellows).

- In the liquid medium just below the ball (usually between 3 and 8 mm below) there is a reduction – oxidation (redox) front where the oxidation-reduction potential (ORP) changes from a positive oxidative state above to a negative reductive state below. Many aggressive bacteria tend to grow first at this redox front. Normally, this will take the form of cloudy growth that may be very "fuzzy" in form or quite "tight" and form very distinct plate-like structures in the medium. Bacteria growing at these sites are commonly a mixture of aerobic and facultatively anaerobic forms able to utilize, directly or indirectly, the selective medium diffusing up the liquid medium column in the inner test vial.

- Below the redox front in the lower third to half of the liquid medium, the ORP is negative (reductive). This means that only those bacteria able to grow anaerobically (without oxygen) will be active here. Commonly, there are more color reactions at these sites associated with the reductive end products (e.g., blacks and greens). Generally, visible growths are more gel-like (colloidal) and denser.

- Inside the base of the inner test vial, two major events occur. First, the medium crystallized into the floor dissolves and diffuses upwards meaning that the deposited chemicals disappear and, commonly in some of the BART testers, it is possible to see the liquid medium through the base. Second, there are reactions within the inner test vial that cause changes in the color and texture of the basal chemical deposits. These reactions can cause the base to blacken or change to a different color. It should be noted that the occurrence of a white deposit commonly occurs in an IRB-BART but has not yet been assigned as a significant reaction.

The BART has two modes in which it can be used. For the field testing where the BART tests are actually performed in the field, the full BART test should be used in which the outer tube gives the additional advantages discussed above. In the laboratory setting, the outer tube is redundant since the inner test vials are being used in a more secure environment. As a result, well-equipped laboratories with trained technical staff may prefer to use the more economical LAB-BART™ versions of the standard field test.

This test (LAB) is packed in units of fifteen tests rather than the standard BART tester (with outer tubes) that are packed in units of nine tests.

C. Who Should Use BARTTesters?

Gradually, the roles of bacteria in the myriad of natural and engineered events are becoming appreciated. These range from the obvious (e.g., taste, odor, corrosion and slime formation) to the subtle (e.g., bioaccumulation and occlusion). Virtually any management practice involving water could be subjected to the impacts of bacteria and other microorganisms, and the BART testers provide a means to monitor either the state of the microbial aggressivity or the impact of treatment.

For managers of water systems, there is a need to understand the potential and real challenges that can be caused by these nuisance microorganisms. Unfortunately, very often, microbiological fouling of a system (whether the base medium be water, oil or gas) is slow and covert without any obvious signals to show that it is microbial in origin. Often, these degenerative processes are put down to the normal aging of the facility and it is now considered that these processes could be driven by microbes and managed by monitoring the levels of aggressivity in these nuisance bacterial events using the BART testers.

Who should use the BART testers? Anyone who understands that bacteria and other microbes can affect the lifespan of a facility in a very real manner. These effects can range through a whole range of characteristic changes including:

- Corrosion in which the microorganisms corrode the solid structures (e.g., steel or concrete) in such a manner as to severely weaken the structure causing failure.

- Plugging in which the microorganisms form thick biofilm growths (slimes) within porous media which cause significant losses in conductivity (hydraulic or thermal).

- Radical changes in water quality caused by the casual sloughing of the slimes which are loaded with microbial cells and their associated accumulates. This sloughing can cause sudden dramatic changes in the concentrations of some chemicals (e.g., iron and phosphorus) in the water.

- False data generation due to the biofilms within the upstream zone above the site of special interest. These biofilms (or slimes) can accumulate very large concentrations of recalcitrant chemicals that would otherwise have found their way into the sampling site. This is a form of bio-filtration and accumulation which gives a falsely improved water quality until the growths begin to slough. Monitoring wells may be particularly prone to these events when organic pollutants (e.g., BTEX, PAH, VOH) approach the well and are accumulated into the biofouled zone around the well. This biological interface acts as an effective filter until maturation causes the collapse of the biofilm structures.

- Odors can be generated by a whole range of microorganisms with some well known odors being:

  (1) rotten egg (SRB generating hydrogen sulfide),
  (2) fishy (commonly heterotrophic aerobic bacteria and, in particular, *Pseudomonas* species),
  (3) earthy-musty (geosmins generated primarily by the *Streptomyces*),
  (4) septic (generated by various members of the enteric bacteria including the coliform bacteria) and
  (5) vegetable/fruity odors (from a variety of algae and yeast).

One useful tool to aid in the confirmation of the source of odors is that the odors will concentrate between the outer tube and the inner test vial of the BART test when odor-generating microbes have grown in the tester. Loosening the outer cap and cautiously sniffing the gap between the cap and the outer tube will reveal the types of odors being generated by these microbes. Often this smell is coincident with an odor being detected in the sample itself. This can often convince a doubter that it is the microbes in the BART that are capable of causing the odor problem and a focus on managing the problem is now understood.

- Turbidity has often been thought of as simply a chemical event associated with chemical colloids, silts or precipitation. These will cause the sample to go cloudy. More commonly than not, the cloudiness in the sample is a combination of turbulence swirling up sediments into the liquid medium and the growth of microbes within that sample. If the cloudiness is microbial, then it can be expected that the BART testers will detect very aggressive microbial populations.

- Color is most commonly generated by microbes through the accumulation of iron (yellows, browns, reds and oranges) although occasionally pigment can be generated by the microbes themselves as pigments. These pigments are most commonly browns, yellows, greens, blue-greens and reds and are generally more transient.

- Biodegradation is a major industry today as a part of the environment industry. Where there is a biologically driven degradation occurring, there is an inevitable increase in the aggressivity of those microbes in the environments that are associated with an observed degradation. To monitor this aggressivity, the BART testers can be used. Generally, if the degradation is basically aerobic and involves a narrow spectrum of organic pollutants, then the heterotrophic aerobic (HAB), the fluorescing Pseudomonad (FLOR) and the slime-forming (SLYM) BART testers are most likely to detect the increased aggressivity of the degraders. This can then be used as a "benchmark" for the vitality of the microbial consortium causing the degradation. If the degradation is anaerobic, then a different spectrum of bacteria may be the most aggressive. These could include the sulfate-reducing (SRB), the slime-forming (SLYM) and the denitrifying (DN) bacteria.

The BART testers are suitable as a field test for any manager or consultant concerned about managing problems which are likely to be either instigated by, or worsened by, the presence of the various groups of microorganisms detectable using the BART testers. Just who would use the BART testers would depend upon the level of biological activity occurring whether this be biofouling, biofiltration, or biodegradation. Some examples of who would use the BART testers are listed below:

- **Water Well Operators**. Water wells are a "site unseen" operation. The extent of any visible fouling is limited to camera logs down the well or obvious fouling of filters and lines downstream of the well head. Often, the bulk source of all of the biological activity is outside of the well screen and not visible. What is visible is the "tip of the iceberg" which is the colloidal structures floating in the well water column (well snow), encrustations, tubercles and slimes attached to the walls and screens of the well and as deposits in the bottom of the well. Detecting even the most aggressive bacteria under these conditions is not simple. The bacteria often have to be "tricked" by changing the normal operational procedures for the well in order to be able get them into the water so that they can be detected using the BART testers. Most commonly used of the BART testers are:

    IRB (where there are known iron problems);

    SRB (where there are anaerobic, black water and corrosion problems);

    SLYM (where there are slimes forming in and over the well casing, screen or pump); and

HAB (if there is turbidity, odd odors, cloudiness, fluorescence and high organic loadings in the water).

- **Water Treatment Plant Operators.** Water treatment facilities usually involve water that has become aerated, possibly filtered, disinfected, clarified and stored. It should be remembered that the BART testers are proofed against the possible effects of chlorine-based disinfectants by the inclusion of a neutralizer that is effective for concentrations of up to 5,000 ppm of chlorine. In general, apart from the concern for the elimination of coliform bacteria from the water, there is little regulated limitation to the microbial loadings in potable, industrial and recreational waters. Consequently, the need to monitor nuisance microbes is more in the interest of the operator rather than regulatory compliance. Unfortunately, the common attitude that water should be free disenfranchises the ability of the operator to assure a maximum operational efficiency in favor of bulk acceptable water produced at the lowest cost. Biofouling causes many covert (and commonly negative) impacts which often go unnoticed until it is too late to effectively control and then radical "surgery" has to be performed to replace the fouled parts. Common problems relate to massive slime formations (SLYM and HAB are good for checking this), corrosion of equipment (SRB), encrustations in pipes, tanks and filters (IRB and FLOR), and sudden fluctuations in water quality (HAB, SRB and DN). Fluctuating nitrate problems could be related to changes in the biofouling with a greater probability of nitrate expression in waters high in oxygen and low in organics. The organics would trigger a greater rate of denitrification particular under a suppressed oxygen regime. Routine use of BART tests in the ongoing operations of the treatment plant can allow earlier control of potential serious biofouling events.

- **Bottled Water Plant Operators.** Bottled water represent a growing fraction of the consumed water since it reflects a superior product in the minds of the consumer to potable water supplies provided by local agencies. While ozonated and carbonated waters do have the microbial loadings suppressed to varying degrees depending upon the techniques employed, there is still a potential for the water to degenerate as a result of microbiological activity. Most commonly, this will take the form of clouding, deposits, tastes and odors. If these events occur when the product is already with the distributor or final retailer, then this would have serious consequences for the bottling company. Quality assurance and quality control can be achieved using the BART testers to determine that the source water is not fouled with aggressive bacteria and that the ozonation or carbonation has effectively acted as a disinfectant to suppress the nuisance microbes.

- **Environmental Managers.** The largest biomass by far on Earth belongs to the microorganisms. This group is not sitting there passively while the biota (animals and plants) quietly does all of the "work." Microorganisms are ubiquitous and functionally active whether they are in the human body (90% of the cells in the human body are microbial cells), in soils, waters, oil and gas, muds and sedimentary rocks. Environmental managers face the task of "managing" the environment and it is essential that the role of microorganisms in that environment be recognized. The BART testers offer the potential to take "snapshots" of the aggressivity of the various components in the microbial biomass that can have a significant impact on the environment of concern.

- **Sanitary Landfill Operators.** There are a number of microbial challenges faced by sanitary landfill operators simply because of the highly organic nature of the fill materials deposited in the landfill. In going down through a landfill, there are a series of stratified activities predominantly microbial in form. These include (going from the top down):
  - ▸ Surface growths on the redox front dominated by methanogenic bacteria that are able to degrade methane.
  - ▸ Biogas generation zone in which methanogenic bacteria are very active producing copious quantities of methane.
  - ▸ Drainage systems in which bacterial activity causes the generation of thick plugging slimes (dominated often by SRB and SLYM bacteria). Should these growths get too aggressive, then there could be reduced permeability that would lead to the water mounding in the landfill and breaking out through side erosions.
  - ▸ Leachate outflows from the drains. Very aggressive aerobic activity is likely to occur around the redox fronts at these sites leading to radical nitrification (nitrate production) and heavy slime growths (dominated by HAB, SLYM, FLOR and IRB).

Both the functionality and stability of sanitary landfill operations can be severely compromised by aggressive microbial activities. An ongoing monitoring of these nuisance microbial groups using the BART testers can aid in predicting and controlling problems before they become uncontrollable.

- **Operators of Recreational Waters.** These waters range from spas, swimming pools, hot tubs and beaches. With these waters there is a primary concern to reduce the hygiene risks to the users by the routine examination for coliform bacteria. However, there are other problems particularly with hot tubs, swimming pools and spas that are caused by other nuisance bacteria that can be detected using the BART testers. The effects of the nuisance bacteria would fall under the categories of reducing plant efficiencies, reducing water quality, and generating unacceptable slime growths. There are both economic and user acceptability issues involved in the microbial biofouling problems which can be monitored and managed using the BART testers.

- **Irrigation Operators.** Vast volumes of water are used in the irrigation industry. This water is subjected to radical changes in pressures and flow rates often under increasingly oxidative conditions. Such shifts in conditions can cause a focusing of microbial slime growths within the system and nozzles that can radically reduce efficiencies and increase operating costs. Most commonly, the SLYM and IRB are likely to dominate under low iron and high iron conditions respectively. If there is a low oxygen concentration in the water, high sulfates or hydrogen sulfide ("rotten" egg odor, black water), then the SRB may be dominant in the irrigation system. Cleanliness and sanitization of the equipment (confirmed by the routine use of the BART testers) is likely to pay dividends through improved efficiencies and higher quality water for irrigation.

- **Hazardous Waste Site Operators.** While these sites may be very hazardous to humans, the environments created may be very conducive to extensive microbial activity. Such activity can be related to the rates of biodegradation and bioaccumulation activities being generated by the naturally attenuated consortia active at the site. Additionally, the operation of treatment facilities, injection and recovery wells, distribution lines and storage tanks can all become severely compromised. For example, returning treated water via the injection wells often becomes highly aerated. Upon injection, a redox front forms around the well in which bacterial slimes grow causing erratic losses in permeability. For the operator of hazardous waste sites, the BART testers provide a simple monitoring tool to determine the level of bacterial activity occurring when used routinely. Management of the site can subsequently be improved through this routine monitoring of the levels of aggressivity (most simply monitored by the time lags observed).

- **Cooling Tower and Heat Exchanger Managers.** As a matter of routine, water is used as the heat sink in many processes. The heat in that water is removed to the air (e.g., cooling tower) or to a greater volume of water (e.g., heat exchanger). For the heat to move efficiently in the transfer from the water to the receiving medium, there should be no interferences. Biofilms (slimes) forming at these interfaces can severely reduce this heat exchange in several ways. Failure to control these biofilms can be expensive due to losses in the process efficiency that causes equipment to fail to meet specifications. Controlling the biofilms is usually achieved by the application of biocides. By the routine testing of the waters using the BART testers, the effectiveness of the biocides in suppressing the biofilms can be determined conveniently and easily. Increases in the aggressivity can be determined by the shortening in the time lags while the success of a biocide treatment may be seen through lengthening of the time lags. As a rule of thumb, a one-day increase in the time lag reflects a one order of magnitude reduction in the numbers of bacteria in the water.

D. Where To Use BART Testers?

BART testers were primarily developed to determine the aggressivity of different groups of bacteria in water. The reason the BART testers are so suitable for the determination of the types of aggressive bacteria is that so many different environments are presented in such a small volume (15ml). When there is activity, this is recognized by activity within the test vial that may be seen as such events as color shifts, cloudiness, and gassing. These are convenient to observe and so a full laboratory is not necessary in order to conduct BART tests. As a result, the tests themselves can be performed away from the laboratory in an office, a field station, even in a trailer, a tent or even in a hotel room! It should be remembered that the BART format for field use has the outer tube that provides an additional barrier to prevent possible odors or leakage from the inner test vial.

One important question is:"At what temperature should I keep the BART test at while the test is running?" Microbiologists usually refer to the temperature at which the BART test is "running" as the incubation temperature. Commonly, the incubation temperature of choice is room temperature and that can be anywhere from 19°C to 25°C. Samples can range in temperatures from 4°C to 35°C. The ideal would be to operate the BART tests within 5°C of the temperature at which the sample was taken. For waters with temperatures of between 15 and 35°C, room temperature may be fine since the maximum difference between sample and incubated temperature would be 17°C and, commonly, it would be less than 5°C. If the water was sampled from a site where the temperature was less than 15°C, then the types of bacteria that would be aggressive would probably grow better at lower temperatures. These types of bacteria are called "psychrotrophic" and can probably best be grown in a refrigerator set at 8 to 10°C. This is not a very cold setting but would be within the optimal growth range for many psychrotrophic bacteria. In tropical countries where the temperature is close to blood heat year round, then optimal incubation conditions would at 35 to 37°C. This could be undertaken in a room that is not air-conditioned. It should be remembered that the BART tests should always be incubated out of direct sunlight although regular room lighting does not appear to affect the tests. The exception to this rule is the ALGE-BART™ that does require indirect sunlight or continuous daylight fluorescent lights to allow the algae to photosynthesize.

Where water samples have been obtained from source water at temperatures higher than 35°C, there is a concern about where and how to conduct the BART tests. As a rule of thumb, the incubation temperature should be, in these circumstances, within 10°C of the original water temperature (preferably within 5°C above that temperature). To conduct these tests, there would need to be an incubator adjustable to those temperatures or a very warm location would have to be found. It should be remembered that the safe upper limit for incubating the standard BART test is 70°C. Above that temperature, the grade of polystyrenes used in the test vials begins to lose its structural integrity and buckle.

E. When To Use BART Testers?
There are three conditions under which the BART tests may be used:

1. To determine the cause of an abnormal event that may involve microbial activity.
2. To monitor the effectiveness of a treatment designed to control the abnormal event diagnosed as being at least in part microbial.
3. To effectively prevent a recurrence of the abnormal event through an ongoing testing and reactive treatment scenario.

Each of these three circumstances involves a different approach. For condition 1, the type of BART to be used is not certain because an abnormal event has occurred that is thought to involve microbial activity. As a result of this uncertainty, a broad spectrum of BART testers should be used to test the water sample. Commonly, the range of testers that could be used would include (aerobic conditions):

HAB, SLYM, IRB, FLOR and SRB

While under anaerobic conditions, a different spectrum of BART testers may be selected:

SRB, BIOGAS, SLYM and IRB

For condition 1, the sample should be taken from the site where the abnormal event is occurring or just downstream of the event. Remember that for most of the time greater than 90% of the microbes are in slimes (biofilms) attached to surfaces and so these would not even be present in the water sample! A negative BART does not mean a negative problem but simply means that the bacteria causing the problem were not in the water sample being tested (they were in the slimes the water passed over before being sampled).

For condition 2, the circumstances are slightly different in that BART tests have already been conducted and, normally, the time lags would be different, at least marginally, for each of the types of the BART tests used. Commonly, it is the two BARTs™ which have the shortest time lag that may be selected to determine whether a treatment management strategy now being applied to the sample is effective. It should be remembered that BART tests giving longer time lags may also be important. This is particularly true of the IRB-BART™ that can produce complex reaction patterns that reflect the form of the bacterial consortium in the sample (seen as the sequence within which the reactions actually are observed). If these reaction pattern signatures do shift during the treatment, then there is a list of the meaning of each reaction pattern signature (the order in which the reactions occur) in terms of which type of bacteria are dominant.

Essentially in condition 2, the objective is to try to evaluate the success of the treatment strategy applied primarily through the impact on the time lags. A dogmatic interpretation of this would be that:

▸ For each day of additional time lag delay, it can be considered that there would be one order of magnitude reduction in the population for each day's lengthening in the time lag. For example, a water sample contained 100,000 cfu/ml and had a time lag of 2 days before treatment and this lengthened for 4 days after treatment. This meant that the treatment cased a two-day lengthening to the time that would be two orders of magnitude (99% reduction to 1,000cfu/ml).

▸ If the time lag did not increase or decrease, then the treatment applied did not have any effect on the aggressivity of the bacteria being monitored using the BART tests.

▸ If the time lag shortened after the treatment, then not only did the treatment prove to be ineffective, but it created a condition in which the bacteria were able to become more aggressive. This stimulation of the bacteria may have been due to either: (1) the treatment including chemicals that could directly stimulate the bacteria (e.g., ineffective organic biocides, phosphates, organic carriers); or (2) the treatment could cause the release of bacteria into the water that had been attached. Remember that, in the latter case, when the bacteria are in the attached state, they will not be in the water and so essentially could be missed as simply not detected. In this case, the treatment may have worked effectively at dislodging the attached (sessile) bacterial growths, but had not killed, or removed, the cells from the water being tested.

Condition 2 is one that would be used to begin to determine whether the treatment was effective at controlling the microbially driven problem and also which BART testers could most conveniently be used to determine the effectiveness of an elected treatment management strategy.

Nothing lasts forever and so a single treatment of a water problem should not be viewed as ending the problem forever! Condition 3 is an essential part of a preventative maintenance strategy. Here, one or two types of BART testers are used in a routine manner to check to see that the water is still showing the lower bacterial activity level that was achieved by the treatment. If the time lags return to the pre-treatment levels then, clearly, the treatment may need to be repeated to again suppress the bacterial activity. If the time lag begins to shorten, then there is that potential to conduct a lower intensity of treatment to return to time lag to the post-treatment lengths. For water wells, one common scenario is to conduct monthly testing with just one or two BART tests (e.g., IRB and SRB). If the post-treatment time lags for these were 10 and 14 days respectively, monthly testing would show that the recovery was holding if the time lags remained the same. In practice, it may be determined that a time lag of 8 or 10 days respectively was a concern. It may be that the water sample contained some sloughing material that triggered the shorter time lag. Repeat the testing, but this time confirm with duplicate BART tests to determine whether the aggressivity was a result of a chance sloughing or the bacteria becoming more aggressive. If the time lags remain shorter in that duplicates, then a preventative treatment would need to be applied to again suppress the bacteria and get the time lags back to the longer (and more acceptable) levels.

It is not responsible to propose the same time lags as being acceptable for water systems of all the various types that require management. Since each water system or well offers some unique parameters, it is much better that the routine (conditions 1, 2 and 3) be followed and a practical strategy developed that is appropriate to that water. The target should be set in the light of the activity associated with conditions 1 and 2 and then used to support sustainable water management of that system with a minimum of diversions from the established schedule of testing and treatments. Condition 3 would then be used to operate the system with the confidence of an "advanced warning system" and a treatment that has been validated by repeated appropriate application. Should the treatment start to fail, this may be because the microbes causing the problem have adapted to that treatment. It should be remembered that microorganisms are adaptable and they have the ability to adapt to treatments when these are used repeatedly. The history of antibiotic therapy is plagued with failures due to the adaptation of the targeted bacteria to the treatments.

F.   Which BART Testers To Use?

Some would call this the million-dollar question. There are two ways to address this question: (1) define the environments that each of the BART testers can best be used to detect microbial aggressivity; and (2) take each environment and define which BART testers would be most appropriately applied on condition 1 events. This assumes that many of the applications of the BART testers would begin with an imminent or serious problem for which rehabilitation is urgently required. It has to be remembered that bacterial consortia may not be detectable by just one of the BART testers. Sometimes the consortium can cause reactions in more than one of the BART testers. This means that there can be overlap. The range of detection of nuisance microorganisms will be discussed below for each of the major BART testers.

IRON-RELATED BACTERIA, IRB-BART™

Iron is well known to be a critical substance for all life. In animals, it is a common part of the mechanisms for moving oxygen throughout the living body. Because iron plays such an important function in the energy metabolism, there is considerable biological competition for iron. Microorganisms also compete for iron and the use of various types of proteins called siderophores (e.g., hydroxamates and catechols). Additionally, many bacteria can also bind ferric (Fe^{+++}) ions into chelating structures know as ligands. This means that many bacteria are able to bind and hold iron in many forms to make large iron-rich structures that are sometimes seen as encrustations, tubercles and bog iron ore deposits. Little is known of the possible use of this iron to generate electro-motive forces (EMF) as a part of the growth of these iron-related bacteria. There is one group of bacteria, called the magnetotactic bacteria, which actually posses small magnet-like structures (magnetosomes) and are able to sense magnetic fields.

So complex are these various biochemical systems for holding onto iron, the precise nature of these events remain only partially understood. However, there are many bacteria which can continue to accumulate iron to the point that the growth becomes almost saturated with oxidized iron and forms a hardening clog or encrustation. Such mineralizing growths may also incorporate carbonates and sulfides with a high iron content (going from 1% up to as high as 40% dry weight) and reducing organic content (declining to as low as <1% organic carbon). The formation of hardening clogs/encrustations can seriously impair the designed hydraulic characteristics of the infested region, causing degenerated water quality and production capacities.

In using the IRB-BART to examine waters for the presence of iron-related bacteria, it has to be remembered that iron bacteria grow predominantly on surfaces and not directly in the water. When testing water, the BART user has to assume that the IRB have detached, are suspended, and possibly are active in the water. As a consequence of this problem, there is a potential for an IRB-BART to give a "false" negative since the IRB are absent from the water but are present on the surfaces over which the water is flowing towards the sampling site. To get IRB to release and enter the flowing waters, it is necessary to cause a shift in the local environment that will make the conditions more hostile to the IRB. This is easily done by changing the pumping conditions (e.g., turn the pump off for a day if it is an active well) or applying a mild chemical shock using something like a low-dosage hypochlorite.

IRB infestations usually occur in the presence of oxygen and so may be more readily seen as slimes, clogs or encrustations. Over the century, these growths have had two common features: the presence of high concentration of ferric (Fe^{+++}) and of high populations of IRB (either as stalked *Gallionella*, the sheathed IRB or the heterotrophic IRB). The seriousness of these growths in engineered structures has led to the use of the term "Iron Bacteria." Recent research has shown that these bacteria are able to shunt the iron through oxidative and reductive states through ferric (Fe^{+++}) and ferrous (Fe^{++}) forms respectively. The ^{BART} biodetector is designed for the detection of these bacteria and is able to perform both the oxidative and reductive based reactions involving iron. This comprehensive group is known as the "Iron-related Bacteria" (IRB).

The medium selected for the culture of the IRB is based on an original formulation developed by Sergei Winogradsky in which the major form of iron is presented as ferric ammonium citrate. The IRB-BART thus provides the major carbon (citrate), nitrogen (ammonium) and iron (ferric) from the same complex chemical form. When the crystallized pellet in the base of the test vial begins to dissolve after the sample has been added, a complex series of reactions occur. These reactions are influenced by both the chemical and biological composition of the sample and the redox and nutrient gradients created in the BART test. Under sterile conditions, a sample may be expected to cause a gradual dissolving of the nutrients from the pellet with the formation of a colored transparent diffusion front which gradually ascends through the fluid column until all of the liquid medium has a similar color. Where there has not been any major chemical reaction and the sample contains some oxygen (oxidative), the resultant color can generate yellow. If the sample is reductive (devoid of oxygen) and contains a relatively high calcium-magnesium concentration, the diffusion front may become a transparent green color.

RPS (reaction pattern signatures) revolve around a complex pattern of signals which are generated when the IRB in the water sample begin to utilize the nutrients and manipulate the ferric form of iron present in the base of the inner BART test vial. Common events range from:

- gas formation (common where anaerobic conditions exist),
- clouding (commonly at the REDOX (reduction-oxidation) front),
- slime formations (commonly starting at the base or around the FID ball in the test vial),
- color changes (which can pass through various shades of yellow, red, brown, to black, or through shades of green).

Careful QC is employed during manufacturing to ensure that the ferric ammonium citrate yields a consistent reproducible response to the various test cultures.

Iron-related bacteria (IRB) are difficult to enumerate since they are subdivided into a number of groupings (e.g., iron oxidizing and iron-reducing bacteria). These bacteria function under different Redox conditions and utilize a variety of substrates for growth. By the routine (e.g., monthly) testing of water or wastewater using this technique, the levels of aggressivity, possible population and community structure (RPS) can all be determined. The status of an iron-related bacterial population within a given sample can be determined and related to any biofouling in the surrounding environment.

To conduct the test, it is necessary to add 15ml of the sample to the biological activity reaction test biodetector. The ball floats up and restricts the entry of oxygen into the liquid medium. At the same time, components in the modified Winogradsky selective culture medium for IRB begin to diffuse upwards into the sample from a dried medium pellet in the base of the biodetector. Two gradients form within the fluid column: nutrients diffusing upwards, and oxygen diffusing downwards. These gradients form a variety of different habitats in which IRB can flourish. The color displayed by microbial activity may be a result of the form into which the ferric iron becomes modified in the medium.

It should be noted that, in a biologically active BART tester, the ferric form of the iron added with the selective Winogradsky medium will revert to the ferrous form along the reductive (lower) part of the redox gradient. Commonly, where there is a radical reduction of the ferric form to the ferrous during the early phase of an IRB-BART test, the color of the diffusing medium in the bottom of the BART tester may shift from a yellow to a green. This should be considered negative unless this "greening" at the base of the inner test vial is accompanied with clouding.

Reaction Patterns

There is a range of reactions that can occur in the IRB-BART, all of which can be observed. It is recommended that the BART tester be held up to a diffuse light to confirm some of these reactions which may be difficult to see against a dark background. (Note that traditional numbered reactions are shown in parentheses and some reactions have been discounted as unreliable):

BC	Brown Cloudy	(Reaction 4)
BG	Brown Gel	(Reaction 3)
BL	Blackened Liquid	(Reaction 10)
BR	Brown Ring	(Reaction 4)
CL	Cloudy Growth	(Reaction 2)
FO	Foam	(Reaction 5)
GC	Green Cloudy	(Reaction 8 & 9)
RC	Red Cloudy	(Reaction 7)

Each of the reactions has been produced in a unique manner by the various species and consortia of bacteria becoming active in the test. There is, therefore, no specific form of any reaction pattern because these are controlled by the form of bacterial growths. Below is listed the descriptions for each of the IRB-BART test reactions.

CL – Clouded Growth

When there are populations of aerobic bacteria, the initial growth may be at the redox front that commonly forms above the medium diffusion front. This growth usually takes the form of lateral or "puffy" clouding which is most often grey in color. It should be noted that if the observer tips the BART slightly, the clouds will move to maintain position within the tube. Commonly, the medium will be darker beneath the zone of clouding and lighter above.

BG – Brown Gel

In this reaction, a basal, gel-like brown growth forms that maintains structure and position even when gently rotated or tilted. This brown gel can occupy the whole of the basal cone of the inner test vial and also extend up the sidewall of the inner test vial to a height of <15 mm. The solution above the gel is commonly clear and colorless. Over time it is often noticed that the size of the gel mass will grow and later shrink. Detachment sometimes happens so that a single brown gel-like mass can be seen floating in the test vial.

BC – Brown Cloudy

Unless there is a very large population of IRB in the sample, this reaction is normally a secondary reaction (often following reactions CL, FO, or RC) and may be recognized as a dirty brown solution that may have a brown ring around the ball.

FO – Foam

This is a very easy reaction to recognize since gas bubbles around the ball form a foam ring or sometimes the bubbles collect over greater than 50% of the underside of the ball. On some occasions, bubbles will collect on the walls of the inner test vial but this is not significant until the bubbles collect around the ball. The solution usually remains clear but commonly has a yellow or greenish-yellow color. The bubbles can sometimes be seen in the foam to be individually coated with slime that may give the bubbles a color ranging from brown through to orange, yellow or grey. Sometimes when integrated together into a foam, this foam is tough enough to either "lift" the FID out of the liquid solution or submerge the FID below the surface of the liquid solution.

Do not confuse this reaction with the generation of bubbles (usually randomly) when oxygen supersaturates as the sample temperature comes up from a lower temperature (of the sample's source). These bubbles are recognized as being reflective and not bound in any slime and dispersed within the inner test vial under the ball and on the walls. They usually disappear within two days.

This FO reaction is most commonly related to a sample in which many microbes are functioning anaerobically. It can often be "harmonized" with the presence of SRB (reactions BB, BT or BA). In other words, the occurrence of a FO in the IRB-BART can often be followed by a positive detection of SRB in the SRB-BART if that test has been performed on the same sample.

RC – Red, Slightly Clouded

The liquid medium remains a clear to a dark reddish solution. The solution will cloud fairly quickly and shift to a BC reaction generally after a BR has formed around the ball.

BR – Brown Ring

A reddish- brown to dark brown slime ring forms around the ball. This ring is entire and tight and usually <3 mm in width. Generally, the brown slime ring will sit between the liquid surface and the equator of the ball and commonly intensifies over time. On some occasions this reaction possesses an unusual feature in that the slime ring can "bio-lock" the ball to the walls of the test vial. In these cases, when the test vial is turned upside down, the ball remains (glued) in-place and the liquid remains above the ball. What has happened is that the ring has become formed biologically into an hydraulic barrier.

GC – Green Clouded

Solution goes to a shade of green and becomes cloudy without necessarily the formation of defined clouds or gel-like forms. No slime ring is formed around the FID. This cloudiness will gradually increase and often this reaction will shift to a dark green very cloudy solution. As the solution becomes a darker green and cloudier, a BR reaction may form but this is usually fairly thin.

BL – Blackened Liquid

This is commonly a secondary or tertiary reaction rather than an initial reaction. It is recognized as a clear, often colorless, solution surrounded by large blackened zones in the basal cone and up the walls of the inner test vial.

Other reactions not coded are described below. These reactions occur less than 1% of the time in water testing using the IRB-BART.

If "fuzzy" growths form around the ball in the IRB-BART, then it is probable that the   water sample had traveled through a semi-saturated zone where there was fungal activity. These create reaction 13 in which a white, grey or speckled "fuzzy" mat forms around and even over the ball. The upper surface of the mat often forms into a tight mass with an irregular surface. The lower surface of the mat can often be seen to be extending into the liquid medium by thread-like processes 2 to 5 mm in length. These growths may bio-lock the ball to the wall of the inner test vial for a period of time. Solution usually remains fairly clear but globular-like deposits may be present. Solution may cloud over time. This reaction is caused by the presence of large populations of fungal spores in the water.

RPS (Reaction Pattern Signatures)

Because of the complex communities that form the iron bacteria, the reaction patterns can develop some very distinctive sequences. In the last ten years, the meaning of the sequences (RPS) has been determined. The common characterizations are listed below:

- **BC – WB – BR**
  IRB with carbonate deposition and some slime formers present
- **CL – GC**
  Mixed heterotrophic IRB dominated by Pseudomonads
- **CL – BG**
  Mixed heterotrophic IRB with some Enteric bacteria (possibly *Enterobacter*)
- **CL – BC**
  Mixed heterotrophic IRB
- **CL – BC – BR**
  Mixed heterotrophic IRB with some slime formers
- **CL – FO**
  IRB with mixed aerobes and some anaerobic activity
- **CL – BC**
  Where a white deposit forms in the vial. Aerobic IRB with carbonate deposition
- **FO – CL**
  Anaerobic bacteria with some aerobic heterotrophic IRB
- **FO – CL – RC**
  Anaerobic bacteria with some aerobic heterotrophic IRB and Enteric bacteria (possibly *Enterobacter, Citrobacter* or *Serratia*)
- **FO – CL – BC – BR** Mixed anaerobic and Enteric bacteria with some slime-forming IRB
- **FO – BR – BC**
  Mixed anaerobic and IRB with some aerobic slime-forming bacteria
- **FO – GC**
  Mixed anaerobic and aerobic bacteria dominated by Pseudomonads
- **FO – GC – BL**
  Mixed anaerobes, Pseudomonads and Enteric bacteria
- **GC**
  Most of the bacteria present are Pseudomonads
- **GC – BL**
  Pseudomonads dominate with some IRB and Enteric bacteria present
- **RC – CL – BR**
  ·Enteric bacteria dominate

The IRB are generally slow growing and often will display the first reaction as either a foam (FO) or a cloudy plate (CP). The consortium is complex and involves a mixture of stalked and sheathed bacteria along with heterotrophic and slime-forming bacteria. Because of the complex nature of this consortium, it takes longer to become established and is more likely to show a succession of secondary reactions as the consortium stabilizes.

Time Lag (days of delay) to IRB-BART Populations

The populations of IRB can be determined using the time lag to the observation of first reaction. This relationship is shown in Table Four.

Table Four

The Relationship Between Time Lag and the Population For Iron-related Bacteria

Time Lag (days)	Population cfu/ml (log)
1	1,000,000 (6.0)
2	100,000 (5.0)
3	10,000 (4.0)
4	5,000 (3.6)
5	1,000 (3.0)
6	100 (2.0)
7	100 (2.0)
8	100 (2.0)

Risk Potential Assessment

The IRB are a complex of many bacteria that possess a common ability to utilize iron. As a result this test has a complex set of reactions which can be displayed. The shorter the time lag to the IRB displaying a reaction, the greater the aggressivity and the need to treat. Not all reactions are equally important in determining the aggressivity of the IRB (and therefore the need to treat). Below is a list of the reactions described previously and the relative importance in relation to the need to treat. Concern can be expressed through the shortness of the time lag (in days) as:

1-2.   Very aggressive (treatment should be started as early as convenient)

2-4.   Aggressive (treatment should be considered in the near future before the condition degenerates further)

5-8. Moderately Aggressive (treatment may not be required but vigilance through ongoing testing should be practiced)

>8. Normal Background Levels (routine testing is recommended)

Table Five

Relationship Between the Time Lag to the First Reaction in an IRB-BART and the Aggressivity of the Iron-related Bacteria

		Aggressivity (significance)			
		Very	Sign.	Moderate	Not
BC	Brown Cloudy	<2	3	4-8	>8
BG	Brown Gel	<1	2-6	7-8	>8
BL	Blackened Liquid		<2	3-6	7-8 >8
BR	Brown Ring	<1	2	3-6	>6
CL	Cloudy Growth	<0.5	0.5-2	3-4	>4
FO	Foam	<0.5	0.5-1	2-4	>4
GC	Green Cloudy	<1	2-4	5-8	>8
RC	Red Cloudy	<1	2-3	4-8	>8

Some remedial treatments should be considered urgently where the time lag (in days) shows aggressivity to be at the 1 or 2 level. Where there has been a RPS (sequence of reactions to form a signature), then the aggressivity should be considered to be equivalent to the most aggressive of the reactions using the above table.

Hygiene Risk

Four of the possible reactions can indicate a potential hygiene risk. These include:

BG, BL, GC and RC. Where these are found to have a time lag that would project an aggressivity of 1 (very aggressive) or 2 (aggressive), then a fecal coliform test should be performed to ensure that there were no fecal coliform bacteria present. Note that the use of the total coliform test could yield a positive since some of the bacteria causing these reactions could be environmental enterics. If the RPS includes **GC**, a test for the presence of fluorescing pseudomonads should also be performed.

SULFATE-REDUCING BACTERIA, SRB-BART™

Sulfate-reducing bacteria (SRB) are a group of anaerobic bacteria that, as a part of their normal activities, generate hydrogen sulfide (H_2S). This product can cause a number of significant problems. These range from "rotten egg" odors, through to the blackening of equipment, waters and slime formations, and the initiation of corrosive processes.

Detection of these microorganisms is made more challenging because they are anaerobic and tend to grow deep within biofilms (slimes) as a part of a microbial community (consortium). Detection of the SRB is therefore made difficult because SRB may not be present in the free-flowing liquid over the site of the fouling but are growing deeper down in the biofilms. Because of this, the symptoms of SRB fouling may precede their detection using the SRB-BART unless a successful attempt is made to disrupt these biofilms and cause the SRB to come up into the liquid.

The sulfate-reducing bacteria are an unusual group in that they utilize hydrogen rather than oxygen as the basic driver for many of the metabolic activities. As a result of this, the SRB are anaerobic and are inhibited by the presence of oxygen. Sulfate reduction appears to be coupled to the formation of ATP (a major energy driver in metabolism) by a proton motive force (PMF) derived from electron transport. The bottom line is that the sulfate is reduced in a step-wise fashion to H_2S while releasing energy for growth. It is the H_2S which creates the problems through electrolytic corrosion, "rotten" egg smells, bad taste problems and the formation of black slimes.

There is another group of SRB which cause the reduction of sulfur to H_2S but these are not detected using the SRB-BART. Usually, these sulfur-reducing bacteria are less common and, hence, have been discounted in the SRB-BART tester. Upon request, there is a tester for the sulfur reducing bacteria (SURB-BART™) which can be made to special order.

SRB activity in the BART tester is easily recognized since the sulfate becomes reduced to hydrogen sulfide. This product now reacts with the diffusing ferrous iron to form black iron sulfides. This sulfide commonly forms either in the base (as black precipitates) and/or around the ball (as an irregular black ring). In the latter event, the SRB may form a part of an aerobic consortium forming around and on the FID ball. Generally, where this happens, the blackening may be seen as granular structures held within the slime ring that is commonly not totally black.

The SRB-BART uses the short chain fatty acids to provide the substrates for the growth of the SRB. On some occasions, heterotrophic anaerobic bacteria can also become very active in the BART test and often grow faster than the SRB. When this happens, the liquid will tend to go cloudy. Usually, this is seen as a gel-like clouding most commonly in the bottom third of the BART inner test vial and shows that anaerobic heterotrophs are present and active. It should be remembered that these bacteria may not necessarily grow in the SLYM-BART™ since the major organic carbon nutrients are not short chain fatty acids.

Under exceptional circumstances, an SRB-BART may display a blackening very quickly (e.g., less than half an hour). In this case it is likely that the sample being tested contains some residual hydrogen sulfide which has rapidly reacted with the iron in the test vial. Where this happens, it is recommended that the water sample be aseptically aerated to drive off the gaseous hydrogen sulfide from the sample before conducting the SRB-BART test. While the aeration would admit oxygen to the sample, the SRB should survive through being protected by the other bacteria within the slime formations.

Reaction Patterns

BB	-	Blackened Base (Reaction 1)
BT	-	Blackening around Ball (Reaction 2)
BA	-	Blackening All in Base and Ball(Reaction 3)

There are three reaction patterns that are positive for the SRB. Detailed descriptions of these is given below:

BB – Blackened Base
The reaction is recognizable by the formation of a blackened deposit in the basal cone of the test vial. It may be first observed by looking up into the underside of the cone of the inner test vial. Blackening frequently starts as a 2 to 3-mm wide ring around the central peg and gradually spreads outwards. Black specking may also occur on the bottom 15 mm of the walls of the test vial immediately above the cone. The liquid medium should be clear (see reaction CG below) and there should be no slime ring around the ball.

BT – Blackening around the Ball
A slime ring may be viewed around the ball with patches of black specking or zones intertwined in the slime growths. The slime itself is not a characteristic of this reaction but the blackening is. The slime usually is either a white, grey, beige, or yellow color and tends to form on the upper side of the ball. The blackening often begins as a specking which gradually expands to patches within the slime.

BA – Combination of BB and BT

A combination of reactions BB and BT constitute a reaction BA. Blackening occurs both in the base and around the ball although the length of the inner test vial may not be blackened.

The other recognized reaction is a negative for SRB but commonly occurs where there are aggressive anaerobic bacteria present. Often this reaction will precede a positive reaction for SRB (i.e., BB, BT and BA). This negative SRB reaction is:

CG – Cloudy Gel-Like

While not a positive indication for the presence of SRB, this reaction is recognized since it does indicate the presence of anaerobic bacteria and often precedes the generation reactions BB, BT or BA. It is recognized by the appearance of cloud-like structures in the colorless liquid medium. Usually these form from the bottom up and initially at a height of 20 to 25 mm up the side-wall of the inner test vial. This clouded zone may expand to render the liquid medium turbid. These clouds are relatively stable structures and have defined edges.

RPS (Reaction Pattern Signatures)

- BB: Deep-seated anaerobic bacteria dominated by *Desulfovibrio*
- BT: Dominant aerobic slime-forming heterotrophs include SRB in the consortium
- BB – BA: Dominant anaerobic consortium including SRB with a fraction able to function aerobically as slime formers incorporating the SRB
- BT – BA: Aerobic slime formers incorporate SRB and are also able to colonize anaerobic conditions

Note that the SRB-BART includes another common test reaction, which does not relate to the presence or absence of the SRB in the sample under test. This test reaction is recognized by the development of a cloudy growth that often begins close to the base and gradually fills at least 20% of the liquid volume. Often this growth reaction appears almost gel-like and has a fuzzy but distinct edge. This may be admitted as a reaction:

CG-Cloudy Gel-like

This reaction does not mean that SRB are present but that anaerobic bacteria are. Commonly, the CG reaction precedes the blackening or occurs shortly after the commencement of the blackening.

Time Lag (days of delay) to SRB-BART Populations

The common relationship between the time lag measured in days of delay and the population of SRB is given in Table Six. Because the SRB commonly are aggressive as a part of a consortium of different species of bacteria, their numbers may be difficult to determine using some of the standard procedures for SRB. This methodology allows the growth of the consortium in the SRB-BART that consequently initiates greater levels of aggressivity. The populations given in Table Six reflect the higher recovery rates, and comparisons with other tests may show the SRB-BART to be the more sensitive.

Table Six

The Relationship between Time Lag and the Population
For Sulfate-reducing Bacteria

Time Lag (days)	Population cfu/ml (log)
1	1,000,000 (6.0)
2	100,000 (5.0)
3	50,000 (4.6)
4	10,000 (4.0)
5	5,000 (3.6)
6	1,000 (3.0)
7	100 (2.0)
8	100 (2.0)

Sulfate-reducing bacteria (SRB) are a narrow group of bacteria that have the common facility to reduce sulfates to hydrogen sulfide. It is this sulfide which reacts with metals (commonly iron) to form the black sulfides. These black deposits cause an identifiable reaction in the base of the tube (BB) or around the ball (BT); the SRB do function as part of a consortium that is either anaerobic (BB) or aerobic (BT).

Risk Potential Assessment

The SRB are a relatively simple consortium in which the SRB tend to either dominate over a facultatively/strictly anaerobic heterotrophic bacterial flora (BB), or become integrated into an aerobic slime-forming heterotrophic bacterial community growing around the ball (BT). Where a more complex and aggressive form of SRB are present (involving both forms of consortial activity (BA), the SRB are usually very aggressive and the BA reaction occurs without being preceded by either of the other two reactions. The risk potential for the severity of a detected SRB event can be expressed through the shortness of the time lag (in days) as follows:

1. Very aggressive (treatment should be started as early as convenient)
2. Aggressive (treatment should be considered in the near future before the condition degenerates further)
3. Moderately Aggressive (treatment may not be required but vigilance through ongoing testing should be practiced)
4. Normal Background Levels (routine testing is recommended)

Table Seven

Relationship Between the Time Lag to the First Reaction in an SRB-BART and the Aggressivity of the Sulfate-reducing Bacteria

			Aggressivity			
			Very	Sign.	Moderate	Not
BB	-	Black Base	<1	2-3	4-8	>8
BT	-	Black Ball	<1	2-4	5-8	>8
BA*	-	Black All	<2	2-5	6-8	>8

Note that the BA reaction (*) listed above must have occurred without either of the other reactions occurring first. If either the BB or the BT reaction did occur first, then the aggressivity should be based on the first of the reactions that did occur. Some remedial treatments should be considered urgently where the time lag (in days) shows aggressivity to be at the 1 or 2 level.

A non-SRB reaction can also commonly occur in this test when a cloudy gel (CG) forms. This is indicative of the presence of anaerobic bacteria (not SRB). However, on some occasions, these anaerobic bacteria can also become very aggressive and can cause deep-seated plugging. The aggressivity for these bacteria can be judged using the table below:

Table Eight

Relationship Between the Time Lag to the CG Reaction in an SRB-BART and the Aggressivity of the Anaerobic Bacteria

		Aggressivity			
		Very	Sign.	Moderate	Not
CG -	Cloudy Growth	<0.5	0.5-2	3-4	>4

SLIME-FORMING BACTERIA, SLYM-BART™

Slime-forming bacteria (SLYM) is the name given to bacteria that are able to produce copious amounts of slime without necessarily having to accumulate any iron. These slime-like growths are therefore often not dominated by the yellows, reds and browns commonly seen where IRB are present. Some of the IRB also produce slime but it is sometimes denser and has more texture due to the accumulation of various forms of insoluble iron. SLYM bacteria can also function under different reduction-oxidation (redox) conditions but generally produce the thickest slime formations under aerobic (oxidative) conditions. These can develop in the SLYM-BART™ as slime rings growing around the floating ball. Slime growth can also be seen as a cloudy (fluffy or tight plate-like structures) or as gel-like growths which may be localized or occur generally through the body of water medium. Very commonly the gel-like slime growths form from the bottom up in the test vials. One common check for these types of growth is to tilt the BART gently and see that the cloud- or gel- like growths retain their structure and tilt with the tube.

A vast majority of bacteria can produce slime-like growths. The slime is actually formed by a variety of exopolysaccharide polymers that are long thread-like molecules. These extracellular polymeric substances (EPS) literally coat the cells into a common slime-mass within which large volumes of water become clustered and bound. Often 95 to 99% of the volume of slime are actually water. Some bacteria produce an EPS that remains tightly bound to the individual cell. These are called capsules. Other bacteria generate such a copious amount of EPS that it envelops whole masses of cells within a common slime.

The role of the slime appears to be protective. If environmental conditions are harsh (e.g., due to shortage of nutrients), the slime layers tend to get thicker. Not only does the slime act as a protectant to the resident bacteria, but it also acts as a bio-sponge by accumulating many chemicals that could form either a nutrient base or be toxic to the cells. EPS may be produced by enzymatic activity (e.g., dextran sucrase or levan sucrase) on carbohydrates. In addition, EPS may be synthesized within the bacterial cells and released to form an enveloping slime.

Slime-forming bacteria tend to be aerobic and form slimes at redox fronts. In the BART tester, this front may form around the ball causing a slime ring, or deeper down in the liquid medium column to form an observable growth. This growth may be plate-like and appear to float at a specific depth, cloud-like with indefinite edges, form as basal dense slimes in the conical base of the test vial, or be gel-like and maintain its shape even when the vial is tilted. Since slime tends to be formed by bacteria under stress, it is common for the slimes to form after there has been an initial growth that may take the form of a localized or general cloudiness.

Many slime bacteria can produce various pigments that will color the slime. Such growths are usually white, grey, yellow or beige in color. These often darken over time particularly in the presence of daylight. Distinctive colored slimes include red (commonly associated with *Serratia marcescens*) and violet (associated with either *Chromobacterium* or *Janthinobacterium* species). Blackening may also occur particularly after growth. This may be a result of the production of either iron sulfides or carbonates which is commonly associated with the presence of mixed cultures including enteric bacteria in the SLYM-BART.

SLYM-BART can be used as a simple presence/absence (P/A) test capable of indicating to some extent the population size and the types of SLYM organisms present in the water sample. Different microorganisms utilize various sites along the redox gradient under a ball to grow, and regular careful observations are needed to catch the start of growth so that the time lag can be determined.

Slime-forming bacteria cause very serious engineering problems since the slime formation can compromise the engineered specifications into many systems. Primarily, the effects of the slime growths are to reduce hydraulic or thermal conductivity and reduce water quality (generally, the first symptom is increased turbidity followed by taste, odor or color problems). As the slimes slough into the water later during the infestation, it can be expected to see sudden rises in the total organic carbon, increases in aggressivity and reductions in water quality.

Reaction Patterns

DS	Dense Slime (Gel-Like) (Reactions 1 & 4)
SR	Slime Ring around the Ball (Reaction 4)
CP	Cloudy Plates layering (Reaction 2)

CL   Cloudy Growth (Reaction 5)
BL   Blackened Liquid (Reaction 6)
TH   Thread-Like Strands (Rare Reaction not recognized)
PB   Pale Blue Glow in U.V. Light (Reaction 5-PB)
GY   Greenish-Yellow Glow in U.V. (Reaction 5-GY)

Of the above reactions, it is the CL (cloudy) reaction that is by far the most common. Often the CL will be preceded by a CP which will be transient (lasting commonly less than 24 hours). Descriptions of the various reactions are given below:

DS – Dense Slime
This reaction may not be obvious and may require the observer to rotate gently the BART test at which time slimy deposits swirl up. These deposits may swirl in the form of a twisting slime when the tube is gently rotated. This swirl can reach 40 mm up into the liquid column, or it may rise up as globular gel-like masses that settle fairly quickly. Once the swirl has settled down, the liquid may become clear again. In the latter case, care should be taken to confirm that the artifact is biological (ill-defined edge, mucoid, globular) rather than chemical (defined edge, crystalline, often white or translucent). Generally, these dense slime growths are beige, white or yellowish-orange in color.

CP – Cloudy Plates Layering
When there are populations of aerobic bacteria, the initial growth may be at the Redox front that commonly forms above the yellowish-brown diffusion front. This growth usually takes the form of lateral or "puffy" clouding which is most commonly grey in color. Often the lateral clouds may be disc-like in shape (plates) and relatively thin (1 to 2 mm). It should be noted that if the observer tips the BART slightly, the clouds or plates often move to maintain position within the tube. The edges of the plates are distinct while the edges of the "puffy" forms of layering are indistinct. These formations are most commonly observed 15 to 30 mm beneath the fill line. While cloud formations will tend to extend to cause an overall cloudiness of the liquid medium (CL). These plates sometimes appear to divide (multiple plating) before coalescing into a cloudy liquid medium.

SR - Slime Ring
A slime ring, usually 2 to 5 mm in width forms on the upper side of the ball. The appearance is commonly mucoid and may be a white, beige, yellow, orange or violet color that commonly becomes more intense over time on the upper edge.

CL – Cloudy Growth

Solution is very cloudy and there may sometimes be a poorly defined slime growth around the ball. Sometimes a glowing may be noticed in at least a part of the top 18mm of the liquid medium. This glowing is due to the generation of U.V. fluorescent pigments by some species of *Pseudomonas*. The common pigments doing this are a pale blue (PB) or a yellowish green (YG) color. Note that this glowing may not be readily observable unless a U.V. light is used. The occurrence of the glowing in a U.V. light means that there is a probability of potentially pathogenic species of *Pseudomonas* and confirmatory testing is recommended.

BL – Blackened Liquid

This is commonly a secondary or tertiary reaction rather than an initial reaction. It is recognized as a clear, often colorless, solution that is surrounded by large blackened zones in the basal cone and up the walls of the test vial. The BL often parallels the BL reaction in the IRB when the two BARTs are used together to test the same water sample.

TH – Thread-Like Strands

On some occasions, the slime forms into threads that form web-like patterns in the liquid medium. Sometimes these threads which interconnect from the ball to the floor of the inner test vial.

RPS (Reaction Pattern Signatures)

- **DS - CL** Dense slime-forming bacteria producing copious EPS, facultative anaerobes dominate
- **SR - CL** Aerobic slime-forming bacteria (such as M*icrococcus*) dominating with some facultative anaerobes
- **CP - CL** Motile facultatively anaerobic bacteria dominate (e.g., *Proteus*)
- **CL - SR** Mixed bacterial flora including some aerobic slime-formers
- **CL - BL** Slime formers dominated by Pseudomonads and Enteric bacteria
- **CL - PB** *Pseudomonas aeruginosa* dominant member of the bacterial flora
- **CL - GY** *Pseudomonas fluorescens* species group present in the flora
- **TH - CL** Aerobic bacteria dominant which are able to generate slime threads (e.g., *Zoogloea*)

The slime-forming bacteria are amongst the fastest growing aggressive consortia and the medium used in this BART is very enriching and causes a wide variety of bacteria to grow rapidly. However, when the bacteria do not grow quickly, this indicates a very low population of aggressive bacteria. As a result of this, the time lags of between 3 and 6 days show a rapid decline in populations when compared to the IRB- or SRB-BART tests.

Table Nine

The Relationship Between Time Lag and the Population For Slime-forming Bacteria

Time Lag (days)	Population cfu/ml (log)
1	5,000,000 (6.6)
2	500,000 (5.6)
3	50,000 (4.6)
4	10,000 (4.0)
5	5,000 (3.6)
6	1,000 (3.0)
7	1,000 (3.0)

Risk Potential Assessment

The slime-forming bacteria are complex consortia involving many bacteria. These consortia inhabit a common "growth" of slime that acts as a communal chamber. Within these slimes, the bacterial cells are commonly dispersed and occupy only a small part of the total volume (<0.1%). Most of the slime is water bound to the organic polymers that bind the slime together. The SLYM-BART reflects the activities of bacteria that are present in the water as a result of the sloughing from the slime. As a result of this, the test may exhibit a complex set of reactions depending upon precisely which bacterial species are present in the water sample. Like the other BART tests, the shorter the time lag to the SLYM-BART displaying a reaction, then the greater becomes the aggressivity and the more urgent the need to treat. Not all reactions are equally important in determining the aggressivity of the slime-forming bacteria (and therefore the need to treat). Below is a list of the reactions described above and their relative importance in relation to the need to treat. Concern can be expressed through the shortness of the time lag (in days) as:

1. Very aggressive (treatment should be started as early as convenient)
2. Aggressive (treatment should be considered in the near future before the condition degenerates further)

3. Moderately Aggressive (treatment may not be required but vigilance through ongoing testing should be practiced)
4. Normal Background Levels (routine testing is recommended)

<u>Table Ten</u>

Relationship Between the Time Lag to the Reactions in a SLYM-BART and the Aggressivity of the Slime-Forming Bacteria

		Very	Sign.	Moderate	Not
DS	-Dense Gel Slime	<1	2	3-7	>7
SR	-Slime Ring	<1	2-3	4-6	>6
CP	-Cloudy Plates	<0.5	1-2	3-6	>6
CL	-Cloudy Growth	<1	2	3-6	>6
BL	-Blackened Liquid	<1	2-4	5-8	>8
TH	-Threads	<2	3-4	5	>6
PB	-Pale Blue Glow	<1	2-4	5-8	>8
GY	-Green-Yellow Glow	<1	2-3	4-8	>8

(Aggressivity spans the Very / Sign. / Moderate / Not columns)

Some remedial treatments should be considered urgently where the time lag (in days) shows aggressivity to be at the very aggressive or aggressive (1 or 2) levels. Where there has been an RPS (sequence of reactions to form a signature), the aggressivity should be considered to be equivalent to the <u>most aggressive</u> using the above table.

Hygiene Risk

The most significant hygiene risk generated by this test is the BL reaction that indicates that Pseudomonads and enteric bacteria are present. If this reaction occurs within eight days, then a fecal coliform test should be performed on that water to determine the hygiene risk directly. Where PB or GY reactions are observed, this should be confirmed using the FLOR-BART™.

HETEROTROPHIC AEROBIC BACTERIA, HAB-BART™

Some bacteria are able to degrade organics as their source of energy and carbon. These are known as heterotrophic. By far, the majority of these heterotrophs function most efficiently under aerobic conditions. Much of the biodegradation that occurs under aerobic conditions is due to the activities of these heterotrophic aerobic bacteria (HAB, formerly the total aerobic bacteria or TAB). Since these bacteria play a major role of biodegradation and their presence in oxygen-rich waters can be critical to the efficiency of the engineered operation, the HAB-BART™ was developed to detect these bacteria.

The unique feature of this test is the addition of methylene blue that acts as an indicator of respiratory activity. While there remains free oxygen in the water, the methylene blue dye in the water remains blue. As soon as all of the oxygen has been consumed by bacterial (respiratory) activity, the methylene blue shifts from its observable form to a colorless form. In other words, in the HAB-BART tests, when the liquid medium turns from blue to a colorless (non-blue) form, the heterotrophic aerobic bacteria have been sufficiently aggressive to have "respired off" the oxygen. At this time a methylene blue reductase enzyme becomes activated and this reduces the methylene blue to its colorless form.

Microorganisms present at depths in this test are short of oxygen and "look" for alternatives. The blue dye (methylene blue) in this test forms such an alternate substrate. When the aerobic bacteria use this dye, the color is bleached out. This usually occurs from the bottom (bottom up) or the top (top down) of the tester first. This bleaching action (decolorizing the blue dye) is the indicator of a positive reaction. Note that the dye is added to the test by inverting the charged HAB-BART for 30 seconds to allow the methylene blue chemical dried in the cap time to dissolve into the water. When the HAB-BART is returned to its normal state (cap side up), the ball rolls up through the water sample causing the methylene blue to become mixed into the water to form an even blue solution.

Methylene blue is a basic dye that can bind readily to the negatively charged microbial cells. Traditionally, therefore, this dye has been used to stain microbial cells. A feature of methylene blue is that it changes from a blue color in the oxidized state to a clear form in the reduced state. When methylene blue is added to a medium that is actively converting energy due to microbial respiration, the electrons are transferred to the dye causing it to become reduced and the dye changes from a blue to a clear state (the color disappears). The protocol has been based on the methylene blue reductase test that has been used in the dairy industry for decades to determine the potential for bacterial spoilage of milk. In the HAB-BART the objective is for the user to be able to determine the aerobic bacterial population which may be related to various forms of biofouling and bioremediation. Essentially, the methylene acts as an oxygen substitute and its reduction (bleaching) from the blue to the colorless form can be used an indication of the amount of respiratory function of the bacteria in the sample water.

Therefore, this test is an answer to the need to test water and wastewater for the presence of heterotrophic aerobic bacteria as such without trying to determine the particular groups of bacteria that may be present.

The HAB-BART determines the activity of the heterotrophic aerobic bacteria. When these bacteria are present and active, the blue dye in the biodetector becomes bleached (colorless) either from the bottom up or the top down. The faster this happens, the more aggressive are the heterotrophic aerobic bacteria.

Reaction Patterns

UP   -Bleaching moves upwards from base   (Reaction 1)
DO   -Bleaching moves downwards from ball (Reaction 2)

There are only two recognized reactions (UP and DO) and both of these relate to the form with which the bleaching occurs. There are different forms of clouding which follow the bleaching of the methylene blue and these are recognized using the CABART™ system.

UP – Bleaching moves upwards

Blue solution bleaches from the bottom up. The bleached zone may be clear or clouded. In the latter case, the medium tends to have a light to medium yellow color. Rarely does the bleaching extend beyond the equator of the ball so that a blue ring will remain around the ball with a width of 1 to 5 mm.

DO – Bleaching moves downwards

Blue solution bleaches from the top down. The bleached zone is more commonly cloudy. The bleached liquid medium tends to have a light to medium yellow color. Commonly the bleaching does extend up beyond the equator of the ball and any blue ring remaining around the ball is relatively thin with a width of 0.5 to 2 mm.

(*) Note that there is almost always a blue ring remaining around the ball and that the DO reaction will usually leave this ring intact. Furthermore, the test reaction can only be one or the other and so interpretation is restricted to one or other of these two reactions:

- **UP** Strictly aerobic bacteria may be dominant with some facultative anaerobes often present
- **DO** Facultatively anaerobic heterotrophs dominate along with some anaerobic bacteria

The relationship between the time lag (days of delay) to the bacterial population is given in Table Eleven.

Table Eleven

The Relationship Between Time Lag and the Population for Heterotrophic Aerobic Bacteria

Time Lag (days)	Population cfu/ml (log)
1	5,000,000 (6.6)
2	500,000 (5.6)
3	1,000 (3.0)
4	100 (2.0)

The heterotrophic aerobic bacteria, like the slime formers, grow very quickly and are readily detectable because of the reduction of the methylene blue from the blue (oxidative) to the colorless (reductive) state. Essentially, the methylene blue acts as a redox indicator and rapidly shows when respiratory activity is occurring because the test liquids become reductive and the methylene blue decolorizes. This test is one of the fastest of the BART tests as well as being the easiest to read. It functions most effectively when the bacterial consortia in the sample are dominated by heterotrophic aerobes.

Risk Potential Assessment

The heterotrophic aerobic bacteria are subdivided into two major consortial groups in the HAB-BART. These are dominated by either: the strictly aerobic (UP), or the facultatively anaerobic (DO) heterotrophic bacteria. The risk potential for the severity of a detected HAB event can be expressed through the shortness of the time lag (in days) as follows:

1. Very aggressive (treatment should be started as early as convenient)
2. Aggressive (treatment should be considered in the near future before the condition degenerates further)
3. Moderately Aggressive (treatment may not be required but vigilance through ongoing testing should be practiced)
4. Normal Background Levels (routine testing is recommended).

Table Twelve

Relationship Between the Time Lag to the CG Reaction in an SRB-BART and the Aggressivity of the Anaerobic Bacteria

		Aggressivity			
		Very	Sign.	Moderate	Not
UP	-Bleach Up	<0.5	1-2	3-4	>4
DO	-Bleach Down	<1	2-3	4-6	>6

FLUORESCING PSEUDOMONADS, FLOR-BART™

The seudomonads are a very important group of Gram negative bacteria that are found in very substantial numbers in soils, waters and many other natural materials. In association with many plants and animals, the pseudomonads can act as agents for disease. In aerobic bioremediation and biodegradation processes, members of the pseudomonads often play critical roles in the biochemical breakdown of critical organic compounds. These various important aspects have led to the development of FLOR-BART which generate conditions favorable to the growth of the pseudomonads.

One critical aspect of this biodetector is the ability to generate soluble fluorescent pigments when some species of the genus *Pseudomonas* are dominant in the water. These pigments are usually produced after growth has occurred and generally can be detected most easily in the culture medium around the ball (top 20mm of the liquid column). Detection is by the use of an ultraviolet (U.V.) lamp aimed at the top one third of the liquid column in the FLOR-BART. Maximum excitation of these fluorescing molecules is at 400nm. There are two main pigments, pyocyanin and pyoverdins. Pyocyanin is a distinctive pigment that fluoresces with a pale blue to blue color and is most commonly associated with the species, *Pseudomonas aeruginosa*. This species is commonly associated with clinical specimens (wounds, burns, otitis, sepsis, pneumonia, urinary tract infections), a condition known as "blue pus," and is a hygiene concern in recreational waters. Pyoverdins is the name given to a group of other fluorescent pigments generated by different species of *Pseudomonas*. Commonly these pigments are referred to as fluorescens and generally have a greenish-yellow glow. The species *Ps. fluorescens* generates these types of pigments and is commonly associated with the spoilage of foods (eggs, cured meats, fish and milk). The FLOR-BART has been designed to generate these pigments where there is a dominance of fluorescent pseudomonads (hence the prefix, FLOR). If *Ps. aeruginosa* is detected in a water sample and there is a concern for the potential hygiene risk, it is recommended that confirmatory diagnosis be performed in a recognized diagnostic microbiology laboratory using either the positive FLOR-BART or a fresh sample as the source for the diagnosis.

Other pigments are sometimes produced. These are usually insoluble and non-fluorescent in U.V. light. These are commonly yellow, beige or orange in color and tend to be transitory. One species *Ps. stutzeri* sometimes generates a reddish-brown pigment later in the growth cycle that is very distinctive. This pigment may concentrate either in the slime ring around the FID ball or in the base of the test vial.

Microorganisms present around the ball in the FLOR-BART can generate these different pigments in the presence of oxygen. Usually these pigments are generated after a cloudy growth has developed in the liquid medium but before there are intense slime-like growths around the ball (as a slime ring). The fluorescent pigments may be difficult to observe with natural and artificial light, but they can be seen using a typical broad spectrum U.V. light whereupon the pigments glow (fluoresce).

There is often a need to test waters for the presence of Pseudomonad bacteria because these bacteria are often dominant in waters which contain oxygen and are rich in a narrow range of organic pollutants (e.g., gasoline, jet fuel, solvents). When these bacteria are present and active, there are two particular events that may need to be considered. First, the presence of Pseudomonad bacteria may indicate that aerobic biodegradation is occurring and biofouling may also be happening within the system being tested. Second, some of the Pseudomonad bacteria that produce the fluorescent (glowing in U.V. light) pigments may be a hygiene risk. The faster that clouding and fluorescing happens, the more aggressive are the Pseudomonad bacteria.

For the FLOR-BART, there are two U.V. fluorescent pigments which can be recognized as:

Pale Bluish Glow that will last for one to four days and then gradually fade. The glow is normally fairly faint and should be viewed against a darkened background since direct light may make viewing more difficult. One major species bearing this pigment (pyocyanin) is *Pseudomonas aeruginosa*. It is of concern since this species can be associated with a range of opportunistic infections. It is also one of the bacterial species found associated with mastitis in cattle. This species can also be found in a variety of waters.

Greenish-Yellow Glow that may last for two to ten days and then gradually fade away. The glow becomes fairly obvious and is often visible even without using the U.V. light. One major species bearing this pigment (the pyoverdin, fluorescein) is *Pseudomonas fluorescens*. Generally, this species is not as virulent as *Ps. aeruginosa* and is often more abundant in waters and can be involved in specialized aerobic degradation of organic pollutants.

In essence, this test selectively allows the detection of pseudomonad bacteria in the water with the separation of the fluorescent species. Pseudomonad bacteria can cause a range of problems in waters. Problems range through slime formations, turbidity, taste and odor, corrosion and biodegradation through to greater hygiene risks. In recreational waters (such as swimming pools, hot tubs, restricted natural bathing sites), the presence of aggressive fluorescent pseudomonads should be taken as a potential cause for concern since these bacteria may cause a range of skin, eye, ear and urinary tract infections. Occasionally the pseudomonad bacteria will cause skin infections particularly under tight fitting bathing apparel. This is particularly a potential problem in warmer waters and hot tubs where the bathers remain relatively inactive in the waters for prolonged periods.

The pseudomonad bacteria often dominate aerobic biodegradation of organic pollutants and determining the aggressivity and possible population size can often monitor the rates of degradation. If the organic pollutant is being degraded aerobically or in a situation where there is a significant quantity of nitrates to support respiration, there is a potential for the degradation to be dominated by the Pseudomonad bacteria. Monitoring the aggressivity of these bacteria using the FLOR-BART enables the user to monitor the amount of biodegradation occurring.

Pseudomonad bacteria are also sometimes associated with taste and odor problems in water since many of the species produce distinctive odors such as "fishy" or "kerosene-like" which can become very dominant in the water.

Reaction Pattern

PB	-	Pale Blue Glow in U.V. Light
GY	-	Greenish-Yellow Glow in U.V. Light

These reactions are described in more detail below. Care should be taken to follow manufacturers cautionary notices when using a U.V. light source to observe glowing in the BART tests.

PB – Pale Blue Glow

Solution very cloudy (passes through reaction one) and then generates a glowing around FID when ultraviolet light is shone onto the side-walls of the inner test vial. This glowing fluorescence occurs usually in the top 15 to 20 mm around the ball and gives a pale blue glow. This glowing commonly lasts 2 to 3 days

GY – Greenish Yellow Glow

Solution very cloudy (passes through reaction one) and then generates a glowing around FID when ultraviolet light is shone onto the side-walls of the inner test vial. This glowing fluorescence occurs usually in the top 15 to 20 mm around the ball and gives a greenish-yellow glow. This glowing lasts commonly for 4 to 8 days (latter case).

RPS Reaction Signatures

- **PB** *Pseudomonas aeruginosa* likely to be present
- **GY** *Pseudomonas fluorescens* species group likely to be present

Table Thirteen

The Relationship Between Time Lag and the Population For Fluorescing Pseudomonad Bacteria

Time Lag (days)	Population cfu/ml (log)
1	1,000,000 (6.0)
2	100,000 (5.0)
3	10,000 (4.0)
4	5,000 (3.6)
5	1,000 (3.0)
6	100 (2.0)
7	100 (2.0)
8	100 (2.0)

The fluorescing pseudomonads are only a part of typical slime-forming or heterotrophic aerobic bacterial consortia. As such, they have to be in high populations and very aggressive in order to begin to produce the ultraviolet (U.V.) fluorescence that is typical for the species of *Pseudomonas* that are capable of doing this. A time lag of longer than five days may have a small population but only if the U.V. glow produced is pale blue. This would still be a concern if the nosocomial pathogenic bacterial species *Pseudomonas aeruginosa* was present in the sample and confirmatory tests using the traditional microbiological procedures may need to be undertaken as a precaution.

Population Assessment of Fluorescing Pseudomonads using a BART Extinction Dilution Technique

To measure the population of fluorescing pseudomonad bacteria, four dilutions of the original water sample should be used. These dilutions can be achieved using the following seim-quantitative technique:

1.   Dispense 14 ml of sterile water into each of four FLOR-BART™ tests. Label these tubes: "1", "2", "3", and "4."
2.   Charge a FLOR-BART™ with the water sample (15 ml) and label "0."
3.   Withdraw 1 ml of water from tube "0" and transfer into tube "1." Invert and gently shake tube for 10 seconds. Allow to settle (5 seconds).
4.   Withdraw 1 ml of water from tube "1" and transfer into tube "2." Invert and gently shake tube for 10 seconds. Allow to settle (5 seconds).
5.   Withdraw 1 ml of water from tube "2" and transfer into tube "3." Invert and gently shake tube for 10 seconds. Allow to settle (5 seconds).
6.   Withdraw 1 ml of water from tube "3" and transfer into tube "4." Invert and gently shake tube for 10 seconds. Allow to settle (5 seconds).
7.   Observe the tubes for PB or GY fluorescence after three days of incubation at room temperature. Note that this day may be changed if an alternate day is found to display maximum fluorescence.
8.   Refer to table below to semi-quantitatively determine population based upon the tests that exhibit fluorescence in a U.V. light.

Hygiene Risk

If a PB reaction is observed, there is a risk that *Pseudomonas aeruginosa* may be present in the water sample and could cause an infection in humans. These infections can range from pneumonia to skin, eye and ear infections. Where a population is detected and confirmed using the extinction dilution technique described above, the tube "0" FLOR-BART should be submitted to a suitable microbiology laboratory to confirm the diagnosis. If the fluorescence is of the GY type, then a similar precaution should be taken if the population is >2.0 log fluorescing pseudomonads/ml.

Table Fourteen

**The Relationship of Positive Detection of Fluorescence to
The Population**

Tube #	Population Assessment				
"0"	F	F	F	F	F
"1"	F	F	F	F	----
"2"	F	F	F	----	----
"3"	F	F	----	----	----
"4"	F	----	----	----	----
Possible Population:	>5.0	>4.0	>3.0	>2.0	>1.0
	(log . fluorescing pseudomonads/ml)				

DENITRIFYING BACTERIA, DN-BART™

DN is short for denitrification. This activity. is extremely important not only in environmental but also in geochemical terms. The reason for this is that the essentially all of the atmospheric nitrogen (N_2) has been derived from the process of denitrification which is driven by the denitrifying bacteria. It is therefore an extremely important stage in the nitrogen cycle in the crust of planet Earth. There is a distinctive cycle in which nitrogen from the atmosphere is fixed, cycles through the biomass, is oxidized to nitrate by nitrification (see N-BART™) and reduced back to nitrogen gas by denitrification which is controlled by the denitrifying bacteria.

The denitrifying bacteria are therefore an important indicator group for the decomposition of waste organic nitrogenous materials. These denitrifiers reduce nitrate through to nitrite and some continue the nitrification on down to gaseous nitrogen (complete denitrification). In waters, the presence of an aggressive population of denitrifiers can be taken to indicate that there are significant amounts of nitrate in the water. Such waters are most likely anaerobic (free of oxygen) and relatively rich in organic matter. A common use for the presence of aggressive denitrifying bacteria in waters is that these bacteria signal the latter stages in the degradation of nitrogen-rich sewage and septic wastewater. Aggressive presence of denitrifiers in water can be used to indicate that there is a potential for the water to have been polluted by nitrogen-rich organics from such sources as compromised septic tanks, sewage systems, industrial and hazardous waste sites. It is recommended that, where a high aggressivity is determined, the water should be subjected to further evaluation as a hygiene risk through a subsequent determination for the presence of coliform bacteria. In soils, the presence of an aggressive denitrifying bacterial population may be taken to indicate that the denitrification part of the soil nitrogen cycle is functional.

Denitrification serves as the major route by which complex nitrogenous compounds are returned to the atmosphere as nitrogen gas. There are four steps in the denitrification process:

$$
\overset{(1)}{NO_3^-} \ \rightarrow \ \overset{(2)}{NO_2^-} \ \rightarrow \ \overset{(3)}{NO} \quad \overset{(4)}{N_2O} \ \rightarrow \ N_{2\,gas}
$$

Nitrate   Nitrite   Nitric oxide   Nitrous oxide   Nitrogen

Denitrifying bacteria are not necessarily able to perform all four steps in the denitrification process and have been divided into four distinctive groups that can perform one or more of the various steps in the denitrification process. These are listed below:

Group 1   -   step (1) only
Group 2   -   steps (1), (2), and (3)
Group 3   -   steps (2), (3), and (4)
Group 4   -   steps (1) and (3) only

One of the largest groups of denitrifying bacteria are the enteric bacteria which includes the coliform bacteria. All of these bacteria perform denitrification under anaerobic (oxygen-free) conditions in a reductive environment.

Some of the principal genera associated with denitrification are: *Actinomyces, Aeromonas, Agrobacterium, Alcaligenes, Arthrobacter, Bacillus, Bacteroides, Campylobacter, Cellulomonas, Chromobacterium, Citrobacter, Clostridium, Enterobacter, Erwinia, Escherichia, Eubacterium, Flavobacterium, Geodermatophilus, Halobacterium, Halococcus, Hyphomicrobium, Klebsiella, Leptothrix, Micrococcus, Moraxella, Mycobacterium, Nocardia, Peptococcus, Photobacterium, Proteus, Pseudomonas, Rhizobium, Salmonella, Serratia, Shigella, Spirillum, Staphylococcus, Streptomyces, Thiobacillus, Vibrio.*

As can be seen from the list, a very wide ranging number of bacteria are capable of denitrification. Their ability to perform denitrification is controlled, in part, by the availability of the nitrate, nitrite, nitrous or nitric oxide substrates.

The patented denitrifying bacterial activity reaction test biodetector (DN-BART) has been designed to detect the aggressivity of the denitrifying bacteria that will reduce the nitrite to gaseous nitrogen (steps 2, 3 and 4). These bacteria are an important part of the nitrogen cycle in soils and waters. In waters, their aggressivity may be used to signal the fact that there is a significant degradation of nitrogenous material occurring.

Reaction Code

FB   -   Foam around Ball (Reaction 5)

FO – Foam Formation (formerly reaction five, DN-BART)
Solution usually cloudy but the major positive for FO is the presence of very many bubbles collecting over >50% of the area under and around the ball to form a foam around the ball. This shows that complete denitrification has occurred and the denitrifying bacteria are present. Populations can be assessed by the time lag to the foam formation (Table Fifteen).

There is only one reaction recognized in the DN- BART that occurs when the nitrate is completely denitrified to nitrogen gas that collects as foam (interconnected gas bubbles) around the ball. This is more of a presence/absence test and the foaming usually is generated on the second test of testing at room temperature.

Table Fifteen

**The Relationship Between Time Lag and the Population
For Denitrifying Bacteria**

Time Lag (days)	Population cfu/ml (log)
1	5,000,000 (6.6)
2	100,000 (5.0)
3	1,000 (3.0)
4	100 (2.0)

The denitrifying bacteria tend either to be aggressive and cause a rapid denitrification, or to be relatively placid. This test now functions through the detection of the complete denitrifiers. These bacteria reduce the nitrate to dinitrogen gas that appears as a foam ring around the ball. Generally, if the test is still negative after a time lag of two days, the population can be considered to be very small and non-aggressive.

Population Assessment of DN using BART Extinction Dilution

To quantify the numbers of denitrifying bacteria in the sample, a dilution (extinction) technique would need to be used. To measure the population of denitrifying bacteria, four tenfold dilutions of the original water sample should be used. These dilutions can be achieved using the following technique:

1. Dispense 14 ml of sterile water into each of four DN-BART™ tests. Label these tubes: "1", "2", "3", and "4."
2. Charge a DN-BART™ with the water sample (15 ml) and label "0."
3. Withdraw 1 ml of water from tube "0" and transfer into tube "1." Invert and gently shake tube for 10 seconds. Allow to settle (5 seconds).
4. Withdraw 1 ml of water from tube "1" and transfer into tube "2." Invert and gently shake tube for 10 seconds. Allow to settle (5 seconds).
5. Withdraw 1 ml of water from tube "2" and transfer into tube "3." Invert and gently shake tube for 10 seconds. Allow to settle (5 seconds).
6. Withdraw 1 ml of water from tube "3" and transfer into tube "4." Invert and gently shake tube for 10 seconds. Allow to settle (5 seconds).
7. Observe the tubes for **FO** (foam) after two days of incubation at room temperature.
8. Refer to Table Sixteen below to determine population.

Table Sixteen

Interpretation of the BART Extinction Dilution
For Denitrifying Bacteria

Tube #	Population Assessment				
"0"	FO	FO	FO	FO	FO
"1"	FO	FO	FO	FO	----
"2"	FO	FO	FO	----	----
"3"	FO	FO	----	----	----
"4"	FO	----	----	----	----
Possible Population: (log DN/ml)	>5.0	>4.0	>3.0	>2.0	>1.0

Hygiene Risk

Denitrifying bacteria flourish in waters that have sources of nitrate and organics. Such sources may involve wastewater that contain some septic material and could therefore present a potential hygiene risk. A coliform test should be considered to assess this risk where there is a detected population of denitrifiers (FO observed). Where the DN population is >3.0 log DN/ml, a coliform test should routinely be used to determine the health risk.

NITRIFYING BACTERIA, N-BART™

Nitrification serves as the major route by which ammonium is aerobically oxidized to nitrate. There are two steps to nitrification process:

$$NH_4^+ \xrightarrow{\quad(1)\quad} NO_2^- \xrightarrow{\quad(2)\quad} NO_3^-$$

Ammonium   Nitrite   Nitrate

Nitrifying bacteria are divided according to which of the above reactions they are able to perform:

Group 1 - step (1) only   -   Nitrosofiers –*Nitrosomonas*

Group 2 - step (2) only   -   Nitrifiers – *Nitrobacter*

The polarized relationship between the nitrifying and the denitrifying bacteria is a problem in the testing of natural samples since the two groups are either producing or utilizing nitrate respectively. In developing a biodetection system for the nitrifying bacteria in natural samples, the terminal product (nitrate) may not be recoverable because of the intrinsic activities of the denitrifying bacteria which are also likely to be present and active in the sample. It is because of this difficulty that the N-^{BART} restricts itself to detecting the nitrosofiers that generate nitrite. This nitrite will be oxidized to nitrate by the nitrifiers only to reappear when reduced back to nitrite by any intrinsic denitrification occurring in the sample.

The nitrifying bacteria are an important indicator group for the recycling of organic nitrogenous materials from ammonium (the end point for the decomposition of proteins) to the production of nitrates. In waters, the presence of an aggressive population of nitrifiers is taken to indicate that there is a potential for significant amounts of nitrate to be generated in waters which are aerobic (rich in oxygen). Nitrates in water are a cause of concern because of the potential health risk particularly to infants who have not yet developed a tolerance to nitrates. In soils, nitrification is considered to be a very significant and useful function in the recycling of nitrogen through the soil. Nitrate is a highly mobile ion in the soil and will move (diffuse) relatively quickly while ammonium remains relatively "locked" in the soil. In some agronomic practices, nitrification inhibitors have been used to reduce the "losses" of ammonium to nitrate.

A common use for the presence of aggressive nitrifying bacteria in waters is that these bacteria signal the latter stages in the aerobic degradation of nitrogen-rich organic materials. Aggressive presence of nitrifying bacteria in water can be used to indicate that there is a potential for the water to have been polluted by nitrogen-rich organics from such sources as compromised septic tanks, sewage systems, industrial and hazardous waste sites and is undergoing an aerobic form of degradation. Nitrification and denitrification are essentially parallel processes that function in reverse sequence of each other. It is recommended that, where a high aggressivity is determined, waters should be subjected to further evaluation as a hygiene risk through a subsequent determination for the presence of nitrates. In soils, the presence of an aggressive denitrifying bacterial population may be taken to indicate that the nitrification part of the soil nitrogen cycle is functional. Nitrification is fundamentally an aerobic process in which the ammonium is oxidatively converted to nitrate via nitrite. Nitrite produced by the denitrification of nitrate may also be oxidized back to nitrate.

Reaction Patterns

This test is an unusual test in that the presence of nitrifying bacteria is detected by the presence of nitrite in the test vial after a standard incubation period of five days. Nitrification involves the oxidation of ammonium to nitrate via nitrite. Unfortunately, in natural samples, there are commonly denitrifying bacteria present in the water and these reduce the nitrate back to nitrite. If denitrification is completed, this nitrite may be reduced further to dinitrogen gas (under anaerobic conditions). That is why this test is laid upon its side with three balls to provide a moistened highly aerobic upper surface where nitrification is most likely to occur. The reagent administered in the reaction cap detects nitrite specifically by a red color reaction. This test is interpreted by the amount of pink-red coloration generated, and the location of this color.

PP	Pink-red color on roughly half the ball, solution clear or pale yellow (Reaction 1)
RP	All balls are reddened, solution may be pale pink (Reaction 2)
DR	Balls and the solution are reddened (Reaction 3)

This test is different from the other BART tests in that a chemical reagent is added to detect the product (nitrite) after a standard incubation period. The typical reactions are described below:

PP – Partial Pink on the Balls
Clear solution but a pink reaction may be generated on the FID hemispheres indicating that nitrification has just begun and the nitrite detected is in the biofilm on the balls.

RP – Red Deposits and Pink Solution
Reaction causes a light pink solution with red deposits all over the three balls. Nitrite is now present in solution as well as in the biofilms on the balls.

DR – Dark Red Deposits and Solution
Reaction causes dark red solution with heavy red deposits on ball. High concentrations of nitrite have been detected indicating an aggressive level of nitrification has occurred in the test period.

RPS (Reaction Pattern Signatures)
The reaction represents the population size and does not reflect the variety of microorganisms present in the water sample:
- **PB** Small population of nitrifiers ($< 10^2$ nitrifiers/ml) associated with aerobic slime-forming bacteria in a consortium
- **RB** Moderate population of nitrifiers ($> 10^2$ and $< 10^5$ nitrifiers/ml) forming a major component in the bacterial flora
- **RT** Dominant population of nitrifiers ($> 10^5$ nitrifiers/ml)

Hygiene Risk
The presence of an aggressive population of nitrifying bacteria in water is taken to indicate that there is a potential for significant amounts of nitrate to be generated in waters, which are aerobic. This may indicate a potential health risk particularly to infants who have not yet developed a tolerance to nitrates. It is recommended that, where a high population is determined, waters should be subjected to further evaluation as a hygiene risk, through subsequent determination for the presence of nitrates.

MICRO-ALGAE, ALGE-BART™
Micro-Algae (ALGE) is the name given to various plant-like microorganisms that are able to photosynthesize using light as the energy source for growth. The ranges of algae that can grow in this biodetector

include grass green Algae (*Chlorophyceae*), Blue-Green Algae (*Cyanobacteria*), Desmids, Diatoms and Euglenoids. Such growths may be localized or occur generally throughout the body of the culture medium.

This biodetector is distinctly different from the other products in the series because it is designed to recover and culture phototrophic (photosynthesizing) microorganisms that utilize light and release oxygen as a product. To achieve this, a modified Bold's medium is used which does not contain organics, but does contain the basic nutrients for plant growth (nitrogen, phosphorus, potassium, sulfur etc.). Carbon is presented as bicarbonates and the medium is made slightly alkaline (pH, 8.2) to encourage the micro-algae to utilize this form of carbon. The growth of micro-algae tends to be slower than for the heterotrophic bacteria and does require the presence of light (for photosynthesis). This light does not have to be strong (such as in direct sunlight) and most micro-algae can actually utilize quite low levels of light. The most effective manner for maximizing the probability of photosynthesis is to place the charged ALGE-BART™ on its side and set it about 60 cms from a single 40 watt daylight fluorescent light or 100 cms away from a 60 watt tungsten light source.

Other differences in the ALGE-BART relate to the lateral diffusing of the nutrient chemicals along the sample and the presentation of a variety of eco-niches for possible growth. These niches are formed by the pores within the woven material layered around a part of the vial and the semi-saturated and saturated nature of this material. Some micro-algae gravitate towards the semi-saturated material above the culturing sample within the test vial while others grow within the saturated material or within the liquid medium itself. In practice, this test takes a minimum of two days and a maximum of 26 days to detect significant populations of micro-algae.

This ALGE-BART can be used as a simple presence/absence (P/A) test capable of indicating to some extent the population size and the types of micro-algae present in the water sample. Different algae utilize different sites in the biodetector because of the two woven materials and the lateral position of the test vial on its side. Twice weekly observations for four weeks should be undertaken to observe the various forms of algal growth. These form into six possible reaction patterns in the test sample (see Reaction Patterns below). Observations can determine:

(1) Level of activity of the micro-algae (aggressivity) through the TL before a reaction is observed; and
(2) Community composition of the micro-algae present and active in the sample.

While a positive test will be recognized by the appearance of the first identifiable reaction, further information beyond the simple P/A result can be achieved by continuing to observe the test vial on a regular basis for subsequent reactions that may appear. The nature of these shifts forms the RPS. Two useful interpretations of the data can therefore be:

(1) Aggressivity and possible population as log. colony forming units per ml; and
(2) Community structure based upon the RPS generated by the chronologically sequenced reaction numbers observed during the test observation period.

By the routine (e.g., monthly) testing of a water or wastewater using this technique, the levels of aggressivity, possible population and community structure can all be determined and the status of any micro-algal induced fouling determined.

To conduct the test, it is necessary to add 15 ml of the water sample to the ALGE-BART biodetector. Once this has been done and the inner test vial returned to the outer test vial that is then capped, the test can begin. To initiate growth, it is recommended that the biodetector be laid on its side under a light source. Continuous light is preferred. Incubation is at room temperature and the biodetector protected from any excessive heating due to any artificial lights that are used. Note that temperatures in excess of 80°F (30°C) may inhibit algal activity. Under no circumstances should the test be severely agitated or shaken during the test period.

If there is a need to determine the micro-algae population in a soil or a slurry, the technique would need to be modified. This is necessary to reduce the potential detrimental effects that could be caused by a high organic nutrient in the soil. Such loadings could stimulate the growth of heterotrophic microorganisms at the expense of the micro-algae. To correct for this, take 1 g of the soil or slurry and suspend it in 9 ml of sterile Ringer's solution. Agitate by vortexing for one minute to disperse the particles evenly into suspension and break up some of the larger structures. Aseptically withdraw a 1 ml sample from the midpoint of the suspension and transfer to 14 ml of sterile Ringer's solution. Use this 15ml suspension to charge the ALGE-BART and follow the standard procedures. The possible population would be modified by multiplying the log vu (viable units)/ml by x 2.0 to achieve a possible population in vu/g of soil or slurry.

Because the micro-algae tend to grow slowly, the generation of a growth may be difficult to determine initially. Stereo-microscopic examination will give an earlier indication of a positive growth. Many micro-algae may initially start to grow generating a green color since the chlorophyll pigments used for photosynthesis are dominant. But as the growth continues, other pigments such as the xanthophylls may become dominant and change the color of the growth. This color shift may involve several different colors dominating over time (e.g., green to yellow to orange to brown).

A number of habitats are provided in this BART test which can encourage the growth of various micro-algae. These habitats include semi-saturated porous, saturated porous, aquatic, liquid: solid and liquid: air surfaces. Nutrients provided are inorganic nutrients commonly used by the micro-algae that, together with constant illumination, provide a preferential habitat for these microbes. Growth is slower because of longer generation times commonly found in the micro-algae.

Reaction Patterns

To observe the test, gently examine the biodetector for the presence of colored patches. If the test is negative, the woven material should remain white and there should be no colored patches or cloudiness in the water medium. A positive reaction may be recognized when there is one of these: a colored cloudiness, a distinctive patch of color or a colored deposit generated in the test device. Low magnification stereo microscopy can be used to directly observe the types of algae growing in the biodetector.

GG	Green growth at or above the water level
FG	Irregular patches of green growth over the woven material
OB	Patches of red, orange or brown growths below water level
YB	Yellow patches diffuse over the woven material
GF	Green deposits and/or green growth in the woven material
DG	Blue-green or black growth commonly at the water level

GG – Grass Green growth (formerly reaction one, ALGE-BART).
A grass-green growth may be seen through the porous textile medium usually concentrated at the water line or below water line. As this reaction matures, flocculent green growth may also be observed in the liquid medium.
FG – Fuzzy Green Patches (formerly reaction two, ALGE-BART)
Much of the reaction seen here may be above the water line in the semi-saturated porous textile medium. The pore structure restricts the

extension of the growth and so the growth may be seen as intense grass green zones with ill-defined or radial edges.

OB – Red Orange Brown Patches (formerly reaction three, ALGE-BART)

Red, orange or brown patches usually with clearly defined edges are generated both above and below the water line. These growths may gradually change in color as the growths mature.

YB – Yellow Beige Patches (formerly reaction four, ALGE-BART)

Poorly defined light yellow to beige patches of growth occur on the fabric often at localized sites on the porous fabric, generally these growths are initially difficult to observe.

GF – Green Flocculent (formerly reaction five, ALGE-BART)

Grass-green flocculent deposits are abundant in the liquid medium and on the floor of test vial. Tendency to be dense and lay on the floor of the test vial. The porous fabric may also show some green discoloration particularly on the lower side.

DG – Dark Green to Black Patches (formerly reaction six, ALGE-BART)

Predominantly recognized by dark-green, blue-green or black growths at the water line. Often this is a secondary reaction following reactions GG, FG or GF.

RPS (Reaction Pattern Signatures)

- **G - DG**   Cyanobacteria present with possible *Nostoc* dominate
- **FG - DG**   Grass-Green algae with cyanobacteria present
- **FG - OB**   Grass-Green algae maturing
- **YB - OB**   Diatoms or Desmids may be dominant
- **GG - GF**   Grass-Green algae maturing without pigment production
- **GG-GF-DG**   Grass-Green algae dominant but cyanobacteria dominate

The RPS displays the reaction patterns in the order that they were observed. For example, GG-GF-DG signature indicates the order for the reactions observed were firstly, GG; secondly, GF; and thirdly, DG. The signature obtained from an individual sample will provide an initial understanding of the type of algal community present in the sample.

Population counts are based on direct microscopic examination of samples with the range being set at two standard deviations. This population projection is based on the number of viable units (vu) per ml of original sample volume. In this event, each vu is considered to be a single cell rather than some form of multicellular structure.

Time Lag (days of delay) to ALGE-BART Population

Table Seventeen

The Relationship Between Time Lag and the Population For Micro-Algae

Time Lag (days)	Aggressivity	Population (log vu/ml)
4	Extremely High	5.2±2.1
8	Very High	4.3±1.4
12	Very High	4.3±1.4
16	High	3.2±1.2
20	High	3.2±1.2
24	Moderate	1.9±1.4
28	Very Low	1.0±1.9

Nuisance Bacteria in Recreational Waters, POOL-BART™

There is often a need to test recreational waters for the presence of bacteria that can cause biofouling and increase the hygiene risk to the users. The microorganisms that can cause these problems come from a broad range of species. The POOL-BART™ biodetector has been configured to detect a broader spectrum of potential nuisance bacteria. These events have serious operational and health implications that need to be considered. First, the presence of biofouling bacteria may indicate that there is an increased risk of corrosion, slime formation, foaming, unpleasant tastes and odors, equipment failure due to clogging of filters, pumps and/or drainage systems, and cloudy water. Many of these events are unacceptable to the users and increase the costs of plant operation and maintenance. Second, there is a hygienic concern. Some of the bacteria may be the types of pseudomonads that produce fluorescent (glowing in U.V. light) pigments. These are considered as marker organisms for increased hygiene risks. The faster that clouding (CL) or production of thread-like webs (TH) the greater the risk of biofouling problems. By having an ongoing program or routine testing with the appropriate remedial action, biofouling can be suppressed so that the facility can remain fully functional and at an economical cost.

In essence, this test selectively allows the detection of potential biofouling and hygiene risk bacteria in the water and the separation of the fluorescent species of *Pseudomonas* (PB and GY reactions). The bacteria detected using the POOL-BART can cause a range of problems in waters. Problems range from slime formations, turbidity, taste and odor, corrosion, biodegradation (CL and TH reactions) through to greater hygiene risks (PB and GY reactions). The more aggressive the bacteria are, the shorter the time before a reaction appears in the test period. Where a CL or TH reaction is not followed by a PB or a YG, the bacteria may be considered to be not such a significant hygiene risk since the fluorescent pseudomonad bacteria have not been detected (because PB or YG were not observed). This test method is not specifically designed to monitor for the fluorescent pseudomonads. It is recommended that users concerned about this risk employ the FLOR-BART that has been specifically developed to determine the presence of the fluorescent pseudomonads.

To set up the POOL-BART the inner test vial of the biodetector is removed by unscrewing the outer cap. Next, the inner cap is unscrewed and placed down on a clean surface without turning the cap over. 15 ml of the water sample is carefully added to the inner of tube of the POOL-BART until the water reaches the fill line. Notice that the ball will float up with the water. Screw the cap tightly back onto the inner test vial and return the inner vial to the outer vial and cap tightly. Place the charged POOL-BART on a secure surface at room temperature where it can be observed daily without being disturbed. For hot tub water samples, it is recommended that tests be performed at blood heat (35 to 37°C) or at the temperature of the hot tub. Do not immerse the tests in the hot tub as a convenient method of performing this test since it creates a hazard to the user.

Bacteria are also sometimes associated with taste and odor problems in water since many of the species produce distinctive odors such as "fishy," "earthy-musty," "septic" or "kerosene-like"which can become very dominant. To determine whether these odor-generating bacteria have been active in a positive POOL-BART, unscrew the outer cap of the test vials and smell the air in the gap between the inner and outer vials. If there has been odor generated, it is likely to have seeped out of the inner vial and collected in this cavity. Comparison of the odor generated at this site can then be compared to the odors generated in the waters from which the sample was taken. The user may also find that the intensity of the taste and/or odor problem may be related directly to the aggressivity of the bacteria in the test. The more aggressive, the greater the taste and odor problem. This allows the manager of the system to develop a useful treatment strategy to control these problems. Each day the biodetector should be checked for reactions.

Reaction patterns

To observe whether cloudiness (CL) or thread-like growths (TH) have been generated, the tube may be held up against a natural indirect source of sunlight or a light source such as fluorescent lights. Early cloudiness may be seen as either a cloudy zone immediately beneath the ball, as cloudy plates or, as a general loss in clarity in the water.

CL	Cloudy growth
TH	Thread-like growth
PB	Pale Blue Glow in U.V. Light
GY	Greenish-Yellow Glow in U.V. Light

CL – Clouding (reaction one, POOL-BART)
Liquid medium becomes generally cloudy but with no glowing around ball. Slimes may occasionally be observed but the growth generally causes the liquid medium to become turbid.

PB or GY – Pale Blue or Greenish Yellow UV Fluorescence (reaction two, POOL-BART)
Solution very cloudy (passes through reaction one) and then generates a glowing around FID when Ultraviolet light is shown onto the side walls of the inner test vial. This glowing fluorescence occurs usually in the top 15 to 20 mm around the ball and may commonly be either pale blue (PB) or greenish-yellow (GY). This glowing last 2 to 3 days (former case) and 4 to 8 days (latter case).

TH – Thread-like Growths (reaction four, POOL-BART)
Solution commonly clear but thread-like slime growths hang suspended between the ball and the basal cone of the inner test vial. These slimes may form as single globular suspended structures or as interconnected threads scattered through the liquid medium (sometimes this can resemble a "web").

CL-Cloudy Growth
When there are populations of aerobic bacteria, the initial growth may be at the redox front that commonly forms above the medium diffusion front. This growth usually takes the form of lateral or "puffy" clouding which is most often grey in color. It should be noted that if the observer tips the BART slightly, the clouds will move to maintain position within the tube. Commonly, the medium will be darker beneath the zone of clouding and lighter above. Microorganisms present around the ball in this test can generate different growths in the presence of oxygen. Usually these

pigments are generated after the cloudy growth has developed but before there are intense slime-like growths around the FID. Common pigments are produced by a variety of the bacteria and often form as a ring-like or slime structure that often develops around the ball or in the liquid medium itself. These pigments may be easily observed as yellow, beige, orange, red-brown or violet colors. Sometimes the whole of the cloudy growth may become pigmented.

TH-Thread-like Growth

A further form of growth is exhibited in which there are thread-like slime processes that often connect the ball often to the peg in the base of the tube. These slime-like threads are arranged in a web-like manner and remain in place for days or even weeks. This reaction is known as the "TH" reaction and may be related to *Zoogloea* species of bacteria.

The test vials should be observed for fluorescent (glowing) pigment formation that should normally happen within five days of the first observation of clouding. While there may be some glowing visible even in natural or artificial light, this should be confirmed using a regular U.V. light source. This light can be used to determine whether there is a glowing luminescence. Operators should read the restrictions and cautionary notices supplied by the manufacturers of the U.V. lights before operating the equipment. Where a PB (Pale Blue) or a GY (Greenish Yellow) reaction occurs, it may be as a fluorescent glowing being either of the below colors:

Pale Bluish Glow

This fluorescence can last between one to four days and then gradually fade. The glow is normally fairly faint and should be viewed against a darkened background since direct light may make viewing more difficult. One major species bearing this pigment (pyocyanin) is *Pseudomonas aeruginosa*. It is of concern since this species can be associated with a range of opportunistic infections. Of particular concern are infections of wounds, the urinary tract and the respiratory tract. This species can be found in a variety of waters.

Greenish-Yellow Glow

This fluorescence may last between two to ten days and then gradually fade. The glow becomes fairly obvious and is often visible without using the U.V. light. One major species bearing these pigments (pyoverdin, fluorescein) is *Pseudomonas fluorescens*. Generally this species is not as virulent as *Ps. aeruginosa* and are often more abundant in waters and can be involved in specialized aerobic degradation of organic pollutants.

RPS (Reaction Pattern Signatures)

- **CL-TH**   Aerobic bacteria dominant which are able to generate slime threads possible *Zoogloea* species
- **CL-PB**   *Pseudomonas aeruginosa* dominant member of the bacterial flora
- **CL-GY**   *Pseudomonas fluorescens* species group present in the bacterial flora

Time Lag (days of delay) to POOL-BART™ Populations

The time lag to the reactions can be used to assess the aggressivity and project the size of the bacterial possible population. Calculate the time lag (TL) between when the test was set up (charged with water) and when the first clouding (CL reaction) or thread-like growths (TH reaction) occurred. The relative aggressivity and population size can be predicted using Table Eighteen. The population is given as the mean within two standard deviates.

Table Eighteen

Relationship of Time Lag (TL) to Aggressivity and Possible Population, POOL-BART

Time Lag (dd)	Aggressivity	Possible Population (log cfu/ml)
1	Highly Aggressive	5.0 ± 1.5
2	Aggressive	4.2 ± 1.0
3	Aggressive	3.8 ± 1.2
4	Low Aggressivity	3.2 ± 1.2
6	Low Aggressivity	2.5 ± 1.2
8	Very Few	1.5 ± 0.8

Hygiene Risk

The bacteria that can be considered as marker organisms for an increase hygiene risk are the pseudomonads that produce fluorescent (glowing under a U.V. light) pigments. In recreational waters, such as swimming pools, hot tubs, intensive natural bathing sites, the presence of aggressive fluorescent pseudomonads (PB and GY reactions) should be taken as a potential cause for concern since these bacteria may cause a range of skin, eye, ear and urinary tract infections. Occasionally the pseudomonad bacteria will cause skin infections particularly under tight fitting bathing apparel. This is particularly a potential problem in warmer waters and hot tubs where the bathers remain relatively inactive in water for prolonged periods.

Acid Mine Drainage, AMD-BART™

Acid mine drainage is often caused by the bacterial oxidation of sulfides to acidic products such as sulfuric acid. It is these products which cause the drainage waters from the site to be very acidic and corrosive. The bacteria involved are the sulfur oxidizing bacteria commonly belonging to the genus *Thiobacillus*. Not surprisingly, these bacteria can tolerate (and, in fact, require) very acidic conditions in order to become aggressive. Normally, growth occurs attached to surfaces in the form of biofilms (slimes). Water samples may, as a result of this, not contain many cells. In order to test for the presence of these bacteria in acid mine drainage waters, it is necessary to disturb any slime growths that may be visible in order to get a greater bacterial sample size in the water for testing. Physical disturbance of the slime will cause dispersion into the water. If this is not possible, temporarily "damming" the water to cause an environmental "shock" to the slime growths should have the same effect (releasing cells into the water).

Methodology

1. Remove the AMB-BART™ from the foil pouch.
2. Unscrew the cap and lift. Notice that the inner tube (test vial) will lift with the cap.
3. Gently pull the inner test vial out of the cap.
4. Unscrew the cap to the inner vial and place on a clean surface without turning the cap over (the inside of the cap is sterile).
5. Add 9 ml of the water sample containing dispersed slime to inner test vial.
6. Tightly screw the cap back on the inner test vial.
7. Shake the tube for 30 seconds to dissolve the medium into the water.
8. Return inner vial into the socket on the underside of the larger cap.
9. Screw the inner tube back into the outer tube.
10. Lay the AMD-BART on its side for incubation usually at room temperature.
11. Observe the test regularly on a convenient daily basis for signs of a positive reaction in the tube.

Reaction Patterns

This test incorporates an acidic medium which will normally buffer the pH of the water sample to a range of 2.0±0.2. Four white plastic balls in the inner vial provide an extended surface area, on which the sulfur-oxidizing bacteria would grow and form a biofilm. A positive detection of the bacteria would be any one of the following events.

CL	Cloudy Growth
BS	Brown Slime Growth
BD	Beige Discoloration of surfaces

BS-Brown Slime Growth
A brown slime-like growth at the surface of the balls.
BD-Beige Discoloration of surfaces
A beige discoloration of the exposed areas of the balls due to growth of the biofilm. This is the first indication of slime growth. This reaction will continue on to a dense brown slime.
CL-Cloudy Growth
Cloudiness in the liquid medium is the first indication of biological activity within the AMD-BART. This first reaction should be followed by a brown discoloration and slime growth on the surfaces of the balls.

Time Lag (days of delay) to AMD-BART Populations

The aggressivity of the bacteria may be judged by the time lag to the first observation of one of the above reactions. This aggressivity can be judged using Table Nineteen.

Table Nineteen

The Relationship Between Time Lag
and the Aggressivity of *Thiobacilli*

Time Lag (days)	Aggressivity
1	Extremely aggressive
2 to 4	Very aggressive
5 to 14	Aggressive
15 to 30	Mild Aggressivity
>30	Detected

Note:
- Incubation temperatures should reflect the water sample temperatures and be within 10C° of the original temperature if possible.

- The culture medium used in the AMD-BART is very selective for the *Thiobacillus* species which cause acidic mine drainage problems. It is important to shake the inner test vial for 30 seconds to ensure a rapid mixing of the chemicals in the medium with the water.

COLI-BART™

Hygiene risk has long been a concern with water supplies. Over the last century, these risks have been effectively addressed by the using the presence of coliform bacteria as the indicator of an unacceptable hygiene risk. Many methods have been developed to determine the presence of coliform bacteria in water. Present-day technologies include membrane filtration (MF); fermentation and the use of fluorogenic compounds such as MUG (4-methylumbelliferyl-beta-D-glucuronide). The technique employed in the COLI-BART™ is a patented development of the fermentation techniques. These techniques, in essence, detect the coliform bacteria through their ability to ferment the sugar lactose with the production of the gases (carbon dioxide and hydrogen). Traditionally, these gases were collected in inverted glass test tubes (Durham's tube) set into the liquid medium. As the gas was produced by fermentation, some of the gas would become entrapped in this inverted tube as a gas pocket. By observing the gas pocket in the Durham's tube, fermentation of the lactose could be seen by the presence of the gas and the coliforms could be conjectured. The problem with the gas entrapment and observation technique was that it was often difficult to view and did not give an indication of the population size. To compensate, complex dilution strategies were applied to the sample so that the population could be predicted with a multiple tube fermentation technique. This technique was labor intensive and involved a statistical projection of the most probable number (MPN) for the coliform population based upon a statistical appraisal of the data (i.e., which tube dilutions went positive with gas production and which volume dilutions remained negative). While a valid technique, the MPN lost popularity because of the demands on labor and consumables generated by this technique. Other techniques such as the MF and the MUG methodologies were adopted and are now in wide use. There remains, however, a concern that these techniques also have some flaws and the MPN technique remains a standard used often in a confirmatory role. The COLI-BART represents an attempt to capitalize on the original fermentation detection methodology, but avoiding the use of multiple tubes (used in the MPN test) and yet still being able to give an estimate of the population.

The major advantage that the COLI-BART has is a unique patented gas detection system. While the traditional glass Durham's tube entraps a pocket of gas that may, or may not, be observable, the COLI-BART system entraps the gas in a manner that causes the device trapping the gas to relocate (float) upwards. This gas entrapment system consists of a dense plastic slatted thimble-like structure (gas-trap-thimble, GTT) which sinks to the bottom of the 215-ml test vial. The GTT has open slats at the side so the bacteria and chemicals can freely move into, and out of, the GTT. When the gases are produced as a result of fermentative activities within the test vial, some of the gas bubbles float up to become entrapped within the cavity within the top of the GTT. Once the gas has collected in a sufficient volume, the density of the GTT drops and the GTT floats up to the surface. The standard GTT employed in the COLI-BART test will float up when 0.4ml of gas has collected inside the GTT. The repositioning of the GTT touching the surface of the liquid medium indicates that the test has been positive for the detection of fermented gas. The time lag to the GTT elevating to the surface gives an indication of the size of the population of fermenting bacteria (e.g., the coliform bacteria).

The COLI-BART comes in two formats that are based on the use of different media. These media are selective for either the **Total Coliforms (TC)** or the **Fecal Coliforms (FC)**. Standard recognized culture media are used for both of these tests. The volume of liquid medium used is 50ml of a triple strength concentration of the standard liquid medium. When the 100ml of water sample to be tested is added to the COLI-BART test vial, the medium is diluted to the correct strength. Both of the selective culture media are formulated according to the Association of Official Analytical Chemists (AOAC) and the American Public Health Association (APHA), The media selected are:

− For the TC COLI-BART™ - Presence-Absence Broth
− For the FC COLI-BART™ - Brilliant Green Bile Broth, 2%

The presence/absence broth was developed by Clark and first reported in the *Canadian Journal of Microbiology* (5:771) and incorporated into the 16th edition of the *Standard Methods for Water and Wastewater* in 1985. This is a broad-spectrum test for the coliform bacteria. Those authorities have also recognized brilliant Green Bile Broth, 2% as a suitable medium for the detection of coliform bacteria but with a narrower spectrum of the coliforms producing gas. These are referred as the fecal coliforms of which *Escherichia coli* is one of the most important species.

The formulation of the two media is given below:

Presence/Absence Broth
(TC COLI-BART™) per liter, final strength:.

Beef Extract	3.0g
Pancreatic Digest of Gelatin	5.0g
Lactose	7.5g
Pancreatic Digest of Casein	10.0g
Dipotassium Phosphate	1.375g
Monopotassium Phosphate	1.375g
Sodium Chloride	2.5g
Sodium Lauryl Sulfate	0.05g
Bromcresol Purple	8.5mg

Brilliant Green Bile Broth, 2%
(FC COLI-BART™) per liter, full strength:

Oxgall, dehydrated	20.0g
Lactose	10.0g
Pancreatic Digest of Gelatin	10.0g
Brilliant Green	13.3mg

Both of these media contain lactose as the principal sugar substrate for fermentation. Inhibitory agents are employed to allow the selective culture of coliforms. In the TC test, the agent is sodium lauryl sulfate and in the FC medium it is the oxgall and the brilliant green.

Methodology for the COLI-BART™ Tests
Each of these tests come pre-charged with the triple strength medium and the GTT is already in place within the test. To conduct a coliform test, the following steps should be followed using the commonly accepted aseptic procedures found in microbiology laboratories:

A. Unscrew the cap on the COLI-BART test and place it down on a clean surface without turning the cap over.
B. Add 100ml of the water sample to be tested. This should be done using a pipette discharging the water into the middle of the medium.
C   Screw the cap back down tightly onto the COLI-BART test vial.

D.   Invert the test to mix the water sample with the medium. Do not do this violently since foam may form and make the start-up of the test difficult. The GTT should sink at this time to over the cap. The next step is to invert of the test back to the cap-side up with the GTT sinking. This is because any air trapped within the GTT will cause the GTT to float up.

E.   To get the GTT to sink into the liquid medium, gradually turn the vial 180° so that cap is now again uppermost. While doing this, carefully watch the side of the GTT to ensure that it stays below the liquid surface during the period of turning back upright. If the GTT comes above the liquid level, air may enter and cause the GTT to float up.

F.   Once the water sample has been added and the GTT has sunk to the bottom of the medium, the test should be placed in an incubator at blood heat (35 to 37°C). Incubation should be for 48 hours for the TC test and 24 hours for the FC test. A positive detection would be the GTT now floating at the surface meaning that gas had been produced by fermentation.

This test is described as a presence-absence test in which the elevation of the GTT to float at the surface is taken to be the prime indicator that either total or fecal coliforms are present (depending upon the medium selected).

Interpretation of the COLI-BART Tests

The standard procedure described above will allow a presence or determination based upon whether the GTT rose (floated) to indicate coliforms were present, or remained sunk on the floor of the test vial (coliforms absent). Here, the detection revolves around the water sample carrying more one or more active coliforms in the 100ml to create a presence result, or no coliforms that were detectable in the water sample. The test threshold is therefore one or more coliform bacteria per one hundred ml (minimal detection level, 1 coliform bacterium/100ml). This test functions as a 24-hour test with observation at the end of that time. With more frequent observations of the status of the GTT, it is possible to predict the size of the coliform population. This can most effectively be done using the CABART™ that is described in a later section of this document.

Relationship between Time Lag to GTT elevation and the Coliform Population Size

Confirmatory studies have been conducted to evaluate the linkages between time lag (hours) and populations of coliform bacteria in water.

Using Natural water samples and the TC COLI-BART in the standard format at an incubation temperature of 37°C, a linear regression analysis was performed over time (x axis) against the log population of fecal coliforms enumerated by the standard membrane filtration technique as the log population/100ml (y axis). The trial included eleven sample and four replicates of each sample to make a total of forty tests and the regression analysis was:

$$Y = -0.711 X + 12.65$$

Regression correlation, R, was –0.956; intercept on x-axis was at 17.79 hours.

Trials on the TC COLI-BART™ using ATCC strains were conducted. These were subjected to trials using known log populations based on the principles of extinction dilution of known cell population sizes confirmed by spread-plate analysis using *M*-ENDO Agar LES to enumerate the population of the ATCC strains. Linear regression analysis was conducted on the data from twenty trials. Using *Escherichia coli* ATCC strains, statistical analyses were performed using linear regression analysis. The regression correlations (as R^2) for these confirmatory trials are given in Table Twenty below.

Table Twenty

Regression Correlations for ATCC Bacterial Strains Subjected to TC- and FC- COLI™ Testing (Time Lag to GTT Elevation)

ATCC Strain	Regression Correlation as R^2 (sample number in parentheses)			
25922(TC)	0.767 (20)	0.888 (5)	0.961 (8)	0.825 (6)
11229(TC)	0.889 (22)	0.913 (8)	0.825 (7)	0.930 (8)
25922(FC)	0.923 (22)	0.980 (8)	0.956 (8)	0.908 (6)
11229(FC)	0.776 (24)	0.878 (8)	0.833 (8)	0.888 (8)

High correlations were observed between the log populations of *Escherichia coli* inoculated into the water sample and the time lag measured in hours to the time that the GTT floats up to the surface of the liquid medium. Intercepts through the x-axis occurred at around 19 hours for the TC COLI-BART and at 23 hours for the FC COLI-BART.

The COLI-BART test can therefore be used not only as a presence/absence (P/A) test but also as a quantitative test for predicting the population based on the time lag observed to the floatation of the GTT. This has now been built into the CABART™ system using an "averaged" formula based upon the linear regression analyses obtained from the studies using ATCC strains of *Escherichia coli.*

Disposal of the Used COLI-BART Tests

Since there is a probability that at least some of the used COLI-BART tests would contain an active population of potentially hazardous bacteria, disposal should be conducted under the direction of a person familiar with the art of sterilizing microbiological materials. The common technique for sterilization is autoclaving with a minimum steam pressure of 8 p.s.i., a minimum temperature of 110°C and a holding time of at least 20 minutes.

Methanogenic Bacteria, BIOGAS-BART™

The **BIOGAS-BART™** is based upon a patented gas collection system (GTT) described by Cullimore and Alford, 1993. Here, the GTT was designed to entrap some of the gas generated during a coliform test. The GTT is a dense plastic thimble that elevates (floats up) to the surface when 0.4 ml of gas has been collected and the thimble density is reduced to less than that of the surrounding liquid medium.

The selective medium used in the BIOGAS-BART is a broad-spectrum type described by Baresi *et al.*, 1978 for the acetate utilizing methanogenic bacteria. This medium is presented in a crystallized form dried into the base of the test vial. When the test is charged with the 100 ml of the sample, the indigenous microflora in the sample rapidly generate reductive condition suitable for methanogenic activities. The selective medium for the methanogenic bacteria dissolves and diffuses upwards across the reduction-oxidation front in a manner described by Cullimore and Alford, 1990 in the biological activity reaction test (BART, Droycon Bioconcepts Inc., Canada).

The BIOGAS-BART detects the methanogenic bacteria that are acetate utilizing (Mah and Smith, 1981). In this biodetector, there are two phases in the detection process. The first records the elevation of the detection thimble to the surface after 0.4 ml. of BIOGAS (methane/carbon dioxide) has been generated. As the gas continues to be generated, the thimble continues to rise and the second phase occurs when the thimble elevates above the liquid surface after an additional 3.2-ml of BIOGAS has been collected. The rates of BIOGAS production are calculated as the predicted average for the two phases.

USER QUALITY CONTROL PROCEDURES

Two levels of QC are routinely applicable to the BART biodetectors to ensure:

(1) sterility of the inner test vial which holds the ball, medium pellet and the water sample;

(2) reliability of the medium pellet to allow the appropriate forms of activities and reactions for selected strains of bacteria.

The selection of these two principal QC procedures is based on the critical need for the inner test vial to be free from contamination. This could cause interferences with the performance characteristics of the test and the medium performing in an appropriate manner. In performing these QC procedures, it is assumed that the person undertaking this procedure has an adequate understanding of basic aseptic procedures and the necessary skills to safely culture bacteria.

Confirmation of the sterility of the BART Biodetectors

In the normal event of the confirmation of sterility, there is a change in the characteristics of the liquid medium forming in the test vial. The test vial can be considered to be essentially sterile where the liquid remains crystal clear as the nutrient pellet gradually dissolves and forms into a diffusion front which moves up and eventually dissolves completely to form the culture medium. For each BART biodetector, a characterization of medium diffusion of the test vial is given below, along with test methods describing the suitability of selective medium of each type of biodetector.

Sterility QC on the Inner Test Vial

Aseptically remove the inner test vial of the BART by removing the screw cap on the outer vial and lifting out the inner test vial. Screw the cap back onto the outer vial. A QC test can now be performed on the inner vial. The following directions are common to all BART biodetectors and the procedure should be performed on each of them.

Aseptically add 15 ml of sterile distilled or deionized water (which has been kept at room temperature) to the vial. Do not use non-sterilized distilled or deionized water for these tests since the incumbent bacterial population may be sufficient to trigger a reaction in the test vial. Stand the test vial up at room temperature (22 to 24°C) and observe for twelve days. If the test vial is sterile there should be no evidence of bacterial activity or reactions. If evidence is found that the test vial was contaminated, this could have been an isolated event. Repeat the procedure in triplicate. Should a duplicate result suggesting contamination, please contact the QC dept. at (306) 585-1762 or by fax at (306) 585-3000.

Below is a description of the typical events that occur as the dried medium pellet dissolves and diffuses up into the water column. It is based upon the standard QC procedures discussed above. There are two events to determine. First, the form in which the medium pellet diffuses up into the water. This gives a good indication as to the acceptability of the medium. Second, the clarity with which this happens indicates whether or not there has been some microbial contamination. If contamination has occurred from whatever source, clouding, slime formation, abnormal color shifts and gasses are possible consequences and the test vial would be deemed to have failed the test. It should be remembered that if the water used for this test contained a high concentration of various salts, it is possible for some of these reactions to occur. It must therefore be remembered that only sterile (distilled or deionized) water should be used in these QC tests. Additional tables are used to describe the initial diffusion patterns that should occur in these QC tests and also the most common form of expression of bacterial contamination.

A successful QC test would have the typical characteristics reported for the basal zone and the column. There should be no evidence of microbial contamination that would usually be seen in the form described in the "contamination" column. Every effort is made by Droycon Bioconcepts Inc. to produce a sterile product and routine QC practices are undertaken to ensure that the product has a satisfactory quality. Users may request the "Certificate of Analysis" issued for every batch of BART produced.

Assurance of Status of Crystallized Medium by Visual Inspection

To do this test effectively, the BART inner test vial should be removed from the outer test vial and held upside down at an angle which allows the convenient observation of the dried medium pellet through the conical base of the test vial. Note that there is a vertical plastic peg in the middle of the cone which may, or may not, be observable depending upon the test type. After observing the medium pellet through the cone, any medium deposited on the side-walls of the inner test vial should be examined. These observations should then be compared to the standard (acceptable) range of characteristics listed below for each BART product. Where the crystallized medium falls outside the described characteristics, it may be unacceptable.

It is recommended that, to view the base of the test vial, the tube be held with the basal cone facing directly towards the observer. It is best to view this using diffuse light against a neutral background. After viewing the base, examine the lower walls just above the basal cone for any medium deposits. Some of the media will coat the lower walls up to a height of 5mm above the cone. All BART products are examined for splattering of dried medium on the sidewalls of the inner test vial. If there are any splatter particles of greater than 1 mm diameter or a number positioned to disturb the view through the test vials, the tube is rejected (second grade). These seconds are still effective for conducting the tests. Tubes are rejected where there is an acceptably large amount of splatter (>1% of wall area) which affects the viewing of any reaction, or there are fractures in the polystyrene wall of the vials exceeding 5 mm in length and/or have penetrated through the vial wall.

a) IRB-BART™

The iron-related bacteria (IRB) are recognized as being those bacteria that are iron utilizing. The form of this utilization can range from simply a passive bioaccumulation in the slime (biofilm) growths through to active use in metabolism. This test specifically uses the ability of bacteria to use ferric and ferrous iron in many ways to create various reactions, some of which are colored. It is now understood that some of the IRB may be able to derive respiratory or energy functions out of the reduction (ferrous) to oxidation (ferric) manipulations which occur through the activity of the IRB.

IRB-BART Medium

A brownish green opaque crystal-like hardened mass extends outwards from the central peg towards the walls of the test vial. The center of the mass appears darker than the perimeter. The outer edge of the medium pellet is sharply defined and appears to be yellow and contain brown fibers extending out from the central deposit. A transparent film may extend up the walls for 1 to 2 mm and have a yellow color. When charged, the liquid medium remains clear and generates to a greenish yellow color. However, there are sometimes reactions between the chemicals present in a water sample that will change the color generated usually through to a yellow or a green. The green color appears to be darkened in and near the basal cone region of the test vial and will diffuse to a lighter green or yellowish green. These green colors appear to be associated with higher calcium and/or magnesium levels in the water sample. If there is microbial activity in the test, this will be displayed by color shifts, gassing, cloudiness or the formation of slime. These events reflect a reaction involving IRB.

Table Twenty-One
Medium Diffusion in a Sterile IRB-BART
Inner Test Vial to Confirm a Negative Reaction

	Color		
Time (days)	Basal	Lower column	Upper column
0.25	Green-yellow	Clear	Clear
0.5	Green-yellow	Clear	Clear
1.0	Greenish-yellow	Greenish-Yellow	Clear
2.0 - 10.0*	Greenish-yellow	Greenish-yellow	Green-yellow

*Note that these colors are crystal clear and have been generated using sterilized distilled or deionized water. Natural water samples can cause secondary chemical reactions that may be seen through an intensification of the green color in the diffusion front and crystalline deposits forming in the base of the test vial. Water saturated with oxygen stored at low temperatures can, when used in this test, cause bubbles to form as oxygen comes out of solution as the temperature rises to room temperature.

Table Twenty-Two
QC Characterization of Medium Diffusion in a
Sterile IRB-BART Inner Test Vial

	Color		
Time (hr.)	Basal	Column*	Contamination***
0.5**	Diffusion begins	Clear	Clear
2.0	Green-yellow	Clear	Clear
24	Greenish-yellow	Clear	Clouding or gassing
48 - 240***	Greenish-yellow	Green-yellow	Cloudy or Slime

*The column reflects the whole length of the water column and should stay crystal clear even as the diffusion front created by the dissolving medium passes through.

**Incubation is commonly at a (room) temperature of 22 to 2 4°C.

***An acceptable test should remain crystal clear, have no gassing or slime deposits. Generally, evidence of microbial contamination is not immediately observable but may appear within 24 hours commonly as either a clouding just above the diffusion front, or as an excessive gassing. These contaminants will cause turbidity, color shifts and may involve slime formations by 240 hours. Reference to the BART reaction chart may aid in the determination of the causative microbes.

Confirmation of the Selective Media Composition in the IRB-BART

In order to confirm the suitability of the selective medium for the biodetection of the various iron-related bacteria recognized by this test method, it is recommended that the following A.T.C.C. (American Type Culture Collection) strains be applied to the biodetectors to determine the standard reaction patterns (Table Twenty). Each culture should be prepared as a 48 hour broth culture incubated at 30°C to reach the stationary growth phase. Inoculation of the inner test vial should use a cell suspension of 0.1 ml of the broth culture in 15 ml of the sterile Ringer's solution. This inoculum should be taken from the midpoint of the broth culture immediately after the culture had been gently agitated. This inoculum should be applied directly over the ball as the test vial is filled. Do not shake the inoculated inner vial. Incubate at 22 to 24°C for seven days and observe for activities and reactions. Typical results are listed below for the recommended A.T.C.C. strains in Table Twenty-Three.

Table Twenty-Three

Cultural Characterization of the IRB-BART

A.T.C.C.	Genus/species	Characterization	
8090	Citrobacter freundii	GC	Reaction 6
13048	Enterobacter aerogenes	BR	Reaction 4
27853	Pseudomonas aeruginosa	GC	Reaction 8&9
19606	Acinetobacter calcoaceticus	GC	Reaction 8
23355	Enterobacter cloacae	CL-BG	Reaction 2&3
13315	Proteus vulgaris	CL-BC	Reaction 2&4
13883	Klebsiella pneumoniae	RC-BC	Reaction 7&4
25922	Escherichia coli	FO	Reaction 5

Note: Some of these culture tests will shift from one reaction type to another as the growth in the IRB-BART matures. For example, Citrobacter freundii may cause after 5 to 8 days a bio locking of the ball so that when the test vial is turned upside down, the ball remains "glued" into position with the liquid medium held above the ball. The first reaction normally precedes the second reaction. The test using E.coli is performed at 35°C.

b) SRB-BART™

Bacteria that are able to reduce sulfate to sulfide are often associated with corrosion (H_2S electrolytic), odor problems ("rotten" egg) and black water. This test uses a medium selective for the SRB and provides a source of iron so that the black iron sulfides are produced when the SRB are active. To improve the selectivity of this test, the liquid medium is manipulated to generate an anaerobic (reductive) state that restricts growth to a narrow spectrum of anaerobes.

SRB-BART™ medium

Beige crystalline deposits extend from the central peg to the walls of the test vial. The central 5mm of this deposit often appear darker extending outwards to lighter shades of grey. These deposits are very granular. A thin film of white and beige granular deposits extends irregularly up the sidewall of the test vial to a height of up to 6 mm and have an irregular edge. The characterization of the medium is given in Table Twenty-Four.

Table Twenty-Four

Medium Diffusion in a Sterile SRB-BART
Inner Test Vial to Confirm a Negative Reaction

Time (days)	Medium characterization		
	Basal	Lower column	Upper column
0.25	Yellow ppt	Clear	Clear
0.5	Yellow ppt	Clear	Clear
1.0	Beige ppt	Clear	Clear
2.0 - 10.0*	Beige ppt	Clear	Clear

*Note that liquid medium remains crystal clear when generated using sterile distilled or deionized water. Natural water samples can cause secondary chemical reactions that may be seen as an initial cloudiness that clears rapidly and/or commonly crystalline deposits forming in the base of the test vial. ppt stands for precipitate.

Table Twenty-Five

QC Characterization of Medium Diffusion in a
Sterile SRB-BART™ Inner Test Vial

Time (hrs)	Color		Contamination
	Basal	Column*	
0.5	Beige floc	Beige floc	Not determinable*
2.0	Beige floc	Clear	Not determinable*
12.0	Beige ppt	Clear	Cloudy zones
24.0-240	Beige ppt	Clear	Cloudy zones/black ppt

*See note for Table-Twenty-Four.

Due to the floc-like elements which form within the test vial once the water has been added, it is not possible to determine any contamination until at least twelve hours after the test has been set up. The beige floc forms as the water reacts with the medium pellet and the lipid oxygen barrier impregnated into the pellet. Where contamination occurs, growth is usually slow and tends to form cloudy structures above the basal cone. The presence of blackening in the QC test may reflect contamination by SRB.

Confirmation of the Selective Media Composition in the SRB-BART

In order to confirm the suitability of the selective medium for the biodetection of the various SRB recognized by this test method, it is recommended that the following A.T.C.C. (American Type Culture Collection) strains be applied to the SRB-BART biodetectors to determine the standard reaction patterns. Each culture should be prepared as a 48-hour broth culture incubated at 30°C to reach the stationary growth phase. Inoculation of the inner test vial should use a suspension of 0.1 ml of the broth culture in 15 ml of the sterile Ringer's solution. This inoculum should be taken from the midpoint of the Brain Heart Infusion broth culture immediately after the culture had been gently agitated. This inoculum is administered over the ball as the test vial is filled. Do not shake the vial. Incubate at 22 to 24°C for ten days and observe for activities and reactions. Typical results are listed below for the recommended A.T.C.C. strains cultured in Table Twenty-Six. Note that the *Desulfurovibrio desulfuricans* strain DSM1924 should be cultured on Sulfate Reducer API agar in accordance with the recommendations of the API (*Recommended Practice for Biological Analysis of Subsurface Injection Waters*, Volume 38, 2nd. edition, 1965).

Table Twenty-Six

Cultural Characterization of the SRB-BART

A.T.C.C.	Genus/species	Characterization	
13048	*Enterobacter aerogenes*	CG	Reaction 4
27853	*Pseudomonas aeruginosa*	CG	Reaction 4
13315	*Proteus vulgaris*	CG	Reaction 4
DSM1924	*Desulfovibrio desulfuricans*	BD-BA	Reaction 1&3
A + A	*Ps. aeruginosa + D. desulfuricans*	BU-BA	Reaction 2&3

Note that some of these cultural tests will shift from one reaction type to another as the growth in the SRB-BART matures.

c) SLYM-BART™

Some bacteria that are not IRB also do form copious slimes. They can cause many of the problems commonly associated with IRB but tend to generate white, grey, beige or black forms of slime rather than the browns. Under the right cultural conditions, many aerobic bacteria will generate slimes. The selective medium used in the SLYM-BART specifically encourages the formation of slimes. These are commonly seen as condensed cloudy/plate-like growths suspended in the liquid medium, gel-like rings (around the ball) or globular/swirl forms of slime in the basal cone. This group is known as the slime-forming bacteria.

SLYM-BART™ Medium

A brown hardened gel-like mass with a darker inner circle that extends fully up the cone to the walls of the test vial. On the sidewall of the inner test vial above the basal cone there is a transparent film that extends approximately 3 mm up the walls. This film is also brown in color and contains gel-like deposits. The edge is ill defined and there is normally a thin ring of salt deposits visible around the wall at 6 to 8 mm above the basal cone. The characterization of the medium is given in Table Twenty-Seven.

Table Twenty-Seven

Medium Diffusion in a Sterile SLYM-BART
Inner Test Vial to Confirm a Negative Reaction

Time (days)	Color		
	Basal	Lower column	Upper column
0.25	Brown-yellow	Clear	Clear
0.5	Dark brown	Light brown	Clear
1.0	Light brown	Light brown	Light brown
2.0 - 10.0*	Very light brown	Very light brown	Very light brown

*Note that the liquid medium is crystal clear and has been generated using sterile distilled or deionized water. Natural water samples can cause minor chemical reactions that may be seen through an intensification of the color in the diffusion front and crystalline deposits forming in the base of the test vial. These crystalline deposits can be differentiated from a basal slime since the crystalline deposits swirl up and have a defined edge, do not have a gel-like appearance, and settle rapidly to the base after shaking. Water saturated with oxygen stored at low temperatures can, when used in this test, cause bubbles to form as oxygen comes out of solution as the temperature rises to room temperature. Therefore do not use water taken directly from a refrigerated or cold source but allow the water to rise to room temperature before beginning the test to ensure any surplus saturation of the water with oxygen to have adjusted by the venting off of the surplus oxygen.

Table Twenty-Eight

QC Characterization of Medium Diffusion in a Sterile SLYM-BART Inner Test Vial

Time (hrs)	Color		Contamination
	Basal	Column*	
0.5	Brown-yellow	Clear	Clear
12.0	Dark brown	Light Brown	Clouding
48-240	Very light brown	Very light brown	Clouding/slimes

*See note for Table Twenty-Eight

Generally, where contamination does occur in the SLYM-BART, it will initially be seen as a clouded zone floating above the diffusion front. These clouds may be plate-like or diffuse in form. After a period of time, these growths may condense into a basal gel-like mass that will either settle in the basal cone of the test vial, or form a slime-like ring around the FID. Colors will generally lighten except around the FID where any slime growths may cause colors such as beige, yellow, red, or violet to be generated. For the QC to be acceptable, there must be no evidence of contamination in 240 hours and the color generation should follow the pattern given in the above table.

Confirmation of Selective Media Composition in the SLYM-BART

In order to confirm the suitability of the selective medium for the biodetection of the various slime-forming bacteria recognized by this test method, it is recommended that the following A.T.C.C. (American Type Culture Collection) strains be applied to the SLYM-BART to determine the standard reaction patterns. Each culture should be prepared as a 48 hour broth culture incubated at 30°C to reach the stationary growth phase using Brain Heart Infusion broth. Inoculation of the inner test vial uses a suspension of 0.1 ml of the broth culture in 15 ml of the sterile Ringer's solution. This innoculum should be taken from the midpoint of the broth culture immediately after the culture had been gently agitated. This inoculated solution is applied directly over the ball as the test vial is filled. Do not shake the vial. Incubate at 22 to 24°C for five days and observe for activities and reactions. Typical results are listed below for the recommended A.T.C.C. strains in Table Twenty-Nine.

<u>Table Twenty-Nine</u>

Cultural Characterization of the SLYM-BART

A.T.C.C.	Genus/species		Characterization
8090	*Citrobacter freundii*	CL	Reaction 5
13048	*Enterobacter aerogenes*	CL-BL	Reaction 5 & 6
27853	*Pseudomonas aeruginosa*	CL-PB	Reaction 5 to PB
12228	*Staphylococcus epidermidis*	DS	Reaction 1
23355	*Enterobacter cloacae*	CP-CL	Reaction 2 to 5
13315	*Proteus vulgaris*	CP-CL	Reaction 2 to 5
13883	*Klebsiella pneumoniae*	SR-CL	Reaction 4 to 5
25922	*Escherichia coli*	CL-BL	Reaction 5 to 6

Note that some of these culture tests will shift from one reaction type to another as the growth in the SLYM-BART matures.

d) HAB-BART™

There is a wide variety of bacteria that are aerobic and they share a common feature of the ability to respire using oxygen. One substitute for oxygen that is used in this BART test is methylene blue. This is utilized by bacteria instead of oxygen and it becomes bleached from blue (oxidized form) to a colorless (reduced) form. This test incorporates methylene blue and the activity of the aerobic bacteria is reflected in the speed and form in which the methylene blue is bleached. Occasionally, bacteria will bioaccumulate the methylene blue before it begins to be used for respiration. As a result, the medium may tend to darken where there is a dominance of bacterially initiated bioaccumulation. The major feature of this test is, however, the color shift from blue to colorless (bleaching). The other features are secondary.

HAB-BART Medium

A brown hardened gel-like mass with a darker inner circle extends fully up the cone to the walls of the test vial. On the sidewall of the inner test vial above the basal cone there is a transparent film that extends approximately 3mm up the walls. This film is also brown in color and contains gel-like deposits. The edge is ill defined and there is normally a thin ring of salt deposit visible around the wall at 6 to 8 mm above the basal cone. The characterization of the medium is given in Table Thirty-One.

HAB-BART CAP

On the inside of the cap, there are blue crystalline deposits around the outer edge in a complete circle, or in an incomplete arc. Flakes of the blue crystalline material may cover 20 to 50% of the inside area of the cap itself. These flakes are firmly held to the cap surface but excessive vibration is shipping may cause some of the smaller crystals to detach. If this happens, the water sample, when added to the inner test vial, may show some small areas of blue coloration. Occasionally, also, these flakes may be seen on the ball or attached to the walls of the test vial. Their presence does not interfere with the test procedure since they dissolve rapidly once the sample is added to the test vial.

Table Thirty

**Medium Diffusion in a Sterile HAB-BART
Inner Test Vial to Confirm a Negative Reaction**

Time (days)	Color		
	Basal	Lower column	Upper column
0.25	Greenish-blue	Blue	Blue
0.5	Greenish-blue	Greenish-blue	Blue
1.0	Greenish-brown	Greenish-blue	Blue

*Note that the blue color is crystal clear and has been generated using sterile distilled or deionized water. Natural water samples can cause secondary chemical reactions that may be seen through an intensification of color at the diffusion front. Water saturated with oxygen stored at low temperatures can, when used in this test, cause bubbles to form as oxygen comes out of solution as the temperature rises to room temperature. Water samples with higher salt concentrations may cause the color of the methylene blue to shift from a blue to very greenish shade of blue. This color shift does not, however, affect the sensitivity of the test.

Table Thirty-One

QC Characterization of Medium Diffusion in a

Sterile HAB-BART Inner Test Vial

Time (days)	Color	Contamination***
	Mid-point color	
1.0	Blue	Bleached*
4.0	Greenish-blue	Bleached**
8.0	Greenish-brown	Bleached***

A satisfactory QC would involve the left hand column of events with no signs of cloudiness, or localized concentrations of the blue color. Generally, contamination would be first observed by a bleaching in which the blue color turns to a colorless form. Commonly where this has happened in less than one day (*) there is only a slight clouding. If the bleaching occurs after four days (**), then the liquid will usually also be cloudy. In the event that bleaching did not occur until eight days (***) this would indicate either a very low contamination with aerobic microbes or that the contamination was dominated by anaerobes. Where anaerobic microbes dominate, the blue color may become locally concentrated (e.g., near the basal cone of the test vial).

Confirmation of the Selective Media Composition in the HAB-BART
In order to confirm the suitability of the selective medium for the detection of the various HAB recognized by this test method, it is recommended that the following A.T.C.C. (American Type Culture Collection) strains be applied to the HAB-BART to determine the standard reaction patterns. Each culture should be prepared as a 48 hour Brain Heart Infusion broth culture incubated at 30°C to reach the stationary growth phase. Inoculation of the inner test vial should be using a 1.0ml suspension of the broth culture in 15 ml of the sterile Ringer's solution. This inoculum should be taken from the midpoint of the broth culture immediately after the culture had been gently agitated. This inoculum should be applied directly over the ball as the test vial is being filled. Do not shake the vial. Incubate at 22 to 24°C for seven days and observe for activities and reactions. Typical results are listed below for the recommended A.T.C.C. strains in Table Thirty-Two.

Table Thirty-Two

Cultural Characterization of the HAB-BART

A.T.C.C.	Genus/species	Characterization	
27853	*Pseudomonas aeruginosa*	DO	Reaction 2
25922	*Escherichia coli*	UP	Reaction 1

The HAB (or heterotrophic aerobic bacteria) BART test is very specifically designed for the determination of aerobic bacteria that are heterotrophic. This means that this test is designed to determine the activity of aerobic bacteria that will possess the methylene blue reductase enzyme system and not function well anaerobically. If the biofouling problems are thought to be generated by aerobic microbial activities, the HAB-BART™ performs a role as the "scout" for aerobically generated problems. Commonly, the HAB-BART can therefore be used as the initial test that is then followed by tests using other BART testers. Some of these patterns of potential usage are given in HAB Table Thirty-Three:

Table Thirty-Three

Possible Further Testing of Waters Giving an Aggressive Diagnosis Using the "Scout" HAB-BART Test Method

Problem	Recommended Testing Strategy using BART				
	IRB-	SRB-	SLYM-	FLOR-	ALGE-
White slime	4		4		4
Grey slime	4	4		4	
Black slime	4	4		4	
Brown slime	4		4		4
Green slime	4		4		4
Turbidity			4		4
Taste	4		4		4
Odor		4	4	4	4
Color	4	4			4
Corrosion	4	4	4		
Hygiene risk*	4		4	4	

* Waters that have a possible hygiene risk should be tested for the presence of coliform bacteria.

e) FLOR-BART™

The fluorescent pseudomonad bacteria are a group of bacteria that have created some level of hygiene risk particularly in hospital and recreational environments. This BART utilizes a selective medium that encourages the growth of species of Pseudomonas which are able to generate fluorescent pigments particularly in the upper part of the inner test vial around the ball.

FLOR-BART Medium

A yellowish semi-transparent hardened film covers the basal cone and may extend as a more transparent film up the sidewalls of the test vial to a height not exceeding 3mm. The texture of this deposit is roughened granular and some lighter linear crystalline deposits may be seen radiating out from the central peg in the basal cone. Diffusion characteristics of the FLOR-BART using sterile water is given in Table Thirty-Four.

Table Thirty-Four

Medium Diffusion in a Sterile FLOR-BART
Inner Test Vial to Confirm a Negative Reaction

Time (days)	Color		
	Basal	Lower column	Upper column
0.25	Light yellow	Clear	Clear
0.5	Light yellow	Clear	Clear
1.0	Very light yellow	Very light yellow	Clear
7.0*	Very light yellow	Very llight yellow	Clear

*Note that the liquid medium is crystal clear and has been generated using sterile distilled or deionized water. Natural water samples can cause minor chemical reactions that may be seen through an intensification of the color in the diffusion front and sometimes crystalline deposits may form in the base of the test vial. These crystalline deposits can be differentiated from a basal slime since the crystalline deposits swirl up and have a sharply defined edge, do not have a gel-like appearance, and settle rapidly to the base after shaking. Water saturated with oxygen stored at low temperatures can, when used in this test, cause bubbles to form as oxygen comes out of solution as the temperature rises to room temperature. Therefore do not use water taken directly from a refrigerated or cold source but allow the water to rise to room temperature before beginning the test. This will ensure that any surplus saturation of the water with oxygen will be vented off and not interfere with the test.

Table Thirty-Five

QC Characterization of Medium Diffusion in a Sterile FLOR-BART Inner Test Vial

Time (days)	Color		Contamination
	Basal	Mid- column	
1.0	Very light yellow	Clear	Clouding
7.0	Very light yellow	Clear	Turbid or slimes

The FLOR-BART should remain very clear with only a very light yellow coloration near the base of the test vial. If there is contamination, this is usually caused by pseudomonads or aerobic Gram positive bacteria. These growths usually develop as a general cloudiness that then intensifies to make the medium turbid. Slimes sometimes are formed.

Confirmation of the Selective Media Composition in the FLOR-BART

In order to confirm the suitability of the selective medium for the biodetection of the various pseudomonad bacteria recognized by this test method, it is recommended that the following A.T.C.C. (American Type Culture Collection) strains be applied to the FLOR-BART to determine the standard reaction patterns. Each culture should be prepared as a 48 hour broth culture incubated at 30°C to reach the stationary growth phase using Brain Heart Infusion broth. Inoculation of the inner test vial should be using a 0.1 ml suspension of the broth culture in 15 ml of the sterile Ringer's solution. This inoculum should be taken from the midpoint of the broth culture immediately after the culture had been gently agitated. This inoculated solution should be applied directly over the ball as the test vial is filled. Do not shake the vial. Incubate at 22 to 24°C for five days and observe for activities and reactions. Typical results are listed below for the recommended A.T.C.C. strains in Table Thirty-Six.

<u>Table Thirty-Six</u>

Cultural Characterization of the FLOR-BART

A.T.C.C.	Genus/species		Characterization
13048	*Enterobacter aerogenes*	CL	Reaction 1
27853	*Pseudomonas aeruginosa*	CL-PB	Reaction 1 with PB
12228	*Staphylococcus epidermidis*		No growth
19606	*Acinetobacter calcoaceticus*	CL	Reaction 1

f) DN-BART™

This test detects bacteria that can reduce nitrate (NO_3) to dinitrogen gas (N_2) by the observation of gassing which occurs when the nitrate has been completely denitrified. While nitrite is an intermediate in the denitrification of nitrates (with dinitrogen gas being the terminal product), it does not remain resident for a significant period of time particularly where nitrifying bacteria are active. Where there is an aggressive population of denitrifying bacteria present in the sample, the liquid sample will show a variable amount of cloudiness but there will be a generation of (dinitrogen) gas bubbles. These usually collect around the ball in the form of foam. This foam will last one to three days and usually not show any color. The presence of a foam or (less commonly) gas bubbles under and around the ball covering at least 50% of the submerged area) cloudiness with gas formation represents a positive detection of denitrifying bacteria.

DN-BART Medium

Opaque beige crystalline deposit extends out towards the walls of the test vial. It has a defined but irregular edge and appears to darken somewhat towards the central peg. Occasionally, crystalline deposits may also be seen at the wall conical base interface and may extend up the sidewall of the inner test vial up to 3mm. Table Thirty-Seven gives the characteristics that would be expected due to medium diffusion into sterile water.

Table Thirty-Seven

Medium Diffusion in a Sterile DN-BART
Inner Test Vial to Confirm a Negative Reaction

Time (days)	Color		
	Basal	Lower column	Upper column
0.25	Pale yellow	Clear	Clear
0.5	Pale yellow	Clear	Clear
1.0	Pale yellow	Clear	Clear
2.0	Pale yellow	Clear	Clear

Note that the liquid medium is crystal clear and has been generated using sterile distilled or deionized water. Natural water samples can cause minor chemical reactions that may be seen through an intensification of the color in the diffusion front and crystalline deposits may form in the base of the test vial. These crystalline deposits can be differentiated from a basal slime since the crystalline deposits swirl up and have a defined edge, do not have a gel-like appearance, and settle rapidly to the base after shaking. Water saturated with oxygen stored at low temperatures can, when used in this test, cause bubbles to form as oxygen comes out of solution as the temperature rises to room temperature. Therefore do not use water taken directly from a refrigerated or cold source but allow the water to rise to room temperature before beginning the test to ensure any surplus saturation of the water with oxygen has vented.

Table Thirty-Eight

Characterization of Medium Diffusion in a
Sterile DN-BART Inner Test Vial

Time (day)	Color		Contamination
	Basal	Mid-column	
1.0	Pale yellow	Clear	Cloudy, may be gassing
2.0	Pale yellow	Clear	Turbid, may be gassing
4.0	Pale yellow	Clear	Turbid, commonly gassing

The medium used in the DN-BART can encourage the growth of a range of facultatively anaerobic and nitrate respiring bacteria. If there has been contamination of the test vial, then this test will commonly exhibit a cloudy medium that gradually becomes more turbid with time. If any of the contaminants are complete denitrifiers, gassing may occur. Incubation for this test is normally at 22 to 24°C but using blood heat (35 to 37°C) can speed up the growth of many contaminants.

Confirmation of the Selective Media Composition in the DN-BART

In order to confirm the suitability of the selective medium for the biodetection of the various bacteria recognized by this test method (see text above), it is recommended that the following A.T.C.C. (American Type Culture Collection) strains be applied to the DN-BART to determine the standard reaction patterns. Each culture should be prepared as a 48 hour culture incubated at 35°C to reach the stationary growth phase using Brain Heart Infusion broth. Inoculation of the inner test vial should be with a suspension of 0.1 ml of the broth culture in 15 ml of the sterile Ringer's solution. This inoculum should be taken from the midpoint of the broth culture immediately after the culture had been gently agitated. This inoculated solution should be applied directly over the FID ball as the test vial is filled. Do not shake the vial. Incubate at 22 to 24°C for one day and observe for activities and reactions after applying the reactant cap following the standard procedure. Typical results are listed below for the recommended A.T.C.C. strains in Table Thirty-Nine.

Table Thirty Nine

Cultural Characterization of the DN-BART

A.T.C.C.	Genus/species	Solution
13048	*Enterobacter aerogenes*	++, clouding,
27853	*Pseudomonas aeruginosa*	++, slight clouding
12228	*Staphylococcus epidermidis*	-, clouding,
19606	*Acinetobacter calcoaceticus*	++, no clouding,
25922	*Escherichia coli*	++,clouding

* Gassing or Foaming (++) is considered the prime test for complete denitrification that can be recognized as a foam ring or intense bubbles under and around the ball. Clouding is not a confirmation of denitrification and should be considered negative.

g)   N-BART™

This test detects the nitrifying bacteria that are able to oxidize ammonium (NH4) to nitrite (NO_2) and on to nitrate (NO_3). This test uses a selective medium for the bacteria able to oxidize ammonium to nitrite by examining chemically for the nitrite product. The additional two balls used in this test provide a larger solid: medium: air area on the upper hemispheres of the three balls. This encourages nitrification in the liquid film over the balls. In the early stages, the first (product) nitrite is detected at these sites. A reactant cap is used to detect the presence of nitrite that is generated during the early stages of nitrification. If the sample being tested also contains denitrifying bacteria, nitrite may again be created by the reduction of nitrate (denitrification). This test method has been developed in consideration of the greater likelihood of nitrite being detectable rather than the (product) nitrate. Note that this test cannot function in water samples with a natural nitrite level of greater than 3.0 ppm. Water samples with greater than 28 ppm of nitrite will automatically turn the liquid medium to a yellow color when the reaction cap test is applied.

N-BART medium

Circular white crystalline opaque deposit remains clustered around the central peg in the basal cone. This extension may be 5 to 8 mm in radius with a defined largely smooth edge. In the normal event of the confirmation of sterility, there is a change in the characteristics of the liquid medium forming in the test vial. These are detailed in Table Forty below.

N-BART Reaction Cap

This cap is a small screw type white plastic cap which can be screwed down onto the inner test vial. This cap contains rough porous paper disc fitted within the inner flange. When viewed from the under side, this disc is colored. It varies from a solid pink to a yellowish pink center with a darker pink perimeter. This color is generated by the reactants used to detect nitrite in the test medium. While there is a variation in the color, it has been found that this does not affect the accuracy of the test method.

Table Forty

Medium Diffusion in a Sterile N-BART
Inner Test Vial to Confirm a Negative Reaction

Time (days)	Color		
	Basal	Lower column	Upper column
0.25	Grey	Clear	Clear
0.5	Clear	Clear	Clear
10.0	Clear	Clear	Clear

*Note that the liquid medium is crystal clear and has been generated using sterile distilled or deionized water. Natural water samples can cause minor chemical reactions which may be seen through an intensification of the color in the diffusion front and occasionally crystalline deposits may form along the floor of the test vial.

Table Forty-One
Characterization of Medium Diffusion in a
Sterile N-BART inner test vial

Time (days)	Color		Contamination
	Basal	Mid-column	
1.0	Clear	Clear	Cloudy
10.0	Clear	Clear	Turbid, possible gassing

The N-BART is a relatively specialized test in which the medium will not support a wide range of contaminants. Where there is contamination, the initial expression of growth is a light clouding (day 1) which gradually intensifies causing the medium to go turbid. If there are any complete denitrifiers among the contaminants, gassing may occur.

Confirmation of the Selective Media Composition in the N-BART

In order to confirm the suitability of the selective medium for the biodetection of the various bacteria recognized by this test method, it is recommended that the following A.T.C.C. (American Type Culture Collection) strains be applied to the N-BART biodetectors to determine the standard reaction patterns. Each culture should be prepared as a 7 day culture incubated at 25°C to reach the stationary growth phase. The cultural techniques to be used are referenced in Verhagen et al. (1993) "Effects of Grazing by Flagellates on Competition for Ammonium between Nitrifying and Heterotrophic Bacteria in Soil Columns" in *Journal of Appl. and Environmental Microbiology*. Inoculation of the inner test vial should be with a 0.1ml suspension of the broth culture in 15 ml of the sterile Ringer's solution. This inoculum should be taken from the midpoint of the broth culture immediately after the culture had been gently agitated. 7.75 ml of this inoculated solution should be applied directly to the test vial. Do not shake the vial. Incubate at 22 to 24°C for five days and observe for activities and reactions after applying the reactant cap following the standard procedure. Typical results are listed below for the recommended A.T.C.C. strains in Table Forty-Two.

Table Forty Two
Cultural Characterization of the N-BART

A.T.C.C.	Genus/species	Characterization/solution
25391	*Nitrosomonas winogradski*	clouding, red reaction, nitrite +
27853	*Pseudomonas aeruginosa*	-, no reaction, nitrite negative
19718	*Nitrosomonas europae*	clouding, red reaction, nitrite +

QC Procedure for the N Reactant Cap

Prepare or obtain analytical grade solutions of sodium nitrite stock solution and dilute standard sodium nitrite solution (5mg and 50mg N-NO$_2$/L). These analytical solutions should be labeled nitrite 5, and nitrite 50 and should be freshly prepared using sterile distilled water and aseptic techniques. The protocol for testing the reactant cap would be similar for all solutions. This protocol would involve the following steps:

A. Sodium nitrite stock solution (**solution A**)
    Dissolve 100mg NaNO$_2$ in 50ml distilled water. Transfer it to a 200-ml volumetric flask. Add distilled water to mark line. Final concentration is 500mg/L.

B. Dilute standard NaNO$_2$ solution:

    Solution B1: Pipet 0.5ml **solution A** to a 50-ml volumetric flask. Add distilled water to mark line. Final concentration is 5mg/L.

    Solution B2: Pipet 5.0ml **solution A** to a 50-ml volumetric flask. Add distilled water to mark line. Final concentration is 50mg/L.

Prepare the solutions before use.

C. Prepare three tubes. Put one white ball in each tube. Pipet 15mls of **solution B1** into each tube. Prepare another three tubes. Put one white ball in each tube. Pipet 15mls of **solution B2** into each tube. Screw on D-RX cap. Turn over onto caps for 30 minutes. Turn back and observe the color change after 3 hours. For the tube with **solution B1**, the solution color will be pink. For the tube with **solution B2**, the solution color will be yellow.
    Refer to Table Forty-Three for the interpretation of the observations.

Table Forty-Three

Reaction Confirmation for the QC on the N-BART Reactant Cap

Solution	Color	Absorption
Nitrite, 5	pink	0.36 @500nm
Nitrite, 50	yellow	0.30 @580nm

Note: In weak reactions for the nitrite solutions, a weakened reaction giving lower color generation and absorption values suggests a degenerating reactant cap.

d) ALGE-BART™

This biodetector is distinctly different from the other products in the series because it is designed to recover and culture phototrophic (photosynthesizing) microorganisms that utilize light and release oxygen as a product. To achieve this, a modified Bold's medium is used which does not contain organics, but does contain the basic nutrients for plant growth (nitrogen, phosphorus, potassium, sulfur etc.). Carbon is presented as bicarbonates and the medium is made slightly alkaline (pH, 8.2) to encourage the micro-algae to utilize this form of carbon. The growth of micro-algae tends to be slower than for the heterotrophic bacteria and does require the presence of light (for photosynthesis). This light does not have to be strong (such as in direct sunlight) and most micro-algae can actually utilize quite low levels of light. The most effective manner for maximizing the probability of photosynthesis is to place the charged ALGE-BART on its side and set it about 60 cms from a single 40 watt daylight fluorescent light or 100 cms away from a 60 watt tungsten light source.

ALGE-BART Medium

White opaque crystalline deposits extend from the central peg towards the walls. Commonly, the deposits may not reach out to the side-wall of the inner test vial and exhibit an irregular form but being more granular towards the center. A transparent crystalline film may extend for 2 mm up the sidewall but this is obscured by the textile walls of the test vial. In the normal event of the confirmation of sterility, there is a change in the characteristics of the liquid medium forming in the test vial. These are detailed in Table Forty-Four.

<u>Table Forty-Four</u>

**Medium Diffusion in a Sterile ALGE-BART
Inner Test Vial to Confirm a Negative Reaction**

Time (days)	Color	
	<u>Basal</u>	<u>Above weave</u>
0.25	Clear	Clear
28*	Clear	Clear

*Note that the culture medium should remain crystal clear and have been generated using sterile distilled or deionized water. Natural water samples can cause secondary chemical reactions that may be seen through crystalline deposits forming on the floor of the test vial.

Table QC Forty-Five

QC Characterization of Medium Diffusion in a
Sterile ALGE-BART Inner Test Vial

Time (days)	Color		Contamination
	Fabric	Solution	
7	Off white	Clear	Cloudy
28	Discolored	Cloudy	Cloudy or slime deposits

The ALGE-BART has a very low organic nutrient load in this QC test and so any contaminant will take a period of time to grow. This test should be done under constant illumination at 22 to 24°C and bad contamination may occur, and be visible, in seven days. The woven material may discolor slightly under these circumstances. If this discoloration includes the formation of black or grey specks, then it is possible that the contaminants are molds particularly if this occurs above the water line.

Confirmation of Selective Media Composition in the ALGE-BART

There is a need to confirm the suitability of the selective medium for the biodetection of the various algae. This need is recognized by this test method (Table Forty-Six). It is recommended that the following Culture Collection of Algae and Protozoa (CCAP) strains be applied to the BART to determine the standard reaction patterns. The CCAP 5th edition was issued as ISBN 1 871105 01 3 in 1988. Each culture should be prepared according to the protocols described in the 5th edition CCAP catalog. The cultures should be used when they have reached the stationary growth phase. Inoculation of the inner test vial should be using a suspension of 0.5 ml of the culture in 15 ml of the sterile isotonic salt solution. This inoculum should be taken from the midpoint of the micro-algal culture immediately after the culture had been very gently agitated. Do not shake the vial. Incubate under continuous light at 22 to 24°C for twenty-eight days and observe twice a week for activities and growth reactions. Typical results are listed below for the recommended CCAP strains in Table Forty-Six.

Table Forty-Six
Characterization of the ALGE-BART

CCAP	Genus/species	Characterization
211/62	*Chlorella* sp.	GF
11/77	*Chlamydomonas baca*	GG to GF
276/21	*Scenedesmus quadricauda*	GF to YB
678/4	*Spirogyra* sp.	GF to DG

Note that some of these cultural tests will shift from one reaction type to another as the growth in the ALGE-BART matures.

i)   POOL-BART™

This test uses a selective medium that encourages the growth of bacteria that commonly foul pools and hot tubs. It is broader spectrum than the FLOR-BART and causes more bacteria to generate a reaction one. There is another reaction not seen in the other referenced test where the bacteria form thread-like structures within the liquid medium particularly between the ball and the basal cone. Essentially, this test has been developed to determine and monitor pool and hot tub fouling bacteria.

POOL-BART Medium

Yellowish, semi-transparent, gel-like mass hardened around the central peg within the basal cone. The edge, where visible, is defined and clearly observable. There may be some extension of the film of deposits up the sidewall of the vial to a height not exceeding 3mm. In the normal event of the confirmation of sterility, there is a change in the characteristics of the liquid medium forming in the test vial. These are detailed in Table Forty-Seven.

Table Forty-Seven
Medium Diffusion in a Sterile POOL-BART
Inner Test Vial to Confirm a Negative Reaction

Time (days)	Color		
	Basal	Lower column	Upper column
0.25	Pale yellow	Clear	Clear
0.5	Pale yellow	Clear	Clear
1.0	Pale yellow	Pale yellow	Clear
7.0*	Pale yellow	Pale yellow	Clear

*Note that the liquid medium is crystal clear and has been generated using sterile distilled or deionized water. Natural water samples can cause minor chemical reactions that may be seen through an intensification of the color in the diffusion front. Occasionally crystalline deposits may form in the base of the test vial. These crystalline deposits can be differentiated from a basal slime since the crystalline deposits swirl up and have a defined edge, do not have a gel-like appearance, and settle rapidly to the base after shaking. Water saturated with oxygen stored at low temperatures can, when used in this test, cause bubbles to form as oxygen comes out of solution as the temperature rises to room temperature. Therefore do not use water taken directly from a refrigerated or cold source but allow the water to rise to room temperature before beginning the test to ensure any surplus saturation of the water with oxygen has adjusted by dissolution of the surplus oxygen.

Table Forty-Eight

QC Characterization of Medium Diffusion in a Sterile POOL-BART Inner Test Vial

Time (hrs)	Color		Contamination
	Basal	Column	
1.0	Very light yellow	Clear	Clouding, slimes
7.0	Very light yellow	Clear	Turbid, threads or slimes

The POOL-BART should remain very clear with only a very light yellow coloration near the base of the test vial. If there is contamination, this is usually caused by aerobic bacteria. These growths usually develop as a general cloudiness but occasionally slimes or thread-like growths may also be observed. These intensify with time to make the medium turbid and slimes may be generated around the FID, in the base or as thread-like structures hanging suspended in the medium.

Confirmation of the Selective Media Composition in the POOL-BART

In order to confirm the suitability of the selective medium for the biodetection of the selected bacteria recognized by this test method (see text above), it is recommended that the following A.T.C.C. (American Type Culture Collection) strains be applied to the POOL-BART to determine the standard reaction patterns. Each culture should be prepared as a 48 hour broth culture incubated at 30°C to reach the stationary growth phase using

Brain Heart Infusion broth. Inoculation of the inner test vial should be using a suspension of 0.1 ml of the broth culture in 15 ml of the sterile solution. This inoculum should be taken from the midpoint of the broth culture immediately after the culture had been gently agitated. This inoculated solution should be applied directly over the ball as the test vial is filled. Do not shake the vial. Incubate at 22 to 24°C for five days and observe for activities and reactions. Typical results are listed below for the recommended A.T.C.C. strains in Table Forty-Nine.

Table Forty-Nine

Cultural Characterization of the POOL-BART

A.T.C.C.	Genus/species	Characterization
13048	*Enterobacter aerogenes*	CL
27853	*Pseudomonas aeruginosa*	CL to PB
12228	*Staphylococcus epidermidis*	No growth
19606	*Acinetobacter calcoaceticus*	CL
25922	*Escherichia coli*	CL

EXPIRY DATE QC

The BART test products are protected from degeneration during storage by the use of an aluminum foil package as a moisture barrier, aseptic manufacturing procedures, crystallization of the dried medium pellet, and post-production QC procedures. Presently the recommended expiry date is set at three years after packaging. There remains however a need to remain confident that the stored BART test hasn't degenerated in any manner that could affect its performance. Recommended QC test procedures for the various BART tests are described above. One feature of the BART test products to be remembered is that they are extremely sensitive to the presence of bacteria in the sample used. In the event of a QC procedure, it is therefore essential to use sterilized (distilled or deionized) water for the test and to use aseptic procedures. It is important to insure that the water is sterilized (commonly by steam sterilization using an autoclave or pressure cooker). It has been found that regular distilled or deionized water commonly do contain an active population of bacteria which is sufficiently large to cause activities and reactions to occur in the BART under test. In the Droycon Bioconcepts Inc.QC laboratory, commercially available distilled and deionized water frequently generate reactions with a time lag (days of delay to reaction) of 2 to 4 days. The use of such nonsterile waters is therefore likely to give unacceptable results. To ensure that the BART test is still suitable for use as a test procedure, the following steps need to be followed:

1. Assurance of the status of crystallized medium by visual inspection.
2. Addition of sterile water to the BART inner test vial and determine forms of diffusion from the basal nutrient medium pellet in the base.
3. Determine whether there has been any microbial activity arising from contamination of the BART test during storage. This is done through looking for clouding, slime formation abnormal color generation and/or gas bubble formation during the test procedure listed as 2 above.

All of the BART test products are subjected to this QC procedure to be assured that there has not been any deterioration of the products over time.

RECORDING AND INTERPRETING DATA FROM BART

BART Data Definitions

There are three clusters of data information obtained from the BART testers. These are defined as:

(1) Time Lag
The time lag is the time lapse period from setting the test up to the observation of the first reaction. The longer the lapse, the less aggressive are the bacteria that have been detected. A successful control treatment would be expected to extend the time lag (commonly registered as "days of delay") and, in simple terms, each additional day of delay can be considered to have reduced that bacterial population by one order of magnitude. For example, if the time lag had been 2 days before treatment and 3 days after treatment, then it could be said that the treatment reduced the population by one order of magnitude (i.e., by 90%). Had the time lag been 4 days after treatment, then the treatment would have been effective at two orders of magnitude. This would mean that the population would have been reduced by two orders of magni-

tude or by 99%. A five-day lag time would have meant a three-day additional time lag delay that would have been three orders of magnitude (i.e., 99.9%). Time lags before and after a treatment determined by the BART testers can therefore give an indication of the success of a treatment.

(2) Reaction Pattern Signatures (RPS)

It is very rare for a water sample to contain a single species of microbe. Normally, there are a number of species in water which commonly range between eight and sixty. Ironically, the greater the number of species, the less "pollution" or eutrophication (intense growth) is occurring. These species do not all "burst into life" at the same time in the BART testers but do so sequentially. This, along with the maturation of the microbes already growing, causes the reaction pattern observed to change. The reaction pattern signature (RPS) is the chronological sequence with which those reactions are observed. If a treatment severely impacts on a part of the microbial community, this impact can be reflected in a change in the RPS. Any shift in the characteristics of the RPS whether it be a change in the reaction codes observed or in the sequence that the reaction codes observed indicates that the treatment has had an impact.

(3) RPS Time Length

This parameter is determined by the length of time between the observation of the first reaction pattern and the last. For example, if the first reaction was observed on day 3 and the last on day 7 then the RPS time length would be 4. Like the time lag, the RPS time length is measured in days. If a treatment has been effective (e.g., at increasing the time lag) but did not impact on the RPS observed, then the impact of the treatment may be determined by the changes in the RPS time length.

There are, therefore, three prime techniques to determine whether a treatment has been effective using the BART testers. In order of priority (from the highest to the lowest):

- time lag gives the an idea of the aggressivity of the bacteria in the water sample,
- RPS gives an idea of the treatment impact on the community (consortium) of bacteria in the treated water, and

- RPS time length shows the treatment impact on that community.

It has to be remembered that the water samples have to be taken at appropriate times both before and after treatment. In both cases, the samples should be taken from a time well before and after the treatment itself under repeatable conditions. It has to be remembered that a common objective in a treatment is to minimally suppress the microorganisms causing the perceived problems. This means that the treatment will, as a part of the process, disrupt and kill some of the microorganisms within the zone of biofouling. This means that large numbers of viable microbes may be released from the slimes (biofilms) into the water during, and immediately after, treatment. If the water is sampled at this time, very highly aggressive populations are likely to be recovered which would give the false impression that the treatment has failed when, really, it has not. When this happens, the time lag and the RPS time length may be expected to be smaller.

Testing to determine the effectiveness of a treatment on a biofouled water system therefore needs careful planning. This should include the following considerations:

- The pre-treatment water sample should be taken under conditions that can be repeated after the treatment has been applied.
- Ideally, the treatment should be sequenced in just after a routine testing of the water has been applied. Normally, this routine testing should be daily, weekly or monthly basis and show an observable trend.
- After treatment, there should be a sequence of sampling and testing immediately after the treatment to determine the direct impact of that treatment. This would involve a sequenced set of samples being taken to monitor the kill (very long time lags), the sloughing of survivors (very short time lags) and the stabilization of the treated biozone (hopefully with longer time lags).
- Post-treatment long-term impact. Returning to the routine sampling and testing on the water using BART testers should allow the long-term impact of the treatment to be determined. Here, lengthening of the time lag and RPS along with changes in the RPS would indicate that the treatment had been successful.

Long term monitoring is an essential part of managing a biofouling system. It should not be thought that a single treatment can continue to be effective for the potential life time of the system. Biofouling will return over time and preventative maintenance (i.e., routine reactive monitoring) can keep the potential problems in check.

Recording BART Data

Each BART generates a distinctive reaction that signals which communities of bacteria are present in the sample. The time at which the reaction is first observed gives an indication of the time lag. This time lag is commonly measured in days and the shorter the time, the more aggressive the bacterial population. In the BART sample data record sheet, the columns across the page represent the time lag in days from left to right. The horizontal rows represent each BART type that could be used on that sample. Each cell represents a particular BART and a specific time lag day. In each cell on the sheet the possible log population (PLP expressed in log cfu/ml units) is given should the reaction be observed and recorded on that day by writing the correct observed reaction code into the cell. Reference to the log population gives an indication of the population size while the background shading gives the aggressivity. In essence, a single row for one BART can show the time lag (by the position of the first reaction recorded in the row) and the reaction pattern signature (RPS) by the subsequent reactions which are observed.

The reactions observed can be different between the various BART tests, and may also vary within a single BART type depending upon the form and aggressivity of the bacteria within each selected sample. Below, the reactions that signal a positive will be described for each BART and should be entered as a Reaction Code by the appropriate letters (e.g., CL). It should be remembered that the bacteria may not stop growing simply because a reaction has occurred and frequently additional reactions may be seen. The letter code appropriate to that reaction should also be entered into the reaction code box immediately below the appropriate time lag box. In viewing the sequence of observed reactions in a particular BART, it becomes possible to understand more concerning the types of bacteria aggressively present in that water sample. Remember that there are two major items of data to be gathered:

1. Time Lag (recorded in days of delay to sighting the reaction)
2. Reaction Code (a two letter code reflecting the type of reaction observed entered into the cell immediately under the day box appropriate to the reaction being observed)

Upon entering this data, the italicized PLP number in the cell into which the code was entered gives the possible log population in log form for the colony forming units/ml. If the arithmetic form is preferred as cfu/ml directly then conversion rows are provided on the Sample BART Data Record Sheet. Additionally, the reaction code box is shaded to represent the likely level of aggressivity that can be expected. A 25% grey shade in the box would mean that the bacteria would have a "HIGH" aggressivity. If the shade is a very light grey (12.5%), this would indicate a "MEDIUM" aggressivity while no shade (open box) would mean that the bacteria had a "LOW" aggressivity. The shade in the reaction code box into which the reaction code has been entered therefore gives a direct indication of the aggressivity for that particular bacterial population in the sample that caused the reaction.

Selection of the appropriate time lag to enter the reaction code should include the following ground-rules:

- Where the data cannot be collected for certain days (due to weekends, etc.), reaction codes should be entered in for the day that the reaction codes are first observed.
- Day "1" extends to 24 hours after the test was first started. Reactions occurring after that time and before 48 hours would constitute day "2" data. Similarly day "3" data would go from 48 hours to 72 hours. Technically, therefore, reactions collected after the test had run for 47 hrs would be considered day "2" data while data collected at 49 hours would be day "3" data.
- Where a time lag day has been missed because there was no opportunity to observe the tests, then an "X" should be put in the upper time lag box to show that data for that day is missing.
- Where several reaction codes are entered at subsequent time lags for the same BART test and sample, the PLP shown in the reaction code box into which the data was first entered reflects the population for the bacteria causing that particular reaction code to occur. Note that descriptions of the community bacterial structures causing the different reaction codes are described for each of the BART types.
- The length of time (RPS time length) between the first reaction code being observed and the last (terminal) reaction is observed has some value in assessing the effectiveness of treatments. The longer the time it takes for the first reaction code to shift to the terminal

reaction code gives an index of the extent of the treatment. The longer the time difference, the less the impact of the treatment (see the section below on the use of BART tests in evaluating the effectiveness of treatments)

Note that each BART type has a distinct set of possible reaction codes based upon that double letter code series. It should be noted that there is a more comprehensive interpretation that has been traditionally used. This traditional coding uses numbers and may be found on the Internet site at: **http://www.dbi.sk.ca**. The coding is listed by BART type below.

BART Reaction Codes Definitions

The BART tests generate a set of reactions which reflect primarily the microbial composition of the water and, secondarily, any chemical reactions which may occur between the medium pellet in the basal cone of the inner vial and chemicals which would be in the water sample itself. This second interaction makes for occasionally more complex reactions. A third parameter that does influence the reactions is the water sample. This is very unlikely to consist of a single species of microbe but rather a consortium (or a community). Species within this consortium may compete within the BART test as it proceeds to cause the displayed reactions to change as the consortium matures. Needless to say this renders every water sample unique and the reaction patterns that are described below represent the most common reaction patterns encountered.

Where reactions do change over time, these form into a reaction pattern signature (RPS) which represents a "snapshot" of the community which should be repeated in subsequent water testing if the consortium has not changed. Genera most commonly associated with the different reactions have been identified for many of the reaction patterns using the MIDI and the API identification systems. This work was undertaken from 1988 through to 1994.

Each reaction set for an individual ᴮᴬᴿᵀ will begin with opening comments on the mechanisms employed to generate reactions and this will be followed with a description of each reaction. Readers should note that the reaction numbers have been changed to a two-letter coding to allow easy understanding of the reaction being observed. For example, UP means bleaching in a HAB-BART has gone upwards, and BR means a brown ring in an IRB-BART test.

Each reaction is described proceeded by a letter code that best fits that description. The letter code is limited to a two letters, which should be entered on the BART data record sheet in the correct (by day and BART type) cell.

Example of how to use the BART Data Record Sheet

Each time the BART tests are observed (normally daily until at least the first reaction is observed), any changes such as the generation of a new reaction (code) should be added to the data sheet. Place two lettered reaction code for that reaction into the reaction box under correct time lag column. A typical set of reactions are recorded in the example and the interpretation is given below:

IRB-BART™, iron-related bacteria detected
- **FO**, time lag of 3 means anaerobic IRB at a PLP of 4.0 (10,000 cfu/ml) High Aggressivity
- **BR**, time lag of 5 mean aerobic slime-forming IRB at a PLP of 3.0 (1,000 cfu/ml) Medium Aggressivity
- **BC**, time lag of 7 means facultative anaerobic IRB at a PLP of 2 (100 cfu/ml) Medium Aggressivity
- The RPS is FO – BR – BC with the RPS time length of 4 days (7 minus 3)

SRB-BART™, sulfate-reducing bacteria detected
- **BB**, time lag of 2 means deep seated anaerobic SRB at a PLP of 5 (100,000 cfu/ml) High Aggressivity
- **BA**, time lag of 4 means mixed aerobic/anaerobic SRB at a PLP of 4 (10,000 cfu/ml) High Aggressivity
- The RPS is BB – BA with the RPS time length of 2 days (4 - 2)

HAB-BART™, heterotrophic aerobic bacteria detected
- **UP**, time lag of 3 means aerobic heterotrophs at a PLP of 3 (1,000 cfu/ml) Medium Aggressivity

DN-BART™, denitrifying bacteria detected
- **FO**, time lag of 2 means denitrifying bacteria at a PLP of 5 (100,000 cfu/ml) High Aggressivity

SLYM-BART™, slime-forming bacteria
- **CL**, time lag of 2 means slime-forming bacteria at a PLP of 5.6 (500,000 cfu/ml) High Aggressivity
- **BL**, time lag of 5 means pseudomonad & enteric mixed bacteria at a PLP of 3.6 (5,000 cfu/ml) Medium Aggressivity
- The RPS is CL – BL with the RPS time length 3 days (5 minus 2)

FLOR-BART™, fluorescent pseudomonad bacteria
- **PB**, time lag of 7 means possible *Pseudomonas aeruginosa* at a PLP of 2 (100 cfu/ml) Medium Aggressivity.

Considerable information can be obtained from the various BART tests and the time lag to the appearance of the reactions. A sequence of reactions, particularly in the IRB-BART, can help to identify which bacterial groups are aggressive in that particular water sample.

It should be noted that confirmatory tests are recommended, particularly the coliform test, where there are concerns about the levels of aggressivity being observed. Note that the BART test is more sensitive to many more bacteria than the standard heterotrophic agar spread-plate test because of the greater variety of environments provided which can support growth. It is therefore not uncommon to have very aggressive bacterial counts using the BART system but not get any colonial growths on the agar plate technique.

Multiple BART Enumeration of Bacterial Populations

In the standard BART test using a single test vial, the population can be predicted based upon the observed time lag. The premise is that the larger, and the more aggressive, the bacterial population, the shorter the time lag would be. In this scenario, there is no direct enumeration of the population as such. In the BART test, the first reaction observed may reflect the most aggressive microbial species as much as larger populations of less aggressive species. The advantage of the BART tester is its "keep it simple stupid (KISS)" approach. Therefore it is most reactive to whichever group of bacterial species is able to generate the first (and subsequent) reactions. Extinction dilution techniques can be used to quantify the populations with more precision. The next section covers the protocols for determining the microbial populations in water samples using the BART testers. It is recommended that the LAB-BART™ testers be used for these studies and performed by persons familiar with the basic range of microbiological techniques.

Extinction Dilution Techniques Application to BART Testers

The BART testers all use a 15ml base volume. This base volume was selected because early trials revealed that a smaller volume (e.g., 10ml) did not generate a satisfactory blend of a redox gradient and an ascending nutrient front which are both important parts of the BART test. Selecting this volume means that tenfold dilutions would have to involve the transfers of 1.6ml of one diluted sample into 15.0ml of the sterile isotonic diluant. The recommended diluant is a phosphate buffer, pH 7.2. The formula per liter of distilled water is:

- Potassium Dihydrogen Phosphate   26.2g
- Sodium Carbonate   7.8g

To prepare the stock solution, dissolve the salts into 1 liter of water. Dispense 15.0ml of the solution into clean glass bacteriological tubes and cap. Sterilize the tubes by autoclaving. Store the tubes under refrigerated conditions until being selected for use. Allow the tubes to recover to room temperature before being used. If the dilution tubes are taken straight out of the refrigerator, the dilution of the water sample into the diluant could cause a "cold shock" which may slow down the recovery and growth rates of the intrinsic microbes in the sample.

To conduct the extinction dilution of the water sample, the 15.0ml aliquots of phosphate buffer are arranged in a row and labeled to allow a tenfold dilution series to be prepared from the water sample. To conduct a dilution series from the water sample, the water sample should be gently shaken for one minute to ensure as much mixing as possible to create a composite sample. Note that if the water sample has been stored at a lower temperature, the sample should be left to sit at room temperature for long enough for the temperature of the water to have reached that of the room (19 to 25°C). Using aseptic techniques, transfer 1.6ml of the water sample taken from the mid-point of the sample to the first dilution tube that contains 15.0ml of the phosphate buffer. Raise and lower the (2, 5 or 10ml) pipette up and down in the dilution while dispensing the sample to ensure mixing. This first dilution is a tenfold dilution of the water sample (10^{-1}). Using the sample pipette and following standard microbiological aseptic practices withdraw 1.6ml from the 10^{-1} dilution and transfer to the next dilution tube containing 15.0ml of phosphate buffer. When mixed, this tube now contains one hundred fold dilution (10^{-2}) of the original water sample. The 10^{-1} dilution tube now contains 15.0ml of diluted water sample. At this time, the 10^{-2} contains 16.6ml. To conduct the next dilution, 1.6ml is transferred from the 10^{-2} dilution to the next dilution tube in the series to create the 10^{-3} dilution. There is a possibility to carry on diluting the water sample *ad infinitum* but the normal dilution limitation would vary with the type of LAB-BART™ being used. Recommended dilution practices for the different LAB-BART testers is listed in Table Fifty.

Table Fifty

Recommeded Dilution Sequences for Enumeration Using the Various LAB-BART Formats

LAB-BART type		10^{-1}	10^{-2}	10^{-3}	10^{-4}	10^{-5}	10^{-6}
IRB	(14)	R	R	R	O	O	O
SRB	(14)	R	R	O	O	--	--
SLYM	(14)	R	R	R	R	R	O
HAB	(14)	R	R	R	R	R	O
FLOR	(5)	R	R	O	O	O	O
DN	(3)	R	R	O	O	--	--
N	(5)	R	R	O	--	--	–

R means that it is recommended that this dilution be used in enumerating a population by the extinction dilution technique. O is optional if the sample is thought to contain an aggressive population of that group of bacteria. The – symbol means that dilutions at this level are less likely to detect positively that particular group of bacteria. The numbers in parentheses show the recommended length of the test.

To conduct the LAB-BART tests, transfer the full 15ml of the dilutions to the separate LAB-BART testers. For the greatest dilution, the volume remaining in that tube is 16.6ml and, in this case, only 15.0ml is transferred to that LAB-BART. Aseptically transfer the full 15.0ml dilution of the sample into the appropriate Lab-BART. Do not shake the BART testers after the water has been added so that the gradients can gradually develop. Place the LAB-BART dilution tests in a convenient place and note the reactions that have been produced on the final recommended day of the test period.

Interpretation of the LAB-BART Results

As the dilutions become greater, this should delay the onset of the reactions until such dilutions as are too great to allow the survival of any bacteria capable of producing a reaction. At the end of the recommended test period, the reactions can be observed and interpreted. For some LAB-BART tests, this is simple because there is a limited range of possible reactions. These include the SRB, HAB, FLOR, N and DN. Here, there are narrow spectra of possible reaction types. The population can be calculated crudely by the highest dilution that shows a positive reaction test. For example, if the greatest dilution giving a positive detection reaction was 10^{-3} then the population would be considered to be 10^{-3} (or 1,000) bacteria/ml.

For the IRB- and the SLYM-BART testers, there are a greater variety of possible reactions that can occur. For example, the LAB-IRB-BART™ has the potential to generate the following primary reactions: FO, CL, GC, RC, and BC while common tertiary reactions are: BL and BR. In these cases, it becomes possible to calculate the population sizes that created each of these reactions. However, the easiest approach is to leave the test for 14 days and then interpret the populations by the type of reactions seen at that time only.

Evaluating the Effectiveness of a Treatment using BART Testers

There is always a concern that biofouling will lead to a variety of unacceptable circumstances. These include:

– Unacceptable hygiene risk to consumers and livestock
– Accelerating treatment costs to make the water acceptable
– Rising power costs as water flowing into, and through, the systems meets with more resistance to flow as a result of the biological formation of biofilms, encrustations, slimes, gas pockets and tubercles
– Uncertainty with regards to future production potentials because of the unpredictable nature of the biofouling events
– Potential for the water system to fail completely as a result of a radical plugging which reduces the water production to below the demand (e.g., a water shortage).

In the past, water shortage and hygiene risks have dominated the agenda with regards to the biofouling impacts on water. As overproduction and covert contamination deplete the acceptable surface- and ground- waters, the suitable water reserves will gradually shrink. As this happens, the value of water will rise and more attention paid to the issues of accelerating treatment costs, rising power costs, uncertain production potentials and the risks of catastrophic production failures. All of these events involve a major, if not dominant, microbiological factor. The BART testers allow microbiological events to be factored into the management equation. This section will address the issues of how to determine whether the impact of a particular treatment (whether it be any one, or any combination, of chemical, physical, biological and logistic approaches) can be measured using the BART testers.

Application of the Time Lag in Treatment Evaluation

Time lag commonly measured in days of delay indicates the aggressivity of the bacterial consortium being tested. The time lags for various different degrees of aggressivity are different for each BART (and bacterial consortium). As a result of this, the difference in the time lag before and after treatment can be used to determine the effectiveness of the treatment (long term). The simplest interpretation can be achieved using the formula below:

$$TL_D = TL_{AT} - TL_{BT}$$

In which TL_D is the time lag difference created by the treatment. This would reflect the log population shift resulting from the treatment, TL_{AT} is the time lag (days of delay) after treatment, and TL_{BT} is the time lag before treatment.

A negative TL_D would mean not only that the treatment had been ineffective but that there had been some additional releases of bacteria (from the biofilms into the water) to render the bacteria more aggressive. If the TL_D was within the range of -0.1 to +0.1 then the treatment could be considered to have had no significant impact on the consortium being tested using the BART tester. A significant treatment impact would involve a TL_D of >0.9. It must be remembered that the TL_{BT} should have been performed in a routine manner before any of the treatment was undertaken. Similarly, for the TL_{AT}, that testing should be conducted after the immediate impact of the treatment has passed (e.g., 14 days after the treatment was completed).

The effectiveness of a treatment can be registered by the calculated TL_D. The list below (Table Fifty-One) gives a summation of the effectiveness of a given treatment.

Table Fifty-One

The Interpretation of TL_D Data Calculated From Treatments

TL_D Range	Effectiveness of Treatment
> -1.0	Treatment has failed, the bacteria has become much more aggressive in the water[1]
-0.9 to -0.1	Treatment has failed to reduce bacterial loading in the water[2]
-0.1 to +0.1	Treatment ineffective
+0.2 to +0.9	Treatment marginally effective[3]
+1.0 to +2.9	Effective treatment applied but biofouling likely to commence again[4]
>+3.0	Very effective treatment with >99.9% control[5]

Notes:

1. This increase in aggressivity (due to a large negative TL_D) may be due to the ongoing sloughing of bacteria from the biofouled zone after treatment. It is recommended that the test for the TL_{AT} be repeated. If the time lag still gives a large negative TL_D then it must be presumed that treatment has either stimulated the activity of bacteria in the water or has caused an ongoing sloughing from the regions of biofouling that were impacted by the treatment.

2. TL_D values of between -0.1 and -0.9 signify that the treatment has failed due to the increases in aggressivity in the water after treatment. This range is more likely to be anomalous and a result of post-treatment sloughing. Repeat the BART tests to determine whether the TL_D remains negative. If it does, then some alternative treatment should be tried.

3. Where the TL_D value are between +0.1 and +0.9, testing should be repeated to determine whether the treatment had a delayed impact (time delay before the population of bacteria actually fell significantly in the water samples). There is a probability that this treatment is relatively ineffective. Sometimes, much better results can be achieved by a repeat treatment shortly after the first treatment. Often this second application causes a much larger impact since the microorganisms have been already been stressed by the first treatment.

4. Essentially, TL_D values of between +0.9 and +3.0 indicate that the treatment has been successful but an ongoing preventative maintenance program should be initiated to prevent the biofouling building up again. Often, after achieving these levels of control, a lower level of treatment

applied regularly, or when the time lag start to fall on the routine testing, can help to control the problem.

5. Although the improvements are 99.9% or better, it is not a 100% control and a monitoring schedule needs to be diligently followed to ensure that any recurrence of the problems can be controlled effectively by further treatments as required.

In most cases, TL_D values between +0.9 and +2.0 are the most likely outcome of a reasonably successful treatment. Changes in the time lag before and after treatment is probably the easiest parameter to measure using the BART testers. It should be realized that, where the computer assisted (CABART™) system is employed, the time lag can be determined to within ten to thirty minutes depending on the type of BART. This would provide much more accurate determination of the TL_D. Additional information on the effectiveness of a treatment can be obtained using the RPS information.

Application of the RPS in Treatment Evaluation

RPS is a measure of the types of bacteria forming parts of the consortium causing the biofouling. Treatments can also impact on the consortium. This can be measured by comparing the RPS observed before and after treatment. A number of observations can be made if the RPS does shift. If the RPS remains the same, then the treatment has impacted in a uniform manner over all of the member species of that consortium that are detectable by the particular BART test.

Where the RPS shifts, there has been a change in the bacterial community that was sufficient to affect the RPS. This change in the RPS can usually be recognized by reference to the information on the individual BART type and the interpretation of the reactions. There are a number of trends that may be recognized by the RPS shift. One major one is the shift from aerobic dominance to an anaerobic dominance in the bacteria detected before and after a treatment. This shift can be the result of either the surface (and more aerobic) layers in the biofilm being stripped off by the treatment, or the more aerobic biofilms within the treatment zone being impacted to a greater extent than the more anaerobic biofilms further out from the treated zone. Typical RPS shifts (Table Fifty-Two) that can be associated with this shift from aerobic to anaerobic dominance are:

Table Fifty-Two

RPS Interpretations as to the Form of the Biofouling

BART type	Aerobic Dominated RPS	Anaerobic Dominated RPS
IRB	GC, CL, BR	FO, BC, BL
SRB	BT	BB, CG
HAB	UP	DO
DN	--	FO
FLOR	PB, GY	--
SLYM	CP, CL, PB, GY, TH	BL
N	PB, RB, RT	--

If the treatment has been effective at removing much of the aerobically dominated flora, then the RPS may be expected to shift from some of those reactions shown in the left hand (aerobic) column to those in the right hand (anaerobic) column. Note also that these RPS changes do not all have to occur for this event to happen. These reactions are commonly associated with this aerobic to anaerobic shift.

Application of the RPS Time Length in Treatment Evaluation

The impact of a treatment on the time length over which the RPS was occurred may also be used to evaluate the success of the treatment. The equation to calculate this is given below:

$$TL_{RPS} = (TL_{LRC} - TL_{FRC})_{AT} - (TL_{LRC} - TL_{FRC})_{BT}$$

Where TL_{RPS} is the time length period difference for the development of the RPS from before treatment (brackets shown with a subscript BT) and after treatment (brackets shown with a subscript AT).

In each case, the time lag to the first reaction code observed is given as TL_{FRC} and to the last reaction code observed as TL_{LRC}. In most cases, the treatment may be expected to slow down the rate at which the reaction codes appear to form the final RPS. Where this happens, the TL_{RPS} would be positive indicating that the time length was longer. However, a foreshortening of the TL_{RPS} does not necessarily mean that the treatment has failed since there may be fewer reactions observed during the period that the reaction code shifts were observed. The TL_{RPS} therefore should not be viewed as a critically important factor in the determination of the success or failure of a treatment, but rather, used to gauge the impact of the treatment. Clearly, if the TL_{RPS} is zero, then there has not been an impact. However, if there is a positive or negative TL_{RPS} value this may aid in the evaluation of the treatment. The following is a guideline to the interpretation of the TL_{RPS}:

– TL_{RPS} is >+2.0 means that the treatment had a severe impact on the biofouling community but did not necessarily eliminate any components (unless the RPS values changed significantly from before to after treatment).

– TL_{RPS} is in the range from +0.5 to +1.9. This would suggest that the treatment had a moderate impact on the biofouling community.

– TL_{RPS} of between –0.4 and +0.4 would indicate that the community may have been weakened but the community structure had survived.

– TL_{RPS} more negative than –0.5 would mean that either the treatment had been ineffective and, in fact, stimulated the biofouling community, or much of the community had been destroyed and the survivors were not capable of generating a full RPS.

This information on the TL_{RPS} should be used only in a confirmatory manner with the L_R forming the first level of evaluation and the RPS shifts forming the second level.

Summary of the Use of BART Testers to Validate Treatment Technologies

The data that can be generated and be used in the interpretation of a treatment can be overwhelming. There are three major interpretations of the BART tests conducted before and after the treatment being evaluated. These are the L_R (log reduction in the population caused by the treatment determined by the time lag), shifts in the RPS caused by the treatment, and the changing in the time length over which the RPS generates (TL_{RPS}). In the evaluation of a treatment, the most significant parameter to evaluate is the L_R. This is because of major impact of a successful treatment would be a longer time lag before the first reaction is observed. Where the L_R is greater than +0.9 then the treatment can be considered to be successful at reducing the biofouling. The other two observations (shifts in the RPS and the TL_{RPS}) are of relatively minor significance but would give a better appreciation of the more specific impact of the treatment on the microbial consortia associated with the biofouling.

HISTORY OF THE BART BIODETECTORS

Theoretical

The BART is a patented methodology based upon the coupling of a vertical redox (reduction-oxidation) front with a selective nutrient diffusion gradient in which the nutrient rich-reductive zone is at the bottom diffusing upwards and the oxidative nutrient-depleted zone is at the top. Most microbial activity tends to concentrate at the transitional zone between the reductive and oxidative states where the nutrient diffusion gradient is creeping up the column. Often, the first focus of microbial activity is at this interface as a cloudy or plate-like growth.

Historically, one of the earliest developments of this system was introduced in the Winogradsky column. Here, a vertical water column was established with mud, gypsum and plant material in the bottom of the glass cylinder. These submerged materials triggered microbial activity that took out the oxygen from the lower part of the water column. It became anaerobic (deprived of oxygen) while oxygen diffusing in from the air above kept the top of the column aerobic. Here the aerobic (oxygen respiring) bacteria continued to grow using nutrient diffusing up from the reductive zone beneath. Over time, complex communities of various microorganisms could be built up in the Winogradsky column. The BART is essentially a Winogradsky column with a floating ball which restricts the entry of oxygen into the water and focuses the aerobic activity around the ball.

Today the advances in the BART biodetectors allow the types of bacteria detected and cultured to be restricted to those able to survive and flourish in the selective medium applied as a crystallized pellet on floor of the test vial. The sequence by which the bacterial reaction and activities are generated is as follows:

1. The water sample (or soil suspension) is added to the sterile inner test vial. 15 ml of sample brings the water column up to the fill line.
2. Immediately, the selective crystallized medium in the base of the vial begins to dissolve and form a diffusion front that gradually moves up the column.

3. Any residual oxygen in the sample is rapidly used by the respiratory activities of any active microbes in the sample. The bulk of the column now becomes reductive (anaerobic).

4. There is now a period during which the incumbent microbes now locate and adapt to a suitable environment (having the correct redox potential and concentration of nutrients). Several different focal sites of activity may now be occurring simultaneously within the test vial.

5. A period of accelerating microbial activity (e.g., growth, metabolism) and reactions (i.e., interactions between the microbial activity and the chemical substrates) now causes competition between the various focal microbial communities and sufficient reactions and activities to become recognizable.

6. The first defined visible signal represents the time lag (TL) and can be related to the population (i.e. the larger the population, the shorter the TL).

7. A series of visibly definable changes may occur which will give an indication of the composition of the community in the sample. A series of reactions may be generated into a pattern that gives an indication of the microbial community composition. For example, a reaction of 8 followed by a 9 would indicate that the community was predominantly aerobic heterotrophs belonging to the pseudomonads.

It should be noted that the tests are usually done at room temperature for the convenience of the person doing the tests, but it should be remembered that this may not be the optimal temperature for the microbes in the water and "blood heat" (35-37°C) may be even more detrimental. As a rule of thumb, the test should be performed within 5°C (10°F) of the original sampling site. In many ground water environments, the temperature remains much more stable than on the surface of the planet.

Practical Validation

From the late 1880's, the arrival of agar as a convenient way to grow bacteria into visible colonies became established. The advantage was that these growths could be recognized. Also this would help to identify the bacterial type. In the flood of knowledge which followed this discovery, the agar plate became the most widely used tool for the enumeration of bacteria in clinical specimens, food, drinks, waters and soils. Unfortunately, many bacteria are not able to grow on or in agar culture media and so are essentially missed. This caused low estimates or false negatives (they are there but were not detected), particularly in soil and water samples. The simplest way used today to determine the gross level of biological activity is to detect the ATP (adenosine triphosphate) activity. This molecule is the principal energy storage mechanism and can be "tricked" into converting the energy to light using the luciferase enzyme (the same one fireflies use to glow in the dark). The more ATP present, the more biological activity, and the more light. While ATP does give an indication on the level of activity, it does not define the types of microorganisms present. In today's world, microbes are being recognized as found everywhere (see web page http:// commtechlab.msu.edu/CTLProjects/dlc-me/zoo/ to look at the "microbe zoo"). It is now becoming evident that most bacteria form parts of communities, called consortia, in which the various species work cooperatively to generate a growth such as a clog, tubercle, slime, nodule, or a floating biocolloidal mat. Over the last century, these consortia have gone relatively unrecognized since the attitude was "one species - one problem." This, of course, stems from the medical science findings that diseases are all produced by single pathogenic species. This precludes consortial diseases. In addition, since most of these pathogenic bacteria were found to grow on some form of agar culture media, it was presumed that all microbial activities occur at the species level, will grow on agar culture media under some conditions, and bacterial species are independent not interdependent. The rapid growth of microbial ecology is now supporting the concepts that microbial species cooperate in the production of living structures (consortia) and that they may not be able to grow on agar or even be capable of independent growth.

Since 1971, research has been ongoing at the University of Regina to detect iron bacteria which plague water wells causing discolored water, plugging, taste and odors problems and loss in production. During the next fifteen years, various methods were examined to try to determine the presence and activity levels of these iron bacteria. It became clear that there was little attention paid to methods for determining the presence/absence of bacterial consortia rather than particular species or genera.

In 1984, a range of different tests was developed to try to determine:

- the state of biological fouling in both laboratory mesocosms which simulated plugging water wells, and
- real world situations (water wells, surge tanks, process lines, heat exchangers) where slimes, plugging, corrosion, discolored water and outbreaks of infections can occur.

At the IPSCO Think Tank held in Regina and at the International Symposium in 1986 on aquifer biofouling held in Atlanta, considerable discussions were held on the need for more appropriate microbiological testing methods to determine biofouling.

In 1987, George Alford and Dr. Roy Cullimore began to develop and patent the biological activity reaction test (BART) with the specific aim of developing a biodetection system for the different types of bacterial consortia that cause problems in water and wastewater. Research and development work continued, while parallel studies on the traditional monitoring techniques and, in particular, the agar spread-plate techniques were undertaken.

The initial grouping of bacteria into a form that would eventually be developed as the BART test system began in the mid 1980's. At that time, iron bacteria were thought mostly to be related to the genus *Gallionella* which possess a very long and obvious stalk that can be easily recognized through microscopic examination. In 1987, a proposal was made to reassess the role of bacteria involved in water well biofouling with the two principal groups being the IRB (iron-related bacteria) and the SRB (sulfate-reducing bacteria). It was proposed that broad spectral tests should be developed to be able to detect the various bacteria that do utilize iron in some manner in their growth (IRB) or reduce sulfates (SRB). By 1990, the first water test systems for iron-related, sulfate-reducing, and slime-forming bacteria were being developed and test marketed.

The concept developed over this time period was that the population of bacteria in the water sample could be determined as the time lag (TL or days of delay) to the first signal of a recognized activity or reaction in the BART test vial. Comparative studies showed that the TL link was not linear but **sigmoidal**. Using the MIDI methylated ester fatty acid analysis and the API bacterial identification systems, it was found that the growths in the BART tests usually involved a consortium of bacterial species that may or may not have stratified. The dominant species often changed as the growth in the test vial matured.

Essentially, this test system provided a range of environments from reductive to oxidative, nutrient-rich to nutrient-poor, containing a chemical matrix that would be selective for the activity of the targeted consortia of bacteria. The concept of the BART was originally developed as a biological activity test (BAT) but it was modified to include the word "reaction" since the application of the floating ball (FID - floating intercedent device) caused a stratification of the environments which allowed many more bacterial types to grow.

In 1988, one of the earliest uses of the BART testers was to determine the bacteriological influences on a covert gasoline plume affecting local individual water wells. A combination of these tests, traditional agar spread-plate techniques, laser particle counting and direct microscopic examination showed the data to be most consistent between the laser particle counting and the BART tests. Here, it was revealed that the pseudomonads associated with the biodegradation of the gasoline plume were entering into a biocolloidal phase and shifting downwards into the aquifer in a controlled manner to impact on some of the deeper wells.

Over the period from 1990 to 1995, a compendium of the various genera that could be recovered, the linkages between the TL and the population, and the identification of the consortial type in the BART test became established. This was published in 1993 and an additional BART was introduced for the heterotrophic aerobic bacteria using the methylene blue reductase test. In order to meet the needs to improve quality assurance and quality control procedures in keeping with the ISO 9000 directives, a full support document was published in 1996. By this time the range of test systems had been expanded from three (1990) to four (1992) and up to nine (1995). The additional tests were for nitrifying, denitrifying, micro-algae, pool biofouling and fluorescing pseudomonads. Improved manufacturing procedures have allowed the test shelf to be expanded from 12 months (1992) to 18 months (1994) to two years in 1996. Packaging introduced early in 1998 expanded the shelf life to three years.

There has been an ongoing evaluation of the BART test systems since its inception in 1987. Serious biofouling at some municipal water wells in New Brunswick, Canada in 1990 led to the use of the BART tests in concert with laser particle counting to determine the location and extent of the fouling. Particle counting essentially looking at the suspended particles showed correlations with the BART test data indicating that the bacterial loading in the water was primarily in the suspended particulate state (sizes ranging from 0.4 to 32 microns). High bacterial aggressivity was noted particularly where "spiking" occurred with a high particle number in particular sizes (e.g., 8, 10, 12 micron sizes) for the four water wells. Here, biofouling risks were developed and linkages between some BART reactions and specific bacterial genera determined. In these studies, it was noted that there appeared to be concentric biozones around each water well with periodic sloughing. This meant that the bacteriologic loading of water samples may be affected by the state of sloughing in the well.

The town of Rocanville in Saskatchewan, Canada, was experiencing some biofouling problems and in 1990 a survey was conducted on the wells and distribution lines. Laser particle sizing revealed that some of the lines in the town had very significant particle sizes ranging from 2.5 up to 10 microns or more. At these sites, the IRB-BART registered an aggressive reaction with a **6 going to a 9**. Two species were identified using the MIDI system: *Comomonas acidovorans* and *Pseudomonas vesicularis*. "Spiking" of particle sizes in the 8 to 16 micron was common where these bacteria dominated in the water samples.

Over the year of 1990, considerable progress was made in the understanding of the use of the BART testers to determine bacterial populations and assess biofouling risks. A summary of this work and the application of this test system for water wells was introduced as novel technologies at a Western Canadian water conference. The influence of the biofilms going through expansion, sloughing and compression stages during sloughing was discussed both from the impact that this would have on the bacterial population in the water, but also how the BART testers could be used to determine this. The advent of sequential sampling of water during pump tests was advocated. It was proposed that the biofouling around a water well is in a series of concentric zones that can be determined. Generally, the activity of the aerobic IRB is high in the early samples while the SRB are initially low but become high later in the pump test when water is being drawn from deeper in the formation. It was recommended that sequential sampling of a pumped water supply from a well is essential if the true nature of a plugging is to be determined.

In 1989, linkages were made between the degree of biofouling in water wells that were losing production due to plugging and the bacterial aggressivity reflected in the short TL recorded using a variety of [BART] tests. At the Waverly sites in Tennessee, it was found that the TL on the SRB and the IRB would link to the occurrence of problems (high manganese). As the TL dropped from the 10 to 14 days of delay and the range down to the 4 to 6 day range, the symptoms began to appear. The BARTs therefore provided an "early warning system" which could be done with relatively minimal effort and treatments applied according to the symptoms. This was one of the first examples of the use of the BARTs as a "scouting" tool that was easy to use and interpret.

There was an ongoing concern that the traditional agar-based enumeration methods were not able to successfully enumerate particulate-related bacteria due to the multi-species aggregation within a common biocolloidal structure. The BART, through providing a variety of different environmental niches, would be more able to trigger the growth of various members of the particulate structures.

Over the period from 1989 to 1995, the BART test systems received more and more attention. In the rehabilitation of water wells, Stuart Smith published an extensive evaluation of the suitability of using the BART testers for the diagnosis and monitoring of water wells which may be likely to biofoul.

Mansuy, Nuzman and Cullimore in 1990 reported on one of the first combined uses of the IRB-, SLYM- and SRB-BART on water wells in Kalamazoo, Michigan and Gainesville, Florida (see Selected Bibliography list). A sequence for typical iron bacterial biofouling was described and the use of these tests in a confirmatory manner with treatment procedures were described.

- The relationship of the data from traditional agar spread-plate techniques, laser particle counting (to obtain the total suspended solids down to 0.4 microns) and the BART test systems was addressed in outline in 1990 together with the strategies for detecting the size of plugging biozones around a water well.

BART testers were also used in 1990 to determine the size of plugging biozones around the plugged water well at Armstrong, Ontario. Here, the tests were able to define the types of intensive biofouling that were occurring around that badly biofouled well.

There is a growing acceptance that microbes are present in almost every conceivable environment on this planet where liquid water is present. The potential implications of this in civil engineering was discussed in 1990 at a Federation of European Microbiological Societies conference at the University of Cranfield, U.K. The BART tests offer the potential to provide a simple sensitive test method for maximizing the detection of microbes in these environments.

Over the periods from 1982 to 1991 and again from 1995 to 1997, there has been an ongoing generation of simulated water well mesocosms which have been developed to determine the rates of plugging through the growth of either SRB or IRB. The most recent description of the work was reported in 1997 by Cullimore, Legault, Keevill and Alford.

In association with Ortech International, Toronto, Canada, an extensive project was undertaken to examine the microbial activities which could interfere with the pathways of gasoline diffusion through saturated porous media (in these trials sand was used as the medium). Bacterial activity was monitored indirectly using laser particle counting (LPC) methodologies while the BART tests along with the traditional spread-plate methods were used to determine the bacterial populations. In this study, the BARTs were found to generate a greater level of bacterial activities than were recorded using the spread-plate methodology and also that the TSS (total suspended solids) generated by the LPC techniques showed conformation with the BART data. Relationships were established between the reaction patterns and the types of bacterial species concurrently, or subsequently, grown on selective agar media. The dominant group of bacteria was the pseudomonads with the following genera being the most frequently recovered: *Pseudomonas, Xanthomonas, Agrobacterium,* and *Comomonas.*

Bacterial biofouling became a major problem in the 1980s in golf courses. These infestations caused the development of a dieback condition in the turfgrass and the formation of a black plug layer (BPL) within the (normally) high-sand content green. Over the period from 1987 to 1991, ongoing research defined the cause, developed a successful treatment, and generated a monitoring procedure. It was found that the lateral black plug layer infestations contained a consortium of bacteria involving a minimum of eight species. It was found that the SRB-BART would generate a BA in less than 3 days (commonly 1 to 2 days) where BPL infestations were occurring but would normally take >5 days in a non-infested soil. A linkage was also noted in these studies between the occurrence of a FO (gassing) in the IRB- and BL (blackening) in the SLYM- testers where the SRB-triggered a positive BA in less than three days. This observation was then extended to water wells studies where similar reaction linkages were noted. Correlations were also obtained between the bacterial species identified and the types of BART reactions observed.

The ALGE- and the SLYM-BART were integral parts of a study from 1992 to 1994 to determine the cause of algicidal properties in straws submerged in the water. The straw was coated with various bacterial growths generated using the SLYM- tests and the impact of this, together with nutrient enrichment, was studied on algae (mixed and axenic) cultures introduced to the ALGE-BART either on the straw or as controls. Relationships were uncovered between the bacterial inoculum, nutrient applications and the suppression of both indigenous and inoculated algal growth.

As an experimental part of an ongoing investigation begun in 1993 on the microbial plugging of under-drains in a landfill operation, the BART testers were employed as secondary test methods for the detection of the bacterial consortia associated with the biofouling of the drains. This was performed by at field-scale and using laboratory mesocosms with specific attention being paid to the fate of dichloromethane. In these studies the BARTs were able to detect specific cloistering of various consortia into specific regions within the landfill mesocosms. This study stimulated the development of a new test for methanogenic bacteria (BIOGAS-BART) which is presently undergoing laboratory trials at the Universities of Western Ontario and Regina. Linkages were obtained between the projected populations of bacteria generated by the TL observed on the BART tests at both 17°C and 27°C and the amount of ATP was measured (ng/ml). At

17°C, a population of 2,000 cfu/ml generated 0.2 ng/ml while 100,000 cfu/ml generated 20 ng/ml. At 27°C, the linear relationship was different with 50,000 cfu/ml generating 0.05 ng/ml while >200,000 cfu/ml generated >10 ng/ml. In the studies on the biodegradation of dichloromethane, it was found that the aggressivity of the SRB could be linked to the rates. This study is ongoing and the BIOGAS–BART is now being developed as a part of the study.

As the environmental impact review for the high level nuclear waste disposal in Canada was proceeding, there was a need to address the issues relating to the microbiology of the disposal site in granitic rock as a part of the risk assessment. As a part of that initiative, the Scientific Review Group held a tutorial. Here, the BART testers were employed to determine the range of bacterial activity occurring within the vault environment. The data was compared with the Acridine Orange Method and direct count using spread-plate techniques.

In the municipal district of Kneehills, Alberta, between 1995 to 1997, there was an initiative to determine the extent of biofouling in the extensive network of low producing water wells. This work was spearheaded by the Technical Services division of the Prairie Farm Rehabilitation Administration, Canada Agriculture. The BART test system formed the "backbone" of the microbiological monitoring which formed a part of the survey. This was a three phased initiative that included, in sequence: survey by questionnaire, survey of selected wells by various tests including BART testers, and finally treatment using an ultra-acid-base treatment (UAB™). A summarized report was released in 1997. A summary of these findings was released as a community information document. The SRB-BART was found to detect very aggressive sulfate-reducing bacteria in 67% of the wells as opposed to heterotrophs (17%) and IRB (9%). Video camera logging confirmed the biofouling which had a form that did resemble the growths seen in the test systems. Recommendations from that survey of 275 water well owners and the subsequent testing of 134 wells led to the recommendation of the routine use of the BART tests to monitor the degree of fouling. Some of the most severely biofouled water wells were treated by the UAB™ process in 1997 and the BART testers were used to monitor the efficacy of the procedure. A summary of the survey was also presented at the National Groundwater Association (NGWA) 1997 conference on "The Biological Aspects of Ground Water."

There has not been so much attention paid to the forms of plugging which occur in gas and oil wells. In 1997, a project was undertaken to examine the loading of SRB in oil well samples and the impact of a biocide on the SRB activity levels. Here, the SRB-BART was found to be able to detect the effects of biocide applications by an extension of the TL, usually by several days.

In a 1997 training manual on Iron and Manganese Removal developed for the Saskatchewan Environment and Resource Management department by **AWI** and Reid Crowther, the BART test are specifically addressed as a part of the routine procedures to determine biofouling linked to the IRB, SRB and heterotrophic bacteria.

BART testers have been used on a number of occasions to detect bacterial activities associated with the biodegradation of the RMS *Titanic*. In 1992, an evaluation of the bacteriological activity was performed. The BART tests (IRB and HAB) were modified for direct installation on the bridge deck of the RMS *Titanic* in August 1996. The BART testers were also used on samples of rusticles recovered from the sunken vessel. (This was reported by Hach Company, Ames, Iowa, one of the distributors, in more detail). The BART tests defined a number of community structures within the rusticles that is now being subjected to further investigation. Parallel work at the Nova Scotia Technical University on rusticles has also used the BARTs to successfully isolate a range of iron bacteria. Included in the identification were members of the *Leptothrix/Sphaerotilus* group of sheathed iron bacteria. Specifically, *Leptothrix discophera* was identified as a major component species that had also been reported informally as infesting the rusticles on the *SS Central America.*

By 1997, the BART test systems had seven years of field application and had been sold commercially for the last six years in growing numbers. These test methods present a much more accommodating and diverse series of environments to induce the generation of reactions and activities in the targeted bacterial groups. One major advance that the BART concepts have achieved is the recognition that, as in a biofilm, the water contains complex consortia (communities) of bacteria. It is better to attempt to observe the growth of these community structures as such rather than to devote considerable time and energy to the precise identification of those few bacteria that are able to be easily mono-cultured. There is a very significant probability that the science of identifying microorganisms will move from an almost endless attempt to define mono-cultures at the molecular level ("small picture"). It is probable that the move will be towards a physiognomic approach where the attention is concentrated on recognizing the functional groups of bacteria such as the IRB and the SRB ("big picture"). The BART testers have already allowed that to happen as the relative speeds with which reactions and activities occur are observed. The answer, as always, is the simpler the concept, the more reliable the results.

The BARTs, in taking the "big picture," give a snapshot of what microbial activities are occurring in the water and their significance.

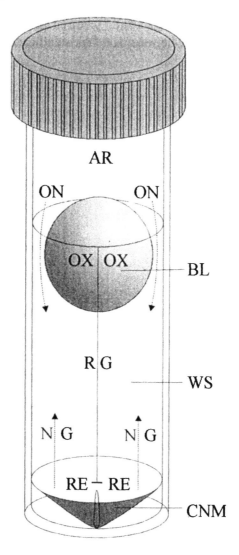

Figure Thirty-Two, The patented biological activity reaction test (BART™) creates two vertical gradients under and around the ball (BL) floating on 15ml of water sample (WS). Oxygen (ON) diffuses down from the air (AR) in the head space of the test vial to create a oxidative (OX) zone above and reductive (RE) zone below to form a redox gradient (RG). At the same time, the crystalized selective nutrient medium (CNM) applied inside the base dissolves into the WS and diffuses upwards to form a nutrient gradient (NG). The BART therefore generates a range of vertically disposed environments that are able to support the growth of a wider variety of microbes than the standard methodologies.

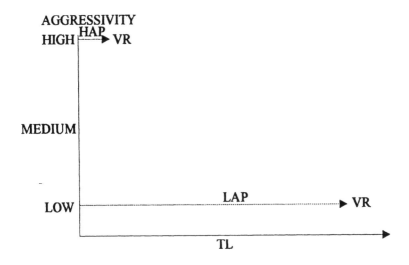

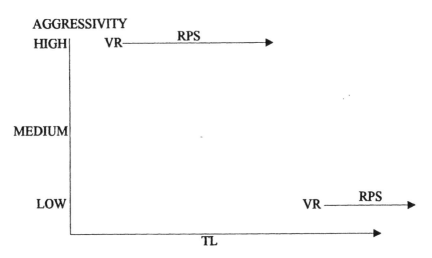

Figure Thirty-Three, Interpretation of the BART™ involves two major observations. The first observation (upper) relates to the time lag (TL) to the first visible positive reaction (VR). Highly aggressive populations (HAP) will tend to display shorter TL than less smaller aggressive populations (SAP). A general rule is that for each day of delay in the time lag, the population would be one order of magnitude smaller. The second observation (lower) is the form of the reactions that occur in the BART. A pattern of reactions occur in the test which reflects the consortial make-up in the sample. These reactions form into a reaction pattern signature (RPS) that gives more information as to which bacteria are present in the consortium.

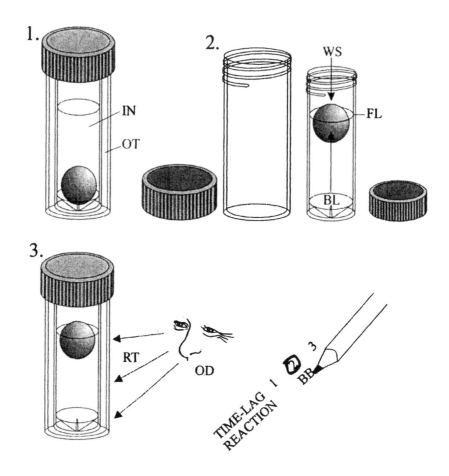

Figure Thirty-Four, Sequence for conducting a standard BART test. There are two test vials, inner (IN) and outer (OT). The OT performs a protective role protecting the user from any leakage and it also accumulates any smells. The IN test vial is used for the test itself. It is filled with 15 ml of water sample (WS) up to the fill line (FL) where the ball (BL) now floats. The test starts as soon as the WS has been added and is usually kept at room temperature (RT) and observed daily (OD). Data can be collated on the BART data interpretation chart (see Figure Fifty).

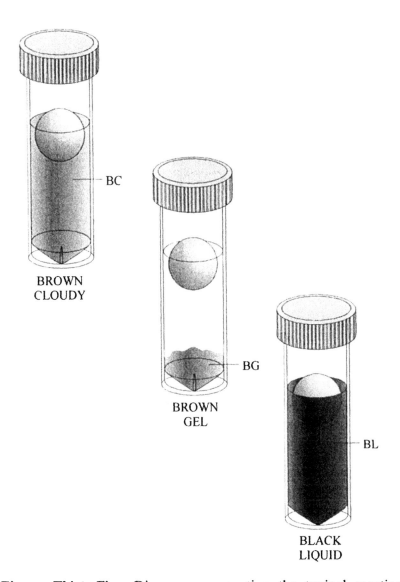

BC

BROWN
CLOUDY

BG

BROWN
GEL

BL

BLACK
LIQUID

Figure Thirty-Five, Diagram representing the typical reaction patterns for the iron-related bacteria in IRB-BART tests: brown cloudy (BC, upper left), brown gel (BG, center) and blackened liquid (BL, lower right). BL and BC are common terminal reaction observed for anaerobic and aerobic bacteria respectively.

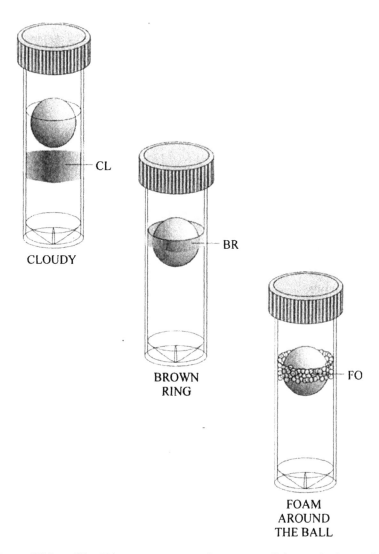

Figure Thirty-Six, Diagram representing some of the typical reaction patterns for the iron-related bacteria in IRB-BART tests: cloudy (CL, upper), brown slime ring (BR, center), and foam around the ball (FO, lower right). The FO or the CL reactions are common first reactions where there are anaerobic or aerobic bacteria present respectively.

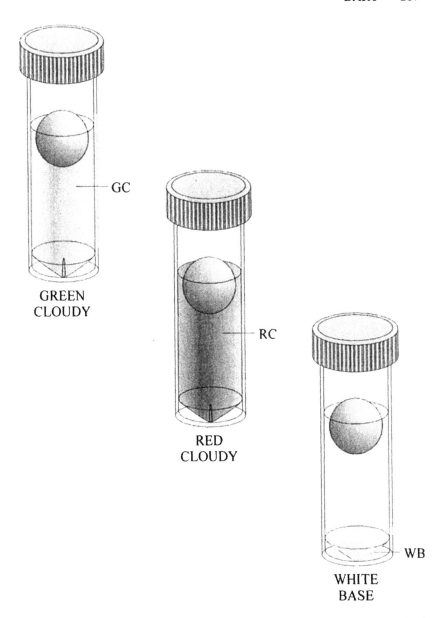

Figure Thirty-Seven, Diagram representing some of the typical
reaction patterns for the iron-related bacteria in IRB-BART tests:
green cloudy (GC, upper left), red cloudy (RC, center) and a whitened
basal region (WB, lower right). GC tends to occur at some stage
where there is a dominance of pseudomonads, while RC dominates
sometimes where the enteric bacteria dominate. WB is a common
reaction for which the significance has yet to be fully established.

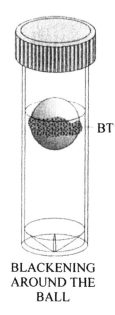

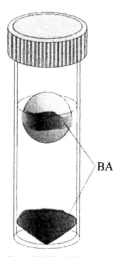

BB

BLACKENED
BASE

BT

BLACKENING
AROUND THE
BALL

BA

BLACKENING
AT BASE
AND TOP

Figure Thirty-Eight, Diagram representing the typical reaction patterns for the sulfate-reducing bacteria in SRB-BART tests: blackened base (BB, upper left), blackening at the top around the ball (BT, center) and blackening around the base and the top (BA, lower right). Most com-monly, a BB reaction precedes a BA in more anaerobic environments while the BT is generated in more aerobic conditions where there are heterotrophic aerobic bacteria active. All three reactions are positive for SRB while the CG (see Figure Thirty-Nine) is negative for SRB but positive for anaerobic bacteria.

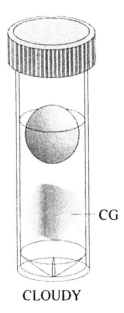

CLOUDY

Figure Thirty-Nine, Diagram representing the typical reaction patterns for the anaerobic bacteria in SRB-BART tests: cloudy gel-like growth often look a little like clouds in the liquid medium (CG). Unlike the other three reactions (see Figure Thirty-Eight for the reactions positive for SRB), the CG is negative for SRB but positive for anaerobic bacteria. CG reactions commonly precede the reactions detecting a positive SRB.

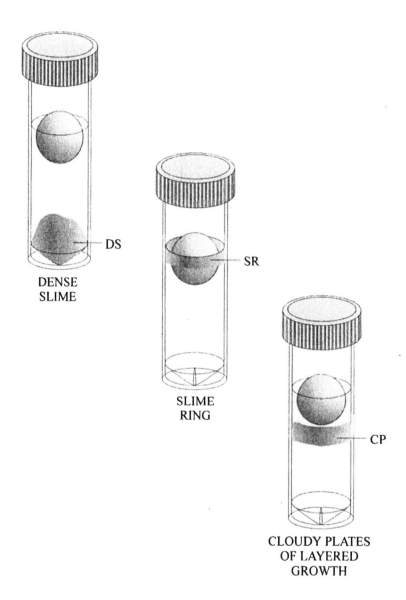

DS

DENSE
SLIME

SR

SLIME
RING

CP

CLOUDY PLATES
OF LAYERED
GROWTH

Figure Forty,   Diagram representing the typical reaction patterns for the slime-forming bacteria in SLYM-BART tests: dense slime in the basal regions (DS, upper left), slime ring around the ball (SR, center) and cloudy plates of layered growth in the water (CP, lower right). The first two reactions (DS, SR) tend to be secondary or final reactions in the test. CP is often an initial reaction.

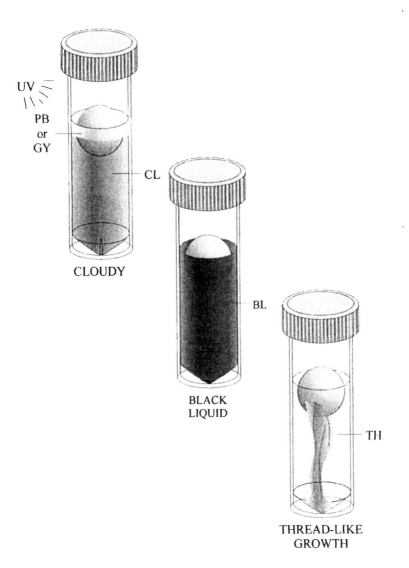

Figure Forty-One, Diagram representing the typical reaction patterns for the slime forming bacteria in SLYM-BART tests: clouded (CL, upper left), blackened liquid (BL, center) and thread-like growth in the water (TH, lower right). Where CL occurs, carefully irradiating the side walls of the BART™ test with a UV light can detect fluorescent pigments (pale blue, PB; and greenish yellow, GY, see also Figure Forty-Three). CL is a very common reaction in this test with BL occur-ring as a terminal reaction where there are a mixed consortium of pseudomonads and enteric bacteria.

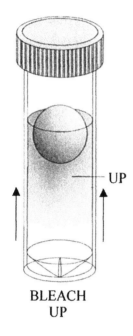

BLEACH
UP

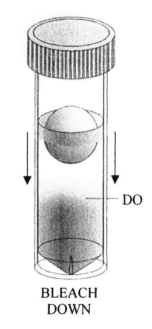

BLEACH
DOWN

Figure Forty-Two, Diagram representing the typical reaction patterns for the heterotrophic aerobic bacteria in HAB-BART tests are: the blue color bleaches out to clear or yellow from the bottom of the test vial upwards (UP, upper left), and the blue medium bleaches from just below the equator of the ball downwards (DO, lower right). The UP reaction tends to occur more often where there are a dominance of aerobic bacteria while the DO tends to be slower having a longer time lag and there are facultatively anaerobic bacteria present in significant numbers.

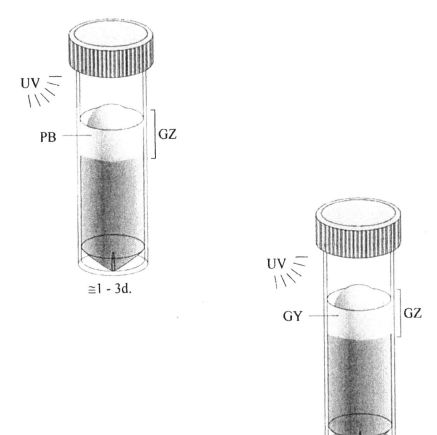

UV

PB

GZ

≅1 - 3d.

UV

GY

GZ

≅2 – 8d.

Figure Forty-Three, Diagram representing the typical reaction patterns for the fluorescent pseudomonad bacteria in FLOR-BART tests that are obtained by carefully illuminating the side walls of the test vial with a UV light. For a positive detection of fluorescence, a glowing zone (GZ) should be visible at the floating ball level and slightly below. If the glow is a pale blue (PB) color, then that could be considered an indicator for the presence of *Pseudomonas aeruginosa*, whereas if the light is greenish yellow (GY), this could be taken as an indicator for the presence of *Pseudomonas fluorescens*. Normally, this glowing will be visible from 1 to 3 days (for PB) and 2 to 8 days (for GY) after the commencement of the clouded growth.

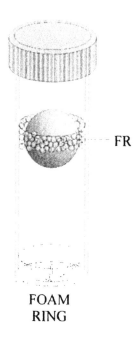

FR

FOAM
RING

Figure Forty-Four, Diagram representing the typical reaction patterns for the denitrifying bacteria in DN-BART tests. In this test, a positive indication is the presence of a foam ring (FO) around the ball usually in 2 to 3 days. It may be accompanied by clouded growth within the test vial (not significant to the test. The gas in the FO is nitrogen (released during complete denitrification of the nitrate) that has become entrapped in biofilms.

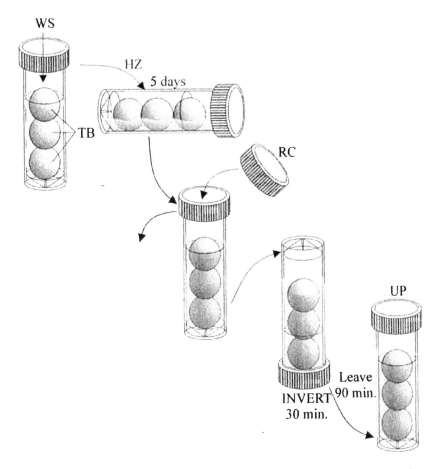

Figure Forty-Five, Schematic for the conducting of a test for nitrifying bacteria using the N-BART. In this test, the BART is set up differently to the other tests. Here, the test vial are filled with the 7.75 mL water sample (WS) laid horizontally (HZ) during the test and there are three balls (TB) in the tube. These balls provide a larger surface area to volume ratio to encourage the aerobic growth of the nitrifying bacteria. Another difference is that test is of a specific length (recommended time, 5 days). In this time, ammonium can be oxidized to nitrite and nitrate with some, or all, of the product nitrate now being denitrified back to nitrite by other bacteria. The indicator product for a positive detection of nitrite is determined by the application of a reaction cap (RC) to the inner test vial that is then inverted for thirty minutes to create the reactions which can be seen within ninety minutes of the test vial being returned to an upright position (UP, see Figure Forty-six).

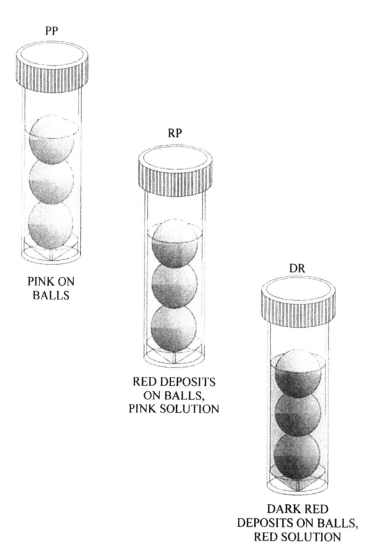

PP

RP

DR

PINK ON
BALLS

RED DEPOSITS
ON BALLS,
PINK SOLUTION

DARK RED
DEPOSITS ON BALLS,
RED SOLUTION

Figure Forty-Six, Reactions recognized as positive for the detection of nitrifying bacteria using the N-BART. Once the reaction has been generated and where nitrite is detected (see Figure Forty-Five), then three reactions can occur where the test vial is now left standing upright (cap side uppermost). These include: a partial pink color generating only on the balls (PP, upper left); red deposits occurring over much of the surfaces of the balls and the solution is now a pink color (RP, center), dark red solution with dense heavy red deposits on the ball (DR, lower right). These reactions represent increasing levels of nitrite being present (i.e., PP < RP < DR).

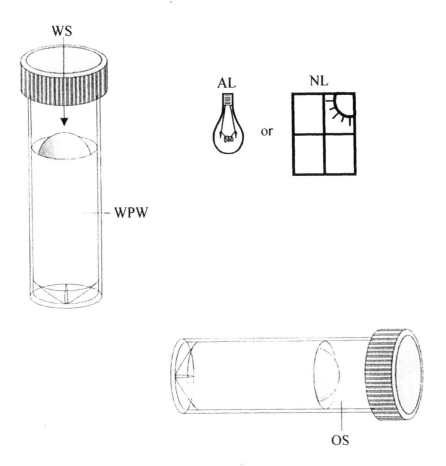

Figure Forty-Seven, Schematic for the conducting of a test for the micro-algae including the green algae and the cyanobacteria using the ALGE-BART. This test is also unique like the N-BART test. Here, the differences are that the test vial is laid upon its side (OS) and illuminated with either a dull natural (NL) or artificial light (AL). The micro-algae need the light to grow and become detectable. The test vial contains a white porous weave (WPW) down the lower half of the test vial. Some of the micro-algae grow in or upon the WPW and become visible as discolorations on the white fabric.

GG

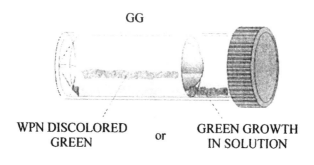

WPN DISCOLORED   or   GREEN GROWTH
GREEN   IN SOLUTION

FG

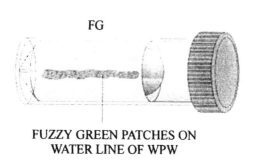

FUZZY GREEN PATCHES ON
WATER LINE OF WPW

OB

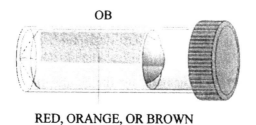

RED, ORANGE, OR BROWN
FORM ON THE WPW

Figure Forty-Eight, Illustration of three of the reactions generated by the micro-algae including the green algae and the cyanobacteria using the ALGE-BART. A general grass-green growth may be either visible in the liquid and/or the WPW may be discolored grass-green (GG, upper), fuzzy green patches may form usually above the water line on the WPW (FG, center), or red, orange and/or brown patches may form both above and below the water line on the WPW (OB, lower).

YB

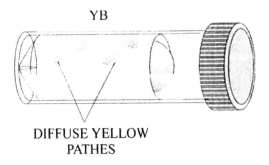

DIFFUSE YELLOW
PATHES

GF

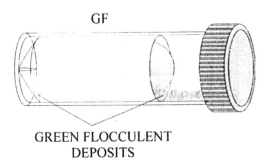

GREEN FLOCCULENT
DEPOSITS

DG

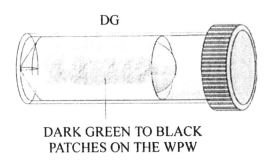

DARK GREEN TO BLACK
PATCHES ON THE WPW

Figure Forty-Nine, Illustration of three of the reactions generated by the micro-algae including the green algae and the cyanobacteria using the ALGE-BART. Diffuse yellow patches form over the WPW (YB, upper left) Green flocculent deposits can be seen settled in the liquid and the WPW may be discolored green (GF, center), and dark green to black patches form predominantly at the water line on the WPW (DG, lower).

BART Type		1	2	3	4	5	6	7	8	9	10
IRB RED CAP	Time Lag (days)	1	2	3	4	5	6	7	8	9	10
	Reaction Codes	6.0	5.0	4.0	3.6	3.0	2.0	2.0	2.0	1.0	1.0
SRB BLACK CAP	Time Lag (days)	1	2	3	4	5	6	7	8	9	10
	Reaction Codes	6.0	5.0	4.6	4.0	3.6	3.0	2.0	2.0	1.0	1.0
HAB BLUE CAP	Time Lag (days)	1	2	3	4	5	6	7	8	9	10
	Reaction Code	6.6	5.6	3.0	2.0	1.0	1.0	1.0	1.0	1.0	1.0
DN GREY CAP	Time Lag (days)	1	2	3	4	5	6	7	8	9	10
	Reaction (FO)	6.6	5.0	3.0	2.0	1.0	1.0	.7	1.0	1.0	1.0
SLYM GREEN CAP	Time Lag (days)	1	2	3	4	5	6	7	8	9	10
	Reaction Codes	6.6	5.6	4.6	3.6	3.0	2.0	1.0	1.0	1.0	1.0
FLOR YELLOW CAP	Time Lag (days)	1	2	3	4	5	6	7	8	9	10
	Reaction Codes	6.0	5.0	4.0	3.6	3.0	2.0	2.0	2.0	1.0	1.0
T-Coli CREAM CAP	Time Lag (hours)	12	18	24	36	48	60	72	84	96	108
	Reaction Code										
Possible Log Population (PLP)		6.6	6.0	5.6	5.0	4.6	4.0	3.6	3.0	2.0	1.0
colony forming units per ml (cfu/ml)		5,000,000	1,000,000	500,000	100,000	50,000	10,000	5,000	1,000	100	10

Aggressivity: | high | medium | low

Reaction Code Summary:

IRB-BART™

BC-Brown Cloudy
BG-Brown Gel
BL-Brown Ring
BR-Brown Ring
CL-Clouded Growth
FO-Foam
GC-Green cloudy
RC-Red, slightly cloudy

SRB-BART™

BB-Blackened Base
BT- Blackened top
BA-Blackened base and top

HAB-BART™

UP-Bleaching from bottom up
DO-Bleaching from top down

SLYM-BART™

DS-Dense slime
SR-Slime ring around ball
CP-Cloudy layered plates
CL-Cloudy growth
BL-Blackened liquid
TH-Thread-like strands
PB-Pale-blue glow (UV)

DN-BART

FO-Foam around ball

FLOR-BART

PB-Pale-blue glow (UV)
GY-Greenish-yellow glow (UV)

Figure Fifty, The standard format for the BART data interpretation sheet frequently used for the recording and preliminary interpretation of multiple tests that have been applied to a water samples.

6

WATER WELL BIOFOULING, DIAGNOSIS

There is a global history of well failure due to the loss in productivity often with the well "going dry." It is generally thought that the well went dry because of the depletion of the ground water in the aquifer or the various chemical and/or physical processes had clogged the well. In the last two decades, there has been mostly experiential and some scientific work relating to the cause of clogging. There remains considerable dispute over the cause, the diagnostic procedures and the effectiveness of management strategies.

LOST PRODUCTION

Causes of Lost Production

Lost production of water from a water well may be due to one of two alternate reasons:

- There is no longer an adequate reserve in the aquifer to meet the demands for water placed on the ground water supply, or
- There remains an adequate supply of ground water in the aquifer but there is impedance to the flow of ground water into the well.

Where there is a loss in available ground water due to the depletion of the aquifer, recovery of production in the well cannot be achieved through well rehabilitation. Consequently these procedures will be directed to the rehabilitation of wells suffering from some form of flow impedance.

Forms of Lost Production

There are a number of terms that have been applied to water wells that have lost production. Traditionally the term "clogged" has been widely used and reflects the concepts that the wells were clogged by the build up of barriers around the well. These barriers were thought to be mixtures of various chemically derived deposits along with clay, sand and silt particles which had moved towards the producing well before clogging up the pore spaces and impeding flow. In more recent times, the term "plugged" or "bio-plugged" has gained acceptance and this condition tends to be applied to water wells where there is evidence of biological activity being instrumental in the clogging process. The biological activity would relate to the generation of site-focussed mineral deposits (such as silicates, carbonates and ferric forms of iron), slimes (biofilms) and casually entrapped materials such as sands, silts and clays. Today, there is some level of confusion between the use of the terms: "clogged" and "plugged."

In general, the term "clogged" is most commonly used to refer to a well condition where the hydraulic efficiency is being marred by a purely physical problem. The entrance velocities being too high might cause this. This would cause a "pulling" of surrounding fines closer to the well screen where they would become entrapped, or by the cone of depression generated around the well casing during pumping being of a type that would force fines downwards towards the well screen and cause clogging. Here, there are purely physical factors causing the problem. Essentially this would involve the "throat" diameter (average size of the gaps) of voids in the porous medium such as a gravel pack being smaller than the average diameter of the fines being entrapped.

The term "plugged" is of more recent origin and has come to reflect essentially the build up of deposits, slimes and accumulates which have been generated as a direct result of biological activity. These three elements of a plug usually fill up the voids within a certain region of the porous media around the well screen. A focused region of plugging usually occurs at the redox (reduction-oxidation) front where biological activity commonly focused in the Eh range from -20 to $+100$ millivolts. To understand the nature of the plug, it is important to understand that it is a biologically derived matrix with three components. These are:

- **Slimes,** this is a high water content protective coating for the microorganisms that live inside this slime. These slimes are complex stratified structures containing various biofilms and water channels in complex communities (consortia). On average, it may be expected that there would be between five and thirty different microbial species living cooperatively within the slime.
- **Accumulates** are deposits which accumulate within the slime matrix as potential useful reserves for the consortia of microorganisms occupying the slime. These accumulates would include organics,

nutrients such as phosphates and various forms of nitrogen, and metallic cations. Commonly, the metallic cations are deposited in oxidized forms. Common groups deposited this way are the ferric oxides, hydroxides and carbonates. The various metallic cations are accumulated at different sites across the redox front with iron, zinc and copper tending to be concentrated at the oxidative end of the front. It is generally though that these metallic accumulates perform primarily protective functions since protozoa will not feed so readily on metal-rich slimes.

• **Mineral deposits**, as the plug ages, the dominant components, by weight, shifts from the slime through accumulates to mineral deposits. Mineral deposits are concentrated as primarily crystalline structures that gradually dominate the plug. These deposits are relatively homogenous structures dominated particularly by specific compounds such as calcium carbonate (e.g., dolomite), ferric carbonates (e.g., siderite, hematite), silicates, and ferric oxides and hydroxides (e.g., geothites). Sometimes these mineral deposits form over a slime-based growth to form a nodule or tubercle.

The process of plugging (some refer to this as bioplugging) inevitably involves microbial activity from the initial formation of the biofilm coating the surfaces, through the generation of the slime, to the hardening of the slimes with accumulates, and the maturation with extensive mineral deposits.

The origins of the microorganisms necessary to cause a progressive plugging have frequently been blamed on contamination of the well environment by organisms introduced from the surface. It has to be recognized that there is a rich and diverse microflora in the aquifer, which can be stimulated by the construction of a well. Such a construction followed by development and production radically changes the ambient environmental conditions to cause much greater levels of microbial activity. This can then lead to a biofouling of the well and plugging. It is more important to determine the environmental conditions around a water well which could make it conducive to plugging (biofouling) rather than try to blame some contaminating surface microbe for the problems that have generated in the well. Martinus Beijerinck in the late 1800's coined the statement: "everything is everywhere, the environment selects." This is true for water wells as it is for other environments.

While the main determination of whether a well is "clogged" or "plugged" remains experiential, the evidence is now suggesting that 80% of the wells with significant losses in specific capacity are affected by plugging which is driven by microbiological processes, while the remaining 20% involve a physical form of clogging.

In the determination of water wells for a biofouling risk-assessment, it is therefore necessary to obtain as complete a background as possible on each well. This would enable the typical environment to be characterized for that well so that potential for clogging and/or plugging can be projected prior to treatment. Remember that "no one size fits all" and "each well represents a unique environment" so that each well should be considered a different and unique challenge.

Information applicable to this environmental plugging risk character-ization includes the following components:

- *Pretreatment*, is the background information on the geology, geographical location, and ground water characteristics of that region.
- *Well Development*, details the manner in which the well was con-structed through to final development.
- *Original Operation*, this includes the initial well production, chemical and bacteriological characteristics of the newly developed well.
- *Historical Operating Data*, reflects the database that has been gathered for that specific well from the beginnings of routine production to the present time. This should give some information supporting the selection of the well as one that needs to be rehabili-tated.
- *Prior to Treatment*, is the data gathered immediately prior to the treatment cycle and would represent the impaired status of the well due to plugging.
- *Treatment Cycle*, is data gathered during the treatment cycle. This can then be used to determine the immediate impact of the treatment(s) applied.
- *Post-Treatment Cycle*, is follow up data obtained to determine the long-term impact of the treatment on the wells production.

Details of each stage of the investigation of an individual well are set out below.

SITE SELECTION, Geological Environment

Although there is considerable variation of fouling potential from one well to another, studies have shown some regional similarities in the form and rate of fouling that occur. These often result from similarities in the characteristics of the water bearing formation. For example, wells in highly transmissive gravels or rock do not tend to plug at the bore-hole wall even if they are significantly biofouled, while aquifers with low transmissivity which have low volume-to-surface area ratios are more vulnerable to plugging. In addition, the formation characteristics can greatly effect the ability of a treatment process to remove clogging material from the area surrounding the well. A treatment process that effectively rehabilitates a well, which receives water from consolidated sandstone, may not be effective in treating wells receiving water from an unconsolidated formation. To make a fair comparison of treatment processes, it is necessary to obtain the following well data:

- Water bearing formation characteristics (e.g., consolidated – sandstone/limestone/granite/dolomite, unconsolidated – silt/sand/ gravel). This determines whether gravel pack is needed (gravel packs make treatment more difficult) and also the influences of the movement of fines to the well.
- Where water is supplied to the well by more than one type of formation. Water quality can change as certain screened areas of the well become plugged and water is drawn from deeper formation.
- Grain size/texture effects hydraulic conductivity.
- Grain size distribution (uniform/non-uniform or well sorted/poorly sorted). Non-uniform formations are prone to mechanical blockage as fines can migrate through pores of coarse material. Materials do not move as readily through uniform porosity of finer material formations so that mechanical blockage is less likely to occur close to the well.
- Transmissivity of aquifer (from grain size, texture, and distribution). This determines the ability of the subsurface environment to transmit water (with nutrients) and therefore support more/less biofouling.

SITE SELECTION, Geographic Environment

Grid maps which pinpoint the exact location of the wells involved in the study and which assign an identification number to each well must be provided. All data collected from the well must be listed under this number. The following data would also be useful:

- Nearby surface waters that may act as recharge to the well. Surface waters can greatly increase organic loading and oxygen levels in a well that increases the rate of plugging
- Point of recharge
- Agricultural activities/industrial activities in region
- Overhead power lines (can cause stray current leading to corrosion)

WELL DEVELOPMENT, Well Construction

Well design and construction can have a significant impact on the rate and extent of plugging. "Wells which are designed to resist corrosion and permit reasonably free but laminar flow (with minimal possible drawdown and oxidation at the intake) are less likely to plug quickly" (Smith, 1995). Poorly installed packs, screens or casings can effect well production and encourage fouling. The following information will help determine the effects of well construction on the treatment processes tested.

- Date of installation/cost of installation
- Materials
- Depth/diameter/volume/static water level
- Screen (opening size/type, position in relation to producing zones)
- Gravel pack (thickness, size gravel used)
- Well seal/annular seal data
- Ground surface sloping away from well?
- Position and length of dead zones
- Well development procedures
- Pump and motor characteristics (voltage, amperage, horse power, energy consumption)

ORIGINAL OPERATIONS

This is taken to be the production phase immediately after the well is developed. The following data can act as a baseline to determine the effectiveness of different treatments:

Well Production
- Original specific capacity (SWL, pumping water level - PWL)
- Hydraulic efficiency
- Entrance velocities profile (variations in entrance velocities with depths)

Water Chemistry
- Total phosphorus, total organic carbon, total nitrogen, potassium
  - Potentially limiting nutrients
- Fe (total and soluble iron), S (total, S^{2-} and SO_4), Mn (total, Mn^{2+}, Mn^{4+})
  - Indications of clogging potential, presence of biofouling, Eh shifts

- Eh (redox potential) and carbon dioxide
  - Indication of microbial activity
  - Direct indication of probable metallic ion states
  - Reflect the ratio of oxidized and reduced species of Fe, S, and Mn
  - Elevated redox potential indicate an environment in which oxides of Fe ($^{3+}$) and Mn ($^{+4}$) are precipitated
- pH
  - Indicates likelihood of corrosion and mineral encrustation
- Hardness (total, Ca, Mg, Carbonate)
  - Encourages scaling, and effects efficiency of cleansing products
- Alkalinity
  - Measures capacity to neutralize acids
- Conductivity
  - Indication of total solids content and a component of corrosivity assessment
- Particle Counts, Size and Density
  - Indication of changes in TSS which may reflect biological activity and whether plugging is occurring
- Turbidity
  - For assessment of changes in particle pumping or bacterial growth
- Silicates

Bacteriology

Although it is unlikely that this data would be available, it is strongly recommended that in the future biological analysis be performed on a well immediately following installation. This would provide a baseline for monitoring the biofilm development.

ORIGINAL START UP TO PRESENT TIME
(Operation/monitoring data)

Any of the data that has been collected during this period of well operation should be added to the database. It is essential that the time of testing be identified. The information listed below would be very useful in this study as operational procedures can influence the fouling process, and changes in water chemistry, bacteriology and well per-formance can provide information on the rate and extent of biofouling.

This may be a problem for the bacteriology since it is possible that no tests were performed on the well at this time. In this case, an owners/operators survey could provide useful information on biological activity in the well. The survey should be used to identify any changes in water quality (e.g., changes in taste, odor and color) and if possible should identify when these changes occurred. It should also ask the well operator if any deposits have been observed on the pump or other well equipment.

Operation Procedures for the Well

Cyclical pumping or long periods of idleness promote biofilm formation and bioaccumulation of recalcitrant materials. Excessive pumping drawdown introduces more oxygen into the system, increasing the rate of abiotic and microbial Fe^{2+}-Fe^{3+} (ferrous – ferric) oxidation.

Symptoms that can be indicative of well problems involving biofouling and plugging would be (effect of plugging is given in parentheses):

- Wellhead pumping rate (decreasing)
- Operating hours (increasing)
- Pump power consumption (increasing)
  - Voltage, amperage draw, occurrence of stray current
  - Variation from manufacturer's specifications.

After Treatment

No water well treatment will effectively prevent some reccurrence of biofouling with the concurrent losses in water quality, production and increasing operational costs. To control these problems, regular monitoring should be performed on each well following treatment. It is essential to know how long it takes for the well to biofoul again after treatment. Some treatment processes can actually leave residual chemicals (such as phosphates) that can actually encourage future microbial activity while other processes do not effectively remove biofilms (leaving the "seeds" for the next crop of plugging). Biofilms left in the aquifer or well can quickly rebound and plug the well. Monitoring tests should be similar to those performed immediately after treatment and should be repeated often enough to identify when changes occur. Common time intervals range from one month (for wells very prone to rapid plugging) to yearly (if the well does not show major symptoms of plugging or other forms of biofouling.

USING THE BART™ TESTS TO DETERMINE PLUGGING/BIOFOULING

It is important to have good record keeping so that the historical occurrence of problems and the success of treatments can be evaluated. This can be done using the BART™ data record and interpretation sheets. Each sheet is designed to allow all of the information to be gathered from the BART™ tests for each particular water sample. It provides the primary data gathering form before entry into the spreadsheets for detailed interpretation. Depending upon the seriousness of the problem, there would be three levels of possible data gathering:

<u>Level One, BART™ Testing</u>

This level would involve the following BART™ tests:

IRB-BART™	-	iron-related bacteria
SRB-BART™	-	sulfate-reducing bacteria
HAB-BART™	-	heterotrophic aerobic bacteria
T-COLI-BART™	-	total coliform bacteria*

All of these tests would be read visually at room temperature except for the T-COLI-BART™ (*) which would be read after 48 hours of incubation at blood heat (35 to 37°C). The test data sheet calls for the BART™ tests (except the coliform test) to be read daily until there is a distinct reaction. These reactions are described below. Once the reaction has started, that period of delay then forms the time lag from which the aggressivity and population size can be determined. In this interpretation, it is considered that the aggressivity relates in an inverse manner to the time lag. In other words, the shorter the time lag, the more the bacteria are aggressive and the greater the population. The time lag should be noted with the observed reaction code being placed in the correct <u>Reaction Codes box</u> when each positive reaction is observed. Refer to the time lag box to determine which is the correct box. Time Lag box marked "1" would reflect a reaction occurring within one day (24 hours) of starting the test. Putting the reaction code into time lag box "1" would therefore mean that the reaction code was observed within the first day of starting the test. Subsequent reactions that may occur can be entered in the appropriate reaction code box for that day of the time lag.

In the left of each reaction code box is an italicized number which represents the Possible Log Population (PLP) which may be converted to colony forming units/ml (cfu/ml) if preferred using the conversion rows shown on the BART™ data collection form. Note that the population in the box is relevant to the population of bacteria causing the reaction observed at that time. Data being interpreted from any water sample may therefore include as many populations as there are reaction codes recorded. If a treatment affects only one part of the microbial flora, then this would be reflected in a reduced population in that part of the population reflecting the extension of the time lag before that particular reaction was observed and entered. Given that the BART™ test may not be able to be read on every day, all reaction codes should be entered for the day (time lag) on which the reaction was observed. To recognize the time lag period when the tests were not observed, an "X" may be entered in the time lag box to show that the test was not observed on that day.

The aggressivity of the bacterial populations can be determined by the shading applied to that particular box into which the reaction code has been entered. If there is a 25% grey shade in the reaction code box, then the potential aggressivity of the associated population may be considered to be "HIGH." If there is a 12% grey shade, the aggressivity risk may be considered to be "MEDIUM." If there is no shade in the reaction code box, here the aggressivity would be considered to be at a "LOW" level.

The T-COLI-BART™ needs only to be read at 48 hours. If the gas thimble has been elevated and is now floating at the surface, this means that total coliforms are present at greater than 1 coliform/100 ml.

Level Two, BART™ Testing

This is a more comprehensive testing program that would include, in addition to the level one test selected, the following additional tests:

DN-BART™	-	denitrifying bacteria
SLYM-BART™	-	slime-forming bacteria
FLOR-BART™	-	fluorescing pseudomonads
E-COLI-BART™	-	fecal coliform bacteria

The first three additional tests are again kept at room temperature and read daily. With the DN-BART™, the objective is to detect denitrifying bacteria that often become very aggressive when there is high nitrogen loading due to organics and nitrates. They often become more aggressive when the ground water is being contaminated with nitrogen-rich wastes such as seepage from septic waste tanks. In addition, the FLOR-BART™

is included to detect the presence of fluorescing pseudomonads. These bacteria are aerobic heterotrophs that often become aggressive when the aerobic degradation of organics is occurring. A U.V. light should be used to check the FLOR-BART™ for any glowing. If a distinctive pale blue glow is seen, this could mean the presence of *Pseudomonas aeruginosa*, which is an opportunistic pathogen. Where this happens, it is recommended that confirmatory test be conducted by a qualified microbiology laboratory to assess the hygiene risk factor. The FLOR-BART™ can be used as a source for isolating these bacteria if the laboratory is conveniently close for the shipment of the tests. Otherwise, the confirming laboratory should work from the original water sample or source. The SLYM-BART™ is included since it is a very broad-spectrum test for many of the slime-forming heterotrophic bacteria that are often dominant in plugging.

The E-COLI-BART™ may be incubated at either blood heat (35 to 37°C) or at the slightly higher temperature of 44.5 °C. If the gas thimble elevates to the surface within 24 hours, the sample would be considered positive for the presence of fecal coliforms (at least one fecal coliform per 100 ml of water). In some regions of the U.S., there are occasions when some coliform bacteria are able to grow within the plugging water wells and therefore generate a false positive. Incubating the test at 44.5 ° C generates temperatures that are too high for these "false" coliforms but not for *Escherichia coli* which is a true indicator of hygiene risk.

Level Three, BART™ Testing

This method employs all of the tests shown at Level Two. The difference is that the technology now incorporates the automatic computer assisted monitoring of many of the tests (test duration on the CABART™ system shown in parentheses):

IRB-BART™	-	iron-related bacteria	(192 hrs)
SRB-BART™	-	sulfate-reducing bacteria	(192 hrs)
HAB-BART™	-	heterotrophic aerobic	(96 hrs)
DN-BART™	-	denitrifying bacteria	(72 hrs)
FLOR-BART™	-	fluorescing pseudomonads	(192hrs)
T-COLI-BART™	-	total coliform bacteria	(48 hrs @35°C)
E-COLI-BART™	-	*E. coli* coliforms	(24 hrs @ 35°C)

It should be noted that an additional test for methanogenic bacteria (BIOGAS-BART™) may also be included in regions of the country where there are significant hydrocarbons entering the ground water.

THEORETICAL ASPECTS, Water Well Biofouling

Water coming from a well head can originate from a number of possible sources, each of which could change the characteristics of that water sample. These sources are listed below in arranged order moving downwards and away from the well head:

Outflow Pipe from the Well

Water in this pipe varies from static (between demand), to turbulent (at onset of demand), to flowing (during demand). During the static period, the line is essentially "dead-ended" during which time microbial growth both on the walls and in the water might intensify. Once the demand affects the static water causing turbulence, sloughing of material and microorganisms from the walls of the pipe is likely to occur dragging up significantly the biological burden in the water. Constant flowing due to a steady demand leads to a condition where the water is influenced gradually less and less by any sloughing from the walls and the water now essentially has the characteristics of the in-flowing water from the pumps.

Pumps

Water passing through the pumping systems is going to be subjected to turbulent and compressive shocks as the velocity of the water is forcefully directed under pressure. Surprisingly, these surges in velocity and pressure actually form sites for microbial growth with the consequent slime, nodule and tubercle formation. When the pump is first started up, there is likely to be sloughing from these sites which means that the water could potentially carry a heavy burden of these sloughed microbes during the start. Once running routinely, sloughing is likely to become a minimalized event.

Static Zones of the Water Column

The water column in a well is elastic in length depending upon the impact of the drawdown on the position and size of this part of the static water head in the well. Therefore, there are two functional zones: an active zone affected by the entry of ground water and the pumping action, and a passive zone where the water remains relatively stagnant above. The static zones are buffered from the pumping activity and so are essentially "dead ended." This static zone may be subdivided into three regions where the form of the stasis could have different effects. These regions are: (1) the drawdown region which is either underwater when there is no demand, or above water when there is a demand; (2) the region between the drawdown water level and the active zone influenced by pumping demand; and (3) the region of static water below the active zone influenced by pumping demand. Each of these regions are likely to have different forms of microbial activity which may be summarized as:

Drawdown region
This is a region where there is likely to be heavy slime growths attached to the walls. This shifting between saturated (no demand) and humid air (on demand) creates conditions that would support the growth of aerobic slime formers that would grow on the walls. Transient sloughing is most likely to occur at the start of pumping.

Region above active region
This is an elastic zone in the water column that interfaces with the drawdown zone above and the active zone beneath. This more resembles the "dead end" characteristics of a distribution pipe except that it is vertical. Often the microbial activity will form into plate-like clouds and grow within only specific zones along the reduction-oxidation (redox) gradient that forms in this zone. The coatings on the wall may also reflect this redox shift. For example, the growths observed when video-camera logging is used might move from (in sequence):

(1) Aerobic slimes (amorphous mucoid bulbous masses),
(2) Hardened plates often rich in red-brown oxides of iron,
(3) A reduced region where the growths on the walls will blacken with iron sulfide and/or iron carbonate deposits.

Video-camera logging can sometimes reveal these forms of biofouling but one common experience in passing through this zone is the danger of the knocking loose some of the attached growths from the walls. These fragments may float in the water or sink to the bottom of the well. In addition to the camera housing knocking off material from the walls, there are often also slime structures floating in the water. These could be called a "well snow" somewhat akin to the "sea snow" often seen during deep-water dives.   This "well snow" is really biocolloidal particles that can range in size from 4 microns to 50 mm. These particles have the same density as the water and so float usually at specific depths in the water column and can form a part of the plate-like clouds of growth.   Particles that get denser than the water sink to eventually enter the active zone beneath and possibly, pass on down to the stagnant zones underneath that.

Region below active region

This zone occurs underneath the active region above the base of the well. It is in this "dead zone" that denser particles collect after descending the water column. In addition, mineral deposits, sands, silts and clays may also collect to forming a biologically active generally very reductive zone. Where there is a significant entry of organics, gassing may occur. If the gas is methane it will collect within these deposits as gas pockets. When the overburden cannot contain the gas any longer, large bubbles (5 to 75 mm diameters) may emerge and rise to the surface. The methanogenic bacteria growing within these dead zone sediments may form this gas. Where hydrogen sulfide is produced and reacts with any iron, it produces black iron sulfides and the bottom deposits will appear to be black.   Generally, this dead zone will gradually collect sediments including dense biocolloids, slime, clays, silts and sand and gradually fill up the space unless it is cleaned out periodically. Often these deposits, when stirred up, can introduce bacteria into the water column which are simply surviving in the sedimentary deposits and are not actively a part of the biofouling.

Active Water Zone in the Well

This is the zone in the well which is directly contributing ground water to the pumped flow. Initially, as the pumping starts up, the first water to be pumped will be water from the column creating the drawdown effect made up with an increasing flow of water entering the well from the surrounding formations. Once the drawdown has stabilized for that pumping rate, the ground water leaving the well will exit from the formation and move towards the pump intake in an increasingly turbulent flow. These compressive and turbulent forces cause microorganisms to attach to sites such as the pump and well screen impeller blades where they grow to form tight masses of slimes or encrustations. In these growths, there is a gradual loss in the efficiency of the pumps, increased resistance to flow due to losses in open slot/grid area, and dramatic shifts in water quality particularly when the pump is first activated.

Microorganisms within this active part of the well water column are subjected to one over-riding event, which is the pump-static cycle. This cycle is created by the pump being activated (pump) and then switched off again (static) in a regular manner. The microorganisms able to function and grow under this cyclic regime gradually harmonize with the cycle. The microbes found in this active zone are a mixture of the natural microbial flora that are fouling up the various surfaces. The microorganisms that are entrancing the water column from the surrounding formations (either as sloughing biocolloids or as individual cells and clumps of cells), and those which are moving between the passive zones above and below and are passing through this active zone. Microbes descending would be in denser structures and would settle in the static zone underneath while ascending would have a low density and be elevating upwards through the zone. Clearly, when the pump is on, these organisms are likely to be taken up with the flow and discharged.

Well Column Entrance Biofouling

There are many different passageways through which water can enter a well. These range from open fractures in a rock bore hole, through to perforations in the casing to engineered vertical or horizontal slots within a screened section of the well. In all cases, the ground water is moving from a region with high surface areas to one with low surface areas (the water well column). The ground water passing through this entrance zone to the well is subjected to sudden changes in turbulence and velocity and may often have to pass through constriction such as the slots in the well screen. Concurrently, the ground water would be subjected to changes in the redox potential (Eh) which may render the conditions more favorable for some of the bacteria. For example, iron-related bacteria tend to become more aggressive and dominant when the Eh shifts to a more oxidative state. As a result of this, the IRB are found to be more numerous in the more oxidative waters close to, and inside, the well water column.

One of the frequent restrictions to the entry of ground water into the well is the width of the screen slots that are designed to prevent the entry of coarser formation material (e.g., silts and sands) into the well. This constriction in slot width coupled to controlled entrance velocities are two principal factors employed to prevent "fines" moving into the well column. Unfortunately, these constrictions also can form sites for the growth of bacteria to form into slimes, encrustations and hardened mineral deposits.

In general, these biological growths which often form into hardening slime deposits, grow the most quickly where there is active movement of ground water into the well through the slots. This growth reduces the flow rate into the well from the infected zone. To compensate for this, ground water may now be extracted from other potential sections of the well that are able to produce. Commonly, these sections are likely to be deeper (involving a higher energy cost to recover) and likely to have different characteristics than the shallower ground water originally extracted. Over time, the shifting in the characteristics of the water being produced can give an early indication that the active entrances to the well itself are being plugged.

Video camera inspections and spinner surveys performed on a routine basis can reveal this form of "tight" plugging and shifts in the zones of the well actively producing water. Treatment of this form of "tight" plugging right around the wells producing zones on the well screen are relatively easy to treat unless these growths have become hardened or have a high phosphorus content. Video camera inspection of the well can reveal whether such growths have been removed by treatment. It must, however, be remembered that the video camera inspection cannot reveal the nature and form of any plugging occurring deeper in the formation and, hence, out of view through the entrances (e.g., slots or fractures) to the well.

Active Turbulent Zone around the Well
The well itself may be surrounded by a pack or be directly inserted into the producing formation. In either case, when there is a demand for ground water, there is a movement of water to the well from the aquifer. An increasing velocity, greater turbulence and changes in the environmental characteristics characterize this movement as the water approaches the well. Of the environmental characteristics, it is probably the shifting in the redox (Eh) potential that controls the shifting in the intensity of the plugging. Often a zone of intense bacterially induced plugging forms around the well at the site where the redox moves from a reductive to an oxidative state. This shifting to an oxidative state occurs generally closer to the well and stimulates

considerable levels of bacterial activity. A concentric zone of enhanced activity therefore is likely to generate around the well and possibly extend into the well entrances and out into the formation.

The form of growths, which occur around the well, will be influenced by the chemistry of the ground water flowing into the site from the aquifer. If this water has a relatively nutrient-rich chemical content, then these enriched zones of growth may extend out into the formation in the direction from which this water is flowing. If the water is already nutrient-rich and has an oxidative redox potential (such as would occur where there is a local river recharge of the aquifer), the growths are likely to extend into the formation away from the well in the direction of the recharge. These types of growths are also likely to be scattered throughout the zone rather than focused. It would be at these sites of biofouling that the ground water flow rates would become retarded as the volume of the slime plugs, accumulates and mineral deposits grow. This would cause diversion of flows to alternate channels prior the infected area plugging up and preventing significant flow. Microorganisms would slough away from these plugging formations in an erratic manner as the slime growths sheer. Where this happens, there could be a sequence of bacterial releases representing the different layers of the plug as each one sheers. Essentially, it is commonly the size and conformation of these plugging zones around the well that are likely to create a form of plugging that is the most difficult to treat.

Background Aquifer Microorganisms
It has now become well established that aquifers are not sterile but contain a rich and diverse, and often slow-growing, microbial flora. Many of these microorganisms, when entering the more active zones around the water well, will respond to the shifting redox potential and other environmental changes by becoming more aggressive. This may lead to their incorporation into the various community structures that are forming the plugging.

SUMMATION OF THE SITES OF MICROBIAL FOULING
There are a series of microbial regions within water wells where complex communities of microorganisms are likely to develop. Most of these are growths attached to surfaces in the form of stratified biofilms and only a few are suspended in the water itself. These regions each reflect distinct biozones occupying regions centered in and around the well itself.

Activating the pumping causes water to flow through the active regions of the well and deliver a product water which microbiologically reflects the origin of the water being pumped to the sampling site at that moment in time. It may therefore be expected that mixtures of suspended and sloughed biocolloids will reach the sampling port in an order related to their travel time to that site. The populations of microorganisms recovered are likely to fluctuate with the passing of microbes from each of these regions in sequence. Needless to say, it can be expected that turbulence both in the pump, the well and the surrounding active formations will cause mixing to occur. Consequently, the sequence of origin for microorganisms arriving at a sampling port set just downstream from the well head would be:

- Distribution line
- Downstream line from pump
- Pump contents
- Active water zone within water well column
- Water passing through openings into the well
- Water moving from the actively plugging regions
- Water from the background aquifer formation

It is only when the pumping has been continued for sufficiently long to ensure that the majority of the water being sampled is actually from the background aquifer formation. Even at this time, there is always likely to be some sloughing from the regions in and around the well that can cause variations in the data being generated. There is therefore a need to ensure that the sampling procedures achieve the objective desired.

The objectives for testing well water for the presence of microorganisms can be viewed as involving three possible concerns:

- Hygiene risk
- Unacceptably high bacterial counts
- Plugging/Biofouling risk

The importance of water well rehabilitation is based upon the ability to recover a biofouled well to its original performance characteristics. Commonly, the loss in performance has been due to plugging events upstream of the well head and so the major concern in rehabilitation relates to the diagnosis of a microbially induced plugging so that an appropriate treatment can be employed to regain performance through rehabilitation. Of primary concern therefore is the assessment of the plugging/biofouling risk while unacceptably high bacteria counts are a secondary concern. It is important to utilize techniques that will be able to demonstrate that biofouling is occurring in a manner and at a site that impacts on the production characteristics of the well.

High bacterial numbers are likely to be a growing and more frequent occurrence as the well plugs but these do not indicate the nature and form of a plugging event. Essentially, bacterial numbers shift from a dribble to downpour to a deluge as the plugging is generated but it remains difficult to determine the nature, location and form of a plugging event merely by recording the size of the bacterial population by counts.

Hygiene risk is generally considered an assessment that is independent of any plugging that might be occurring and relates more specifically to the presence of coliform bacteria. Usually, it is believed that the coliform bacteria travel through the plugging zone so that their presence reflects a potential hygiene risk in the influent water rather than a hygiene risk harbored by the plugging regions of the well. There are experiences where the coliform bacteria can be removed by the microbial activity in the plugging regions and there are also cases where some of the coliform bacteria can become integrated into the community structures forming the plugging. Sometimes, the act of rehabilitation of a water well which is successful in removing the plug formation can emerge with coliform bacteria "appearing" in the water after treatment simply. The reason for this is that the influent water had a coliform population but when it passed through the plugged zone, these coliform bacteria were "filtered" out by the plug. Once the process of well rehabilitation had removed the plug and the surfaces cleaned, there were no longer any active plugging material that could remove these coliform bacteria. The plug had essentially been operating as a biologically active filter that was removing the hygiene risk (i.e., the coliforms) from the water. Once the plug had been removed by rehabilitation, the hygiene risk re-emerged since there was no longer a bio-filtration mechanism in place.

In the management of a well, the emphasis is on obtaining knowledge as to the presence, the position and the scale of any plugging around the water well being tested. Consequently the objective is limited to the determination of the plugging/biofouling risk. To achieve this, conditions have to be generated in the well that will cause at least some of the bacteria that form an integral part of plug to disrupt and, by sloughing, enter the water. Once this disruption has been achieved, then pumping the well can sequentially pull this disrupted material from the well. The procedure therefore calls for a disruptive phase to precede a pumping phase that would involve sequential sampling to entrap the various components from the plugged regions associated with the well.

RECOMMENDED SAMPLING PROCEDURES

From a theoretical perspective there would appear to be no simple manner in which to take a single water sample and expect that single sample to generate a confidence in the data necessary to determine the degree of biofouling is optimistic. The vast range of cloistered microbial communities that may or may not be in a stable attached state means that the microbial loading in the water sample may depend more upon the status of sloughing within the well than any other factor.

One important aspect of sampling, therefore, has to be to determine whether there is any biofouling and what bacteria are associated with that event.

THEORETICAL CONSIDERATIONS, Sampling for Bacterial Testing

To evaluate the aggressivity of the bacteria that may normally be associated with the attached (slimes, nodules, tubercles, and mineral deposits) growths within the well environment at-large, some method has to be incorporated which would cause at least some of the bacteria to detach. Once these organisms have detached, it now becomes possible to recover them in a sample of water pumped from the well. The recommended procedure there includes the following stages:

· Shock the well environment in a manner that is likely to cause at least some of the attached bacteria to break away from the attached growths and enter the free flowing ground water moving towards the discharge zone.

· Since the shock, if successfully applied, would disrupt many of the growth formations in and around the well, these disrupted masses would move towards the well in the same sequence. In other words, water being discharged from the well would begin by discharging in a sequential manner these disrupted masses from the wall.

· Using a sequence of sampling the initial waters of discharge, it now becomes possible to examine each of these waters to determine which bacteria have sloughed from the various zones around the well which have been traumatized by the shock and released these bacteria to the water.

From the data gathered from the analysis of the sequenced water samples, interpretation can reveal the forms of biofouling that are occurring in and around the well.

There are two schools of thought as to where the sloughed bacteria recovered in these sequenced samples came from. After a biofouled water well has been shocked in such a manner as to cause the detachment (sloughing) of some of the biofouling, subsequent pumping causes the sequential releases of these various bacterial groups such as IRB, SRB, HAB and methanogenic bacteria. The peak release times for each of these bacterial groups is often sequenced. With frequent sampling and analysis, these releases can be observed and the time of pumping to the peak release for each group of bacteria can be established. The two schools of thought use different approaches to interpret the data and, on different occasions, one or other may be valid.

The first school of thought (the <u>layered theory</u>) is that the biofouling is tight in one location and that the shock effect causes the various layers of biofilms to sheer away sequentially. This would mean that the biofouling was tightly focused and multi-layered. As each layer sheers off, the next layer is exposed and then that too sheers off into the water which is then discharged. Sulfate-reducing bacteria and biogas forming (methanogenic) bacteria are more likely to be found in the deeper layers of the biofouling and therefore be recovered later than the heterotrophic aerobic bacteria which would tend to be nearer to the surface of the biofouling. The location of the layered form of biofouling can be predicted by the lag time during the pumped discharge to the first maximum microbial release. Where the pump rate, active water well volume and the porosity of the pack and other porous media around the well is known, it becomes possible to project the average distance of the biofouling away from the central vertical axis of the well.

The second school of thought (the <u>concentric theory</u>) proposes that the biofouling occur as a series of concentric biofouling zones (known as biozones) around the well. These ring-like concentric zones develop at points around the well where there are changes in the Eh (redox potential), ground water velocity, turbulence and compression and nutrient availability. Bacterially driven biofouling of these concentric zones now becomes a suitable site for biofouling dominated by more specific groups of bacteria. Here, the innermost zone would normally be considered to be the most oxidative one while the outermost zone would be the most reductive.

Beyond the outermost zone of biofouling is a region in which there is a background level of microbial activity that extends throughout the aquifer.

SHOCK PROCEDURES APPLICABLE TO SAMPLING

A biofouling event commonly matures over a number of years unless it is associated with a biodegradation problem in which case the time scale for biofouling can shift from years to months or even days! Biofouling usually forms in a relatively consistent well environment even though the environment may change between active (pumping) and passive (static) states. Essentially the incumbent microbes adapt to the routine shifting in the environment that accompanies the routine cycling between active and passive states.

.It has to be remembered that the microorganisms associated with the biofouling are adjusted to the natural conditions present around the well. It is not correct to make the premise that a particular temperature, pH, salt concentrations, or the redox potential would not allow biofouling to occur. When conditions become extreme, there is a tendency for the range of microbial species involved in a biofouling event to become limited and yet there always seems to remain some species that will adapt to these extremes and flourish.

The principal factor involved in shocking a well is to radically change the environmental conditions in the well so that the incumbent organisms become traumatized. This trauma will then lead to the detachment (sloughing) of some of the microorganisms from the layers and/or biozones within the biofouled zone. These detached organisms will now be sequentially pumped out from the well.

The critical question is the form of the trauma that would be imposed on the biofouling. There are a number of approaches that can be used depending upon the condition of the well.

Shock Procedure for a Producing Well

A producing well is routinely passing between an active and passive phase or is permanently producing. In the former case, the environmental conditions around the well will be shifting with the redox front moving towards the well (passive phase) and away from the well (active phase). In the latter case, the environmental conditions are stable because the well is constantly producing so that the redox front would remain at a fixed position. In both cases, the biofouling (plugging) would be occurring in a constant manner. To shock these types of wells, the simplest method is to radically change the pumping regime. In other words, turn off the pump for a minimum of 24 hours. This creates a static condition in which the redox front would move out of sequence with the normal activity of the well's environment. As a result of this change from normal, the biofouled zones will begin to destabilize with some of the component microorganisms moving into the water.

It is commonly very unpopular with well operators to request prolonged water well shutdown. Often the argument is made that the well's production is essential of the supply demand. However, if the well's production is going to be an assured supply in the coming years, it is essential to get a reasonable, reliable estimation of the extent and form of any biofouling. In extreme circumstances, a shock can be achieved using shorter time periods if the wells normal operation has been very routine. As a rule, at least a four-hour shutdown should be employed with an overnight twelve-hour shutdown preferred.

Shock Procedure for a Non-Producing Well

A non-producing well is going to be in relatively stable state of stasis with nothing more than the normal flow of the ground water coupled to the influences exerted by the well itself. To cause a shock sloughing of microorganisms from these types of wells, the act of pumping radically changes the environmental conditions around the well and some sloughing will erratically occur during this event. The time and form of this sloughing may be too unpredictable. To compensate for this, the well is pumped (recommend at least eight hours) followed by a 24 hour resting period before beginning the pumping to sequential sample the well. This 24 hour resting period after the pumping provides a time for the microorganisms causing the biofouling to begin to relocate in the destabilized environment. After the resting period, the sequence of sampling can be started.

Shock Procedure for a Well subjected to Organic Loading

The recognition of the impact of organic loadings in a ground water system on biofouling is now becoming accepted. As organic materials (such as JP4, gasoline, and organic solvents) approach a well, they tend to enter into the biofouled zones as accumulates that are degraded over time. This is because of their value of nutrient sources for many of the incumbent microbes. Because of this accumulation followed by degradation, the biofouling becomes dominated by the microorganisms associated with these accumulative and degradation functions. Consequently, the mass of the biofouling grows more rapidly and is dominated by those organisms associated with the degradation function. Plugging may occur much more quickly because of these organics and the biodegraders will dominate the microorganisms within the biofouling.

Delivering a shock to these wells to cause sloughing is generally easier because the rate (metabolic) activity is greater. In most cases, these wells are on very rigid passive-active cycles to which the microorganisms are keenly adapted. Extension of the passive (non-pumping) cycle by preferably one-day should be sufficient to cause enough sloughing to be able to perform the sequential sampling.

Shock Procedure for a Flowing Artesian Well.

A flowing artesian well is under a sufficient hydrostatic pressure that water normally flows from the well without pumping. Here, the biofouling forming the plugging condition is being subjected to two states:

> *An active state* when the water is being drawn from the well into the water system

> *A passive state* where the water may be freely diverted to a surface run-off site or the well is shut off and the well comes under the ambient pressures of formation.

These forms of well are likely to have a more stable and possibly more dispersive form of biofouling. It is clearly more difficult to shock a microbial formation that is under pressure and being subjected to a constant flow (if there is a passive discharge of water during off-demand times). There are two possible approaches to shocking these artesian wells.

The first method for an artesian well is to increase the demand placed on the system by adding additional pump capacity and pump the well at a greater production level than usual (by at least 20% for at least one day). This will cause the environmental conditions to change in the formation immediately around the well and lead to some microorganisms entering the water. A sequential sampling should follow immediately after the enhanced pumping is stopped since the detaching (sloughing) microorganisms are being carried directly out of the well. The second method for an artesian well is to shut down the well for a period of at least 24 hours longer than any normal shut down that may occur during the normal operation of the well. Here, demand by the well is shut off and the flows through the well environment return to the normal ambient conditions for the aquifer. This will cause some level of stress to the microbial fouling and some of the organisms will then detach into the water. It must be remembered that the sequential sampling in this case could involve a rapid surge of detached microorganisms coming out of the well in a virtual "slug" at the beginning and so additional samples should be taken immediately and at 1 and 5 minutes into the sequential sampling.

SEQUENTIAL SAMPLING, SUMMARY

The bottom line in delivering a shock to a well prior to the sequential testing for biofouling (plugging) is to radically change the conditions downhole from that which normally would occur during the production cycle. In the cases where wells have already developed very matured plugging conditions, the success rate for shocking the well to recover some of the biofouling microorganisms goes down. This is particularly the case where there is a high inorganic content (including such materials as carbonates, silicates, ferric oxides and hydroxides, and phosphates along with sand, silts and clays) in the plugging material. In other words, in a well that has significantly lost capacity because of biofouling, the lack of evidence drawn from a sequential sampling does not mean that microorganisms are not involved in the plugging process. It could mean that the microorganisms are deeply embedded within the matured plugging and are not recoverable using the relatively simple shock methodologies described above.

Where it is found that sequential sampling fails to reveal the form of the microbial plugging of the well, this occurrence means that an alternative strategy has to be seriously considered. Such a strategy would involve a more vigorous shock treatment involving the application of a chemical able to disrupt the formation enough to cause the release of some of these incumbent organisms. If well rehabilitation is proceeded with on such a well, there will frequently be very large populations of bacteria in the discharge waters after treatment. These organisms would have been released from the fringes of the plug where the chemical activity had not been intense enough to kill all of the cells.

RECOMMENDED SEQUENTIAL SAMPLING PROCEDURES

Sequential sampling begins after the shock procedure has been applied to the well and it has been left for the recommended and acceptable time period to allow the microorganisms within the plugging zone to slough into the water. It should be remembered that the sampling port should be as close to the well head as possible and flushed to remove any dead-ended microbes that may have been growing within the sampling line and on the outlet itself. The recommended procedure would be:

- Pour boiling water carefully over the sampling port for long enough (e.g., 30 seconds) to ensure that the temperature of the port itself is now hot.
- Flush water through the port for ten minutes. If the water is still cloudy after the ten minutes then continue until the water becomes clear. If the water is not clear after 30 minutes, it can be assumed that this is the background clarity of the water.
- Again pour boiling water carefully over the sampling port for long enough (e.g., 30 seconds) to ensure that the temperature of the port itself is again hot to reduce the risk of "perched" living microbes from entering the water sample from the sampling port.

This practice should reduce the risk of getting false results due to contamination from the sampling line. The volume of water passing through the sampling line will dilute the impact of such residual contamination.

Once the sampling port has been flushed and prepared, then the discharge (by pumping the well or allowing free flow to begin again) can commence. A sequence of samples is taken during this discharge to determine the range and location of the various groups of microorganisms associated with the plugging.

The normal sequence that may be expected to emerge from a well in which the plug is predominantly set on the redox front with the oxidative side towards the well would be (in order moving outwards from the well and later in the sequenced samples):

Closest to the well (oxidative)
Iron-related Bacteria – IRB
Fluorescent Pseudomonads – FLOR
Heterotrophic Aerobic Bacteria – HAB
Slime-forming Bacteria - SLYM
Denitrifying Bacteria - DN
Sulfate-reducing Bacteria – SRB
Biogas (Methanogenic) Bacteria – BIOGAS
Farthest Away from the well (reductive)

These bacterial groups would have arrived at the sampling site for the most part of the various parts of the biofouled regions in a sequence that reflects either their depth in the biofilms (layered school of thought), or their distance from the well (concentric school of thought). In either case, these bacteria arrive at the sampling site in an order that reflects the form and cause of the plugging.

The key to determine the various microbial plugging risk indicators is the appropriate sequencing of the sampling. Not all wells are the same in scale, production capacity and ability to undertake an extensive sequential sampling program. As a result of this concern, three alternative sequences for sampling are listed below in Table Fifty-Three.

Level A Sampling, limits the sequential sampling to only two hours and is, therefore, self-limited to testing just the ground water pumped in from the immediate environment around the well.

Table Fifty-Three

**Recommended Sequential Sampling Times
For the Determination of Biofouling**

Sampling time (Hours)	Level of Sampling A	B	C
10 min.	x	x	x
30 min.	x	x	x
60 min.	x	x	x
90 min.	x	x	x
120 min.	x	x	x
4		x	x
8		x	x
12		x	x
24			x
36			x
48			x

The three level sampling times are displayed above by the positions of the "x" in the tables. Level A would involve five samples, level B would utilize eight and level C would incorporate eleven.

Level B Sampling, utilizes a twelve-hour sequence of sampling to determine the magnitude of the microbial biofouling within the well environment. This test program should reach the outer edge of the plugging in most cases and should give some information on the background microbiology of the ground water.

Level C Sampling, extends the sequence of testing to two days of discharge. This should give enough time to be able to compare the data gathered from samples in which microbial sloughing was occurring through to the regular background flora that would normally be present in the ground water in the aquifer.

As the samples are taken, the ideal circumstance would be to begin testing immediately. This would avoid the possible degeneration of the water samples during any storage and shipment phase. Where the samples have to be held or shipped to another location for testing, there are some rules that should be observed:

- Make sure that the water sample is placed in a sterile container (sealable plastic bag, plastic bottle or glass). Be careful not to contaminate the inside or the neck of the container with hands, breath or non-sterile equipment.

- Do not refrigerate the sample if the testing will occur within eight hours. It is preferable to keep the water samples as close to their original temperature as possible (e.g., in a sealed cooler) in a dark cool place. If the sampling is likely to be delayed more than eight hours, then the samples should be stored over ice bags to prevent the samples heating up to room temperature (where the water samples were collected at lower temperatures). If the samples have to be kept for longer than 24 hours, there is a probability of some degeneration in the microbial population. This can be minimized by not shaking the water too much while keeping the water temperature in the 8 to 12°C range. Lower temperatures than that could cause severe trauma that could delay growth. The only inhibitor that may be used is sodium thiosulfate to inhibit any activity from residual traces of chlorine should this be present in the water. It should be noted that the BART™ tests do contain this inhibitor to suppress any residual chlorine activity. Note that you should always keep the water sample in the dark to prevent any algal activity in the water. Although algae are not commonly found in many ground waters, their presence could cause photosynthesis and the production of oxygen in the presence of light. This could stimulate the aerobic bacteria. There are test methods that can be applied to the water samples in the sequential series for the microbial loading that may either directly or indirectly determine the populations. In the following two sections, these methods will be discussed.

MICROBIOLOGICAL TESING PROCEDURES (Direct)

The methodology selected for the direct determination of the various bacterial groups is based upon the biological activity reaction tests (BART™) because this system is simple to perform in the field as well as the laboratory setting. In addition, these tests cover the range of bacterial groups likely to be associated with a well plugging (biofouling) event. Sequential water samples from all wells included in the project would be subjected to the level one testing. Level one determination would be performed routinely on all of the wells subjected to a sampling level A where selected wells (e.g., 30% of those surveyed) were shown to be plugged with very clear evidences of microbially induced biofouling.

MICROBIOLOGICAL TESTING PROCEDURES (Indirect)

Rapid tests an be performed to determine whether there are any suspended particles (biocolloids) in the water sample which could give an indication that there is biofouling in the well being sampled. Biocolloids in this case are considered to relate to the amount of sloughing that is occurring from the biofilms in the plugged zone. The more sloughing, the greater the amount of suspended particles (biocolloids), the more probable is a plugging event likely to be occurring.

The direct approach to the determination of these particles in the water is to use a laser particle counter to determine the numbers and size of these particles present in the water. The laser particle counter (LPC) for water has opened up a new method for observing particles in water particularly over the range from 0.4 to 120 microns. To achieve this, the LPC shoots an arranged series of bursts of laser light through the water commonly in a concentric pattern. These pulses are detected individually by a photodetector and conform to the pattern generated. If particles interfere with these pulses, a "shadow" is generated in the pattern of pulses being detected by the photodetector. That pattern reflects the size of the particle. Most LPC machines measure particles as ppm of total suspended solids, and then break the particles down by size into bins. This size can be generated as numbers, total volumes or total surface areas for each bin contained particles of that size. Because the LPC is computer controlled, data can be generated very quickly and in copious amounts. The bottom line is that the LPC is able to "see" through the water sample and count all of the particles that interfere with the pulses of laser light.

Because of the ability of the LPC to record all particles within range, the LPC offers considerable potential benefits to the determination of a plugging event and also the effectiveness of a rehabilitation treatment in an indirect manner. Unfortunately, the LPC equipment remains relatively expensive but there would appear to be a good potential to use the LPC to monitor both the generation of plugging and the effectiveness of treatment. For this reason, an LPC methodology has been included and selected wells should be subjected to this form of evaluation. At this time, the methodology would be restricted to the determination of the probability of plugging in which some form of dense deposition is involved.

The methodology involves taking a 60 ml water sample in either a clear 100 ml glass flat-bottomed tube or a BART™ 70 ml outer tube. When filled, the tube is capped and gently shaken for 20 seconds and immediately placed in the laser light pathway of the LPC. Counting is commenced immediately and, within two minutes, the data is collected (shaken). The sample tube is now allowed to stand at room temperature for thirty minutes without being disturbed and then the same testing is repeated and a second set of data (settled) collected. In the thirty minutes between the LPC readings for the shaken and the settled sample, the denser particles have already fallen out of the laser light pathway. For the second reading, only the particles at the same density as the water will be recorded and this does not include the ones that have settled. The difference is the denser particles that have settled out and it is these particles that are more likely to be associated with a plugging event in the well.

The procedure for the interpretation of data generated by these tests (shaken and settled) is given below. Of the data generated, the following are of critical importance to this evaluation:

TSS (ppm)	-	TSS
Mean Particle Size (microns)	-	MPS
% Volume of Particles less than 4 microns	-	PL4
% Volume of Particles greater than 4 microns	-	PG4

These data values are repeated for the shaken and the settled and are shown by the additional initial K or T respectively so that the total data base includes 8 items, four from the shaken (TSSK, MPSK, PL4K and PG4K); and four from the settled (TSST, MPST, PL4T and PG4T).

The interpretation of the data involves a comparison of the shaken (K data) with the settled sample (T data). If there has been no settling then the K data series and the T data series should give the same results ($\pm$ 15%). This would mean that there was not a significant fraction of dense particles in the water and so no dense material from any plugging was recovered. If, however, the T values are down by >15% from the K values for the TSS and/or the MPS values, there is a probability that some of the particles are from a plugging zone which includes dense deposits. Confirmation of this can be achieved using the PL4 and PG4 data. Normally most of the dense particles are of a larger particle size than 4 microns. Where the volume of particles slips significantly between the K and the T test for the PG4 data, then this would confirm that dense particles were present in the water sample and that these probably arose from the dense zones in the plugging. The percentage that came from that source can be calculated using the formula:

$$PDP=((PG4K - PG4S) / (PL4K + PG4K) \times 100)$$

Here, the PDP is the percentage of dense particles in the water. This assumes that the only dense particles that settle out have a diameter of greater than 4 microns. The larger the PDP value above 15%, then the greater is the probability that a plugging event is occurring in that well from which the sample came.

DATA INTERPRETATION (BART™ water testers)

Using the data sheet, it is possible to automatically obtain an assessment of the possible population in cfu/ml as a measure of the aggressivity. The time lag data can be used as an alternate to the cfu/ml particularly where comparisons are being made. Other data generated by the BART™ tests relates to the reaction patterns being generated by observing the tests over time particularly after the first reaction has been observed and the time lag determined by that event. Chapter Four deals with the various common reaction pattern sequences (RPS) in more detail.

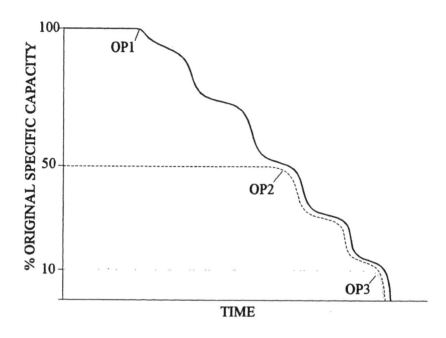

Figure Fifty-One, Theoretical graph depicting typical characteristics of a plugging water well over time (not specified, x axis) using production rate (as the percentage of the original specific capacity, y axis). Three wells are shown. The first (continuous line) well is functioning at the maximum specific capacity of the well. Here, the reduced production relates directly to the amount of plugging. The second well (dashed line) was operated continuously at 50% of the original specific capacity of the well while the third well (dotted line) was operated at only 10% of the original. Note that the first well declines in a slow harmonic manner and then plugs (OP1), the second maintains production at the designed level until the plugging impairs the specific capacity by >50% at which time there is a decline (OP2) while the third well plugs dramatically (OP3) once the specific capacity drops below 10% (the designated production level for the third well.

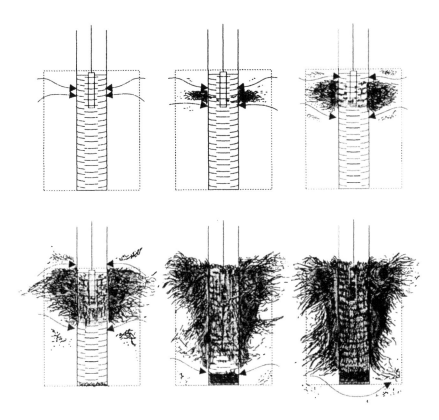

Figure Fifty-Two, Illustration of the movement of the focus sites of biofouling (shaded areas) down the vertical section of a gravel packed well with an irregular (not laminar) flow. Water flow pathways into the well are shown by arrows in each diagram. When the well is newly developed (upper left) there are a few natural focus sites of microbial activity. When the well begins producing (upper central), these growths focus along the water flow paths. These plugging growths cause the flow paths to become diverted around the plug (upper right). The plugs now expand diverting the water flows more severely (lower left) causing water to be picked up from different formations with different qualities (lower center). Virtually complete plugging of the well then occurs (lower right) with no significant water flow into the well but rather flows return to the normal ground water patterns..

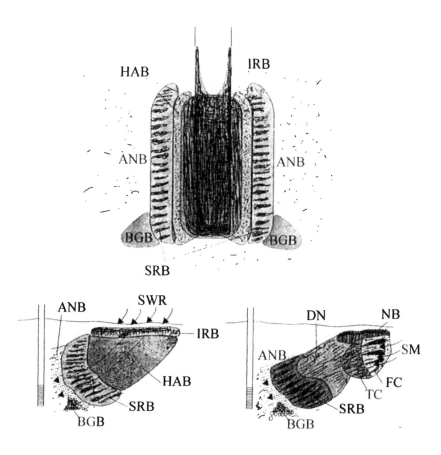

Figure Fifty-Three, Typical sites for the various bacterial groups commonly found associated with plugging around a water well. A vertical profile of a gravel packed water well is shown (upper) with the common location of the various bacterial groups including the iron-related bacteria (IRB), sulfate-reducing bacteria (SRB), heterotrophic aerobic bacteria (HAB), anaerobic bacteria (ANB), and the biogas producing bacteria (BGB). The impact of surface water recharge (SWR, lower left) and leakage of septic materials into the ground water (SM, lower right). In the latter case, fecal coliforms (FC), total coliforms (TC), nitrifying bacteria (NB) and denitrifying bacteria (DN) may also play roles in generating plugging.

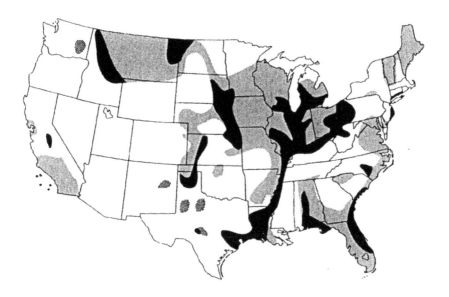

Figure Fifty-Four, Known regions of iron-related bacterial (IRB) plugging in the continental United States of America. The intensity of the shade reflects the severity and frequency with which IRB-induced plugging have been reported to occur in those regions. Unshaded areas on the map can be called the "Great False Negative Land" since IRB are ubiquitous in ground water systems and can "emerge" as components in plugging when the environmental conditions allow.

Figure Fifty-Five, Known regions of iron-related bacterial (IRB) plugging in Canada. The intensity of the shade reflects the severity and frequency with which IRB-induced plugging occurs in those regions. Unshaded areas on the map can be called the "Great False Negative Land" since IRB are ubiquitous in ground water systems and can "emerge" as components in plugging when the environmental condi-tions allow.

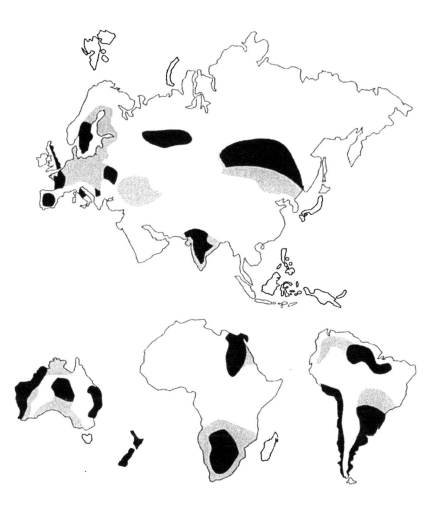

Figure Fifty-Six, Known regions of iron-related bacterial (IRB) plugging in the rest of the world. The intensity of the shade reflects the severity and frequency with which IRB-induced plugging occurs in those regions. Unshaded land areas on the map can be called the "Great False Negative Land" since IRB are ubiquitous in ground water systems. Euro-Asian continents are shown (upper) with the Australian (left), African (center) and South American (right) continents below.

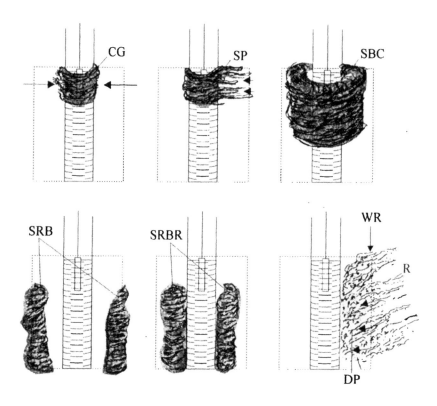

Figure Fifty-Seven, Diagram of the various locations of bacterial plugging that can occur around a water well. Typical IRB infestations occur tight on the well as columnar growths (CG, upper left) that may move out upstream in the direction of ground water flow (SP, upper center) or set back columnar structures from the well on the oxidative side of the redox front (SB, upper right). On the other side of the redox front where it is reductive, the SRB dominate (SRB, lower left) and if the well is not producing, the SRB can move into the well (SRBR, lower center). Local surface water recharges (WR) such as from a river (R) can cause dispersive plugging on a "bacterial front" that can move towards the well (DP, lower right).

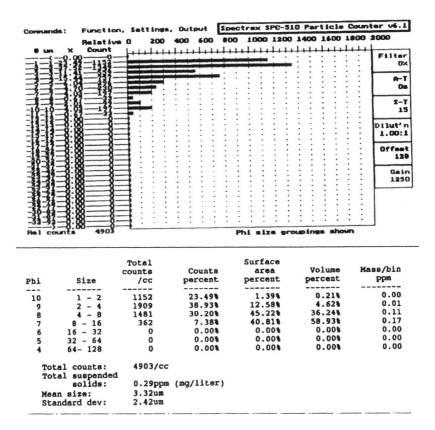

Phi	Size	Total counts /cc	Counts percent	Surface area percent	Volume percent	Mass/bin ppm
10	1 - 2	1152	23.49%	1.39%	0.21%	0.00
9	2 - 4	1909	38.93%	12.58%	4.62%	0.01
8	4 - 8	1481	30.20%	45.22%	36.24%	0.11
7	8 - 16	362	7.38%	40.81%	58.93%	0.17
6	16 - 32	0	0.00%	0.00%	0.00%	0.00
5	32 - 64	0	0.00%	0.00%	0.00%	0.00
4	64- 128	0	0.00%	0.00%	0.00%	0.00

Total counts: 4903/cc
Total suspended
    solids: 0.29ppm (mg/liter)
Mean size: 3.32um
Standard dev: 2.42um

Figure Fifty-Eight, The laser particle counter (LPC) provides an "eye" to look at the biocolloidal structures in water. Measurement is of the diameter and numbers of these particles and can create a calculation of the total suspended solids (TSS). Detection using laser light pulses is commonly achievable in the 0.4 to 120 micron range and so most bacteria and associable biocolloids are enumerable. This is done in the example by histogram (upper) and the data presented in summary tabular form (lower). It is convenient to use the LPC to determine the likelihood of bacteria in a water sample. If no TSS is detected then it can be surmised that, at least in the laser light pathway, there were no bacteria present.

7

BIOFOULING EVENTS IN WELLS, PRACTICAL EXAMPLES

INTRODUCTION

Not everything is learned by the simple rote of reading and understanding "dry" texts that in clinical terms spell out the present state of the art. It has to be remembered that the current passing through the state of the art is always changing that base of knowledge. This, however, does not necessarily apply to the concepts that always observe the basic "Keep It Simple Stupid" rule that it is the simplest form that becomes the most reliable and yet, often, the most sophisticated.

In honor of the flux in the pool of knowledge, the first section of this chapter will be devoted to a series of poems written over a number of years written by me for the *Biofilm Bulletin*. This was a little known, but valuable, anecdotal account of various forms of biofouling as they relate to ground water systems. The bulletin arose out of the exchange of minds and understandings that took place at the 1986 IPSCO Think Tank on Biofilms and Biofouling, University of Regina, Canada. The editor was Dr. Keith Cooksey whose enthusiasm allowed the bulletin to flourish for more than three years. It was found that poetry can sometimes be used to get concepts across very clearly to the reader. For this reason a number of poems are included and the importance of many of the underlying concepts can be more easily understood.

There are six poems in this series. The first one is "Mummy, mummy. The well's run dry!!" The bottom line so often is that water has a zero value until suddenly it is no longer available and then suddenly water rises to the value of gold and beyond, for we can live without gold but we cannot live without water. The child panics and calls to mother who in turn will get father to fix the problem. In the second poem ("A Wandering Plasmid") a cursory look is made of the seemingly intense researches underway on the manipulation of genetic material. As an applied microbial ecologist, I have observed that it is fairly evident that evolution wove into the fabric of life tremendous ability to genetically adjust to changing conditions. This is particularly evident in the Prokaryotes (the microbes). "Suzie in the Slime with Violence" is a poem relating to the research from 1987 to 1993 on the recovery of high sand content golf greens from a lateral black plug layer. This biofouling was controlled and the technology remains in dormancy awaiting the golf industry to realize the extent of biofouling problems which do affect playability!

The last four poems are more theoretical but do give an insight into the mechanisms of biofouling. "Ticking to the Polymeric Clock" covers the probability that like all other organisms, the consortia involved in biofouling do have a pulse-like form of growth. In the fifth poems "Contact" and "Scrubber's Delight", the emphasis is on the nature of the consortium forming a biofilm within a plugging. The last poem "Bella of the Clog" deals with the microbes likely to be found in a plugging event. Each poem will be presented by verse, and the relevance to the biofouling will then be discussed after each poem.

This was published in volume 2(3) of the *Biofilm Bulletin* and includes many aspects of biofouling. Verse one covers the form of many plugs in ground water systems. They are often very large (hence "great plug" and will usually locate at very logical sites (e.g., redox front in a flowing water path). The next lines (2 and 3) deal with the fact that these plugs are not obvious and, therefore, are not a direct concern to the well users.

MUMMY, MUMMY, THE WELL'S RUN DRY!!

Beware that great plug by logic driven
Obscene machine biologic driven
skulking hiding and nowhere seen
a shining sparkling metallic sheen
lusting away for the perfect taste
of vacant pristine interstitial space!
Such slime castles exist pulsing underground
unseen, unknown and making no sound.
They quietly spread an awesome dread
quietly stopping flowing waters quite dead.
Mummy, Mummy, the well's run dry!
There, there my darling do not cry.
Daddy will come and use his gun
to blow away that nasty scum.
Daddy came and quickly took stock
He could not deliver enough of a shock
to that scum bug bent on strangling his well

So he called Big Mister B. C. H. & T.
and told him to come just as fast as could be.
With flashing lights, Mister Big began to detect
the site of that scum bug to next disinfect
He bathed the beast in steaming heat to corrupt
along with soapy acidic feasts to then disrupt.
Assailed on all sides by means so diverse
those slimy clog dogs did howl and disperse.
Mummy, Mummy the well's all clean,
it's the cleanest clean it's ever been!
Yes, thanks to Daddy Darling drink it we can,
So now let's thank Mr. Slimebuster Man.

It would be an "ideal" if there was a technique to point a "camera" down around a well and get a three dimensional image of the size of the "obscene machine" because many resemble the scale of Godzilla in size but they are "nowhere seem." The next three lines cover the fact that the biofilms incorporate a high percentage of metals commonly dominated by iron. This iron can reach 35% (dry weight) in a matured plug. The plug is always growing and looking for that "pristine interstitial space" into which to occupy the vacant voids and cause a plugging. In verse one, the last four lines deal with the nature of the plugging. These plugs now go through a "pulsing event" in a routine manner (expansion, sloughing, compression) which happens without any obvious effect on the ground water flow to the casual observer. This pulsing can be observed by determining daily production using a standard drawdown to trigger pumping. Eventually the plugging occupies so much of the void space within the infested (plugged zone) around the well that conductivity is severely reduced. This may apply to a wide variety of wells including oil and gas wells as well as water, injection and relief wells.

The second verse relates to the panic expressed by a child to the mother because there is no longer any water coming from the well. Mother reassures the child that the father will fix the problem. Father would use his gun to blow away that nasty scum. Violent compressive forces can cause plugging to shear away so that the hydraulic pathways (conductivity) can at least partially be recovered and hence flows. One favorite example of this is at a Canadian forces base where some seasonal water wells were successfully opened up every spring by firing a couples of rounds of .303 directly down the well. This particular practice can have serious long-term impacts on the integrity of the well!

Often, it is not realized that the scale of a plugging event can extend way beyond the well itself into the surrounding porous media. Verse three deals with the patented blended chemical heat treatment (BCHT™, ARCC

Inc., Florida) method for the rehabilitation of water wells. It is important to look at a plugging event as being treatable by a phased approach rather than a single "magic" bullet that will cure all. The BCHT process, like other successful methods, employs a sequence of treatments: (1) shock; (2) disrupt; and (3) disperse. This is, perhaps, similar to the techniques the police would use to control a mob that is out of control. Three elements in this process are the application of heat (chemical reactions occur faster at high temperatures), surfactants (to destroy the polymeric web-like structures which bind the plugs together) and radical pH shifts (in this case, the plugged site is made acidic to destroy the mineralized structures involved within the plug. Dispersion is probably one of the greatest challenges and air surging commonly forms a simple method. It is important to remove "those slime clog dogs" from the well as completely as possible. In practice, it is impossible to remove absolutely all of the plugging microbes from an impacted site. Re-infestation of the well is as inevitable as the sun rising except that the dawn may take longer to come if preventative maintenance is practiced as a normal routine.

In the last verse, the lines are self-explanatory but one of the major features in a successful rehabilitation is that, in addition to the recovery of water flow, the water quality would also improve. This is because the surfaces and voids within the cleaned areas in and around the well are once again being colonized by various biofilms. At this time, these biofilms will act as bio-filters removing nutrients and selected metallic cations from ground water. Concurrently, the microorganisms forming these growths will remain attached so that the product water passing through the well would have low population loadings. It becomes the "cleanest clean it has ever been!"

A WANDERING PLASMID

A wandering plasmid I -
> A thing that induces in snatches,
> In rich nucleic patches,

And Slimy lullaby!
My Molecules are long,
> Through every fashion ranging,
> And into your genes I'm changing

To bloom my clones so strong!

First my apologies to Gilbert & Sullivan for modifying "A Wandering Minstrel, I." from the light opera "The Mikado." This poem was published in Volume 3(1) of the *Biofilm Bulletin*. It deals with the molecular level of biofouling and plugging. Essentially, the plugging can be perceived to have three elements: (1) living microbial cells that commonly form less than

0.1% of the volume; (2) a web-like mass of clustered (bound) water held in a polymeric (EPS) structure; and (3) the mineralized components often dominated by iron, oxides, carbonates and hydroxides. The long molecules are the polymeric EPS that act as protection for the living cells inside. Reality is that changes in the genetic capability created by induction, adaptation and mutation affects the nature of the plugging as a single event. Dominant species or strains within the biofilm reflect the adaptation to that environment. This includes the development of biodegradative and accumulative functions within, an around, the viable cells. In the end it is the clones adapted to, and dominating, that environment that "bloom so strong."

SUZIE IN THE SLIME WITH VIOLENCE

Suzie in the Slime with Violence
 Picture yourself in a slime on a sliver
 with Polkadot ooze and oxidized ions.
 Everybody smiles as they slide down so slowly
 into a world of intoxicated size.

Suzie in the Slime with Violence
 Metchnikoff laughs as he slips out the sliver
 smiles at the ooze with disbelieving eyes.
 how could he know that they grew oh so slowly,
 a quadrifed swirling that high!

Suzie in the Slime with Violence
 Excuse me, but Suzie was watching a black slime
 creeping about while caring to find why
 as it grew and then grew to cancel out time
 on the way to that giant glycocalyx eclipsing the sky.
Suzie in the Slime with Violence

This time my apologies go to Lennon and MacCartney for modifying the words to "Lucy in the Sky with Diamonds" which appeared in Volume 3(2) of the *Biofilm Bulletin*. Each of the three verses deal with a different aspect of biofouling but the theme relates to the development of a management practice to control black plug layer (BPL) in high sand content golf greens. These layers were lateral and competed with the turfgrass causing radical die-back where the BPL dominated.

Verse one deals with the fact that a fully matured plug could be very small. Hence, it is the size of a sliver. "Polkadot ooze" relates to the heterogeneous nature of the BPL. Like most plugs, there are variations in the density of the various components so that passageways are created along with zones where the metallic oxidized cations are concentrated. When moving down into this microscopic universe, the variations and size becomes expanded ("into a world of intoxicated size").

Verse two is a parallel event. Metchnikoff was a microbiologist who believed that the body had a natural immune defense generated by specialized cells within the body (e.g., white blood cells). To demonstrate this, Metchnikoff slipped slivers of wood under his finger tips and observed microscopically the subsequent infection (the "ooze"). Plugging has been believed by many to be circumstantial chemical reactions not relatable to a microbial event. It is not easy to stick a sliver down a well but it is easy to mimic a well in a laboratory as a mesocosm. In those experiments, the role of ground water microbes in the formation of plugging events became self-evident. The "quadrifed" refers to the four main colors for plugging and slime events. "Quadri" denotes four while "fedic" means a surplus of something, in this case that would be occlusive plugging. There are four types of plugging simply split by color into: White (HAB or SLYM) Yellow (SLYM or HAB), Brown (IRB) and Black (SRB).

The final verse is a more lyrical summation of the fact that the rate of growth cannot be readily predicted since there are so many prevailing factors which can alter the aggressivity of the BPL consortium on its way to become the "giant glycocalyx eclipsing the sky." It is only in the last few years that scientists are now coming to grips with the scale of the biomass on this planet that we call Earth. Since the dawn of life, the amount of organic matter generated is now thought to have exceeded the weight of the planet by four to five times. Also, the weight of the microbial cells on this planet equals that of the plants without even taking the "glycocalyx or EPS or slime" into account. If the gross biomass is considered, then the microbial mass outweighs the plant biomass by possibly one or more orders of magnitude.

This next poem, Ticking to the Polymeric Clock, appeared in Volume 3(2) of the *Biofilm Bulletin*. It relates more to the dynamics of development in science rather than to biofouling specifically. We have lost (my personal opinion) much of the freedom to explore in science and technology because of the need for recognition by peers. As a result, new advances can almost be viewed as "fresh" carrion on which the "succeeding" scientist can feed and increase their credibility. Discoveries are often made by the loner seeking to explore using new and revised concepts not "acceptable" by the majority. Remember that science only exists where the errors are consistently constant and, as a consequence, the experiments become repeatable. We need to be aware of those constant errors and investigate them diligently in order to push forward the borders.

The poem "Contact" refers to the early colonization of surfaces where the environment has changed to encourage microbial growth. This appeared in Volume 2(2) of the *Biofilm Bulletin*. One of the most fascinating parts of microbial ecology is the vast mass of "sleeping" microbes out there in the most unexpected sites such as sedimentary rocks, granite, muds and desert sands. These survivors are really ultramicrobacteria (small, dormant, electrically neutral survivors) and are commonly referred to as UMB. In the poem they are called "silent microcells." They are attracted to the "surface smell" of the pristine surface that is really the local environmental conditions. For example, the installation of a well would radically change the surrounding environment to provide favorable "surface smells."

TICKING TO THE POLYMERIC CLOCK
Ticking to the polymeric clock
Sticking to the surface stock
Fusion to melt down tocked on goo
Now here's a problem on which to chew

So diverse the iron rich slime
Always pervasive and always on time
to cause a mean and fast incursion
by means of a subtle and twisted diversion

Publish my perish and perish with publish
Readers will awe, and copy to cherish
They gain with such and ego stride
and wait in recognition not denied

Flocks of readers wait by the knowledge gate
ready to swoop, scoop and through reading rate
your thoughts and experiences as to their needs
to publish so fast that you perish with such speed
that the culture vultures are too fat now to fly
so they leave the pickings over facts quietly to die.

CONTACT
So clean and pristine the surface shore
Waits inviting attachments score
Soon a silent microcell
Stops attracted to that surface smell

Whispery polymers spiralling out
Smacks the surface with many a clout
Reborn and happy, a cell anew

Hunts for nutrients on which to chew
Bouncing and tumbling through reproduction
in a polymeric mattress, site of seduction
Enticed to come in from out of the blue
Innocents arrive into that glistening hue

Now reborn was that vile film
Sometimes known as a biofilm
It swells and pulses to its favourite tune
That one we'll all know pretty soon...

Slime and Sluff
Slime and Sluff
Life gets tough
when you Slime and Sluff

In the second verse, the "spiraling polymers" refers to the fact that the initial anchoring of a microbial cell is done by the cell throwing out strand: of polymers which stick the cell to the surface. Once anchored, the cell nov recovers its former size as a vegetative cell and begins to metabolize and grow. In the third verse, the biofilm is referred to as a "polymeric mattress" and it here that the consortium forms into an efficient community maximiz ing the biodegradative, bioaccumulative and colonization functions. Passing microbes may now either join the consortium or be cannibalized by it. The fourth verse refers to the harmonic nature of the biofilms within the plugged zone as they now grow and mature towards an iron-rich plug. "Slime" refers to the phase of stable growth while the "slough" refers to the intermitten periods of destabilization when the biofilms shear away to be carried on the waters flow.

Well, this next poem was published in Volume 1(2) of the *Biofilm Bulletin* and was based upon a story line using some of the scientists who attended the 1996 IPSCO Think Tank (ITT). The first three lines recognize some of the genera of bacteria commonly found in biofouling wells while the rest deals with the (fanciful) activities of some of the ITT scientists to whom an apology is again extended!

BELLA OF THE CLOG
Now the Bella of the ball is *Gallionella*, so thin and so tall
But trickster *Crenohrix*er's building slime blocks into a wall
While *Pseudomonas* knew so much that things aren't what they seem
was quietly paddling along in the underground big water stream.
Here scientists look down through their clever disguises
and uttered out loud in most raptured surprises.
Cooksey yelled "I wanna looksee"

and D.C. pondered "what could them be?"
while Gaffney muttered "don't ask me,
but this is one heck of a mystery!"

Then the clatter of chopper blades droned into sight
to herald Bill Costerton with his sight on the fight
to recover those beasties, no matter how small
with his spool he would catch them, not one but them all!
A slimed up with rust pick-up roared up at that time,
out jumped Alford the George roaring "that's my slime, mine!"
A crowd gathered round to postulate and profound
When up crept Vlad who dug down with no sound.
A hole he dug deep down via the well casing to go
to microbial depths that no one did know.
The slime did not yield up the rich raw virgin field
but did gradually slowly become silently sealed.

SCRUBBER'S DELIGHT
It's not a crime to clean up slime
It has to be done soon this time
To keep the gleam and lasting sheen
of a sparkling, dancing surface clean
To bring a beguiling smile to each cynic
And make the place as clean as a clinic.

This last poem relates to the scrubbing (in the symbolic sense) of a plugged well in order to rehabilitate and recover production. There has been considerable opposition to well rehabilitation driven by the twin prongs of the economic incentive for the driller of installing a new well, and the lack of full appreciation for the complex nature of the plugging processes and the methods of control. It is surprising that, just in the Canadian prairies, there are 200,000 operating water wells with a capital investment of at least C$1,000,000,000 and yet no policy for making these wells sustainable. The poem, Scribber's Delight, appeared in Volume 2(2) of the *Biofilm Bulletin* and speaks for itself.

THE DETERMINATION OF BIOFOULING

It remains important to determine when, how and where, the BART tests should be used to determine the degree of biofouling, nuisance activities or hygiene risk. To allow this the techniques, application and interpretations to be constructed to maximize the information gained, a series of examples of the use of the BART testers in various applications are listed below.

These examples range from the plugging, biofouling and remediation of ground waters, through to application in surface waters, soils, oil and gas wells, and various potential applications in the food industry.

Ground Water

Ground water by its very nature passes through porous media on the way to reaching the surface waters, becoming locked within deeper aquifers, or returned to the surface as steam. Most of the ground water of interest is in the shallower levels in which the waters return to surface water flows through wells, springs or other forms of discharge. Since the surface environment is oxidative (oxygen dominated) while the ground water systems are oxygen starved (reductive), there is a tendency for microbial biofouling to form at the redox front where the ground water is shifting from a reductive to an oxidative potential. This front often forms the focus for microbial activity.

Within and around the redox front, the form of microbial activities can be varied but inevitably that growth is going to be on surfaces and in the water. Growth on surfaces in the porous media through which the ground water flows will be occupying voids. This void occupancy will reduce the hydraulic conductivity of the infested (biofouled) formation. This will result in a number of impacts depending upon the form and type of attachment and growth which has occurred to the surfaces. These impacts will affect the manner in which bacteria will be present both on the surfaces and within the flowing or stagnant water. Since the BART testers are sensitive only to the bacteria present in the water sample that has been taken, it is important that the water sample be taken in a manner that would reflect the form of microbial activity in and around the site of biofouling.

Before beginning to test for the forms of microbial biofouling, it is important to have a reasonable understanding of where the bacteria are likely to be growing. This will most commonly be on surfaces and plugging voids so that the necessary measures can be implemented to get detached microbes from these sites into the water samples that would then be examined.

One common feature in a biofouling event in ground water is that the microorganisms infesting the sites will have adapted to the environmental conditions present in, and around, the redox front. These conditions would reflect the sequences of directional turbulent water movement (due to pumping) and the more laminar and slower water movement which would occur naturally within the formation when there was no demand created by pumping. Under these conditions, there would be "pulses" of chemicals flowing through the site with each period of pumping. These chemicals could be categorized into three broad groups: nutritional, accumulative and neutral. Nutritional chemicals would be dominated commonly by organics that could be degraded along with, particularly, inorganic forms of phosphorus and nitrogen.

Accumulative chemicals are those chemicals which become accumulated within the growths of the biofouling organisms but would not be utilized in bulk within the microbial cells within the growths. Examples of such accumulates would be non-degradable (recalcitrant) organics, heavy metals commonly dominated by various forms of iron and manganese, and various forms of inorganic carbon dominated by carbonates. Commonly at the redox front, where the most concentrated forms of biofouling can occur, the bulk accumulate is often iron and/or carbonates. These can often form the bulk of the mass being measurable in the range from 5 to 40% of the dried weight.

Neutral chemicals are those chemicals that are not taken up into the biofouling mass but simply pass by. These include many of the anionic ions (such as chlorides and sulfates) and mono-valent cations such as sodium.

These events mean that some chemicals will pass straight through a zone of biofouling (i.e., neutral) while others (nutritive or accumulative) can be removed from the ground water by biofouling microorganisms. Here, these chemicals may be utilized for growth and metabolism (nutritive) or simply continue to accumulate within the biofouling mass.

One other chemical that plays a dominant role in biofouling and is often ignored is water. Water is well documented as existing in three common forms (ice, liquid water and steam). Little attention is however paid to the various forms that liquid water can take and these forms may be very relevant to the type of biofouling and the methods of recognition. The general premise is that liquid water is simply that, a volume of unclustered water molecules that move around in a random manner as essentially "free" liquid water. Water within a biofilm is tightly clustered either by being within the microbial cells or enmeshed in the polymeric "web" (EPS) which forms the bulk of the growing biomass.

Today, the nature of water is becoming more clearly understood in that the bulk of the liquid water is clustered even when it may appear to be "free liquid and unclustered." The polymers generated by microbial growth are ubiquitous in the waters where the growth has occurred. The essential nature of the clustering of the water is that it shifts to a colloidal form, some of which may be filterable (i.e., sub-micron dimensions). At this time, water that has been filtered through a 0.22 or a 0.45micron filter is thought to contain only dissolved chemicals. If there are filterable masses of colloidal material passing through these filters, then the dissolved chemical content may be overestimated.

In essence, water can be viewed as a very important basic substrate that can become a "limiting" nutrient under some circumstances such as a water-starved oil or gas well. Normally, water is never considered to be a limiting substrate to growth although the acquisition of tightly clustered water by a biofouling mass can be a reason for plugging and loss in flow through a porous medium.

By following the movement of water into a zone of biofouling, some of the factors leading to plugging can be better appreciated. Ground water approaching a zone of biofouling would have a low level of clustering since the level of microbial activity would be low and there would, consequently, be relatively little dispersed polymeric material in the water from microbial activity. Within the zone of biofouling, there would be a gradually increasing amount of polymers associated with the biofilms. These would tend to attract the water. First, the water would move into the various conduits (passageways) into the biofilm. Once inside the biofilms, some of the water could become more tightly clustered and held within the "web-like" structures formed by the polymers within the biofilm. Some of this water may now become clustered more tightly and move close to, and into, the microbial cells inhabiting the biofilms. Water that did not become entrapped would move out of the biofilms back into the free-flowing water. Some water would, however, become entrapped in the biofilms (by clustering). A fraction of this could also be released back into the flow within polymeric (colloidal) structures along, possibly with some of the incumbent microbial cells from within the biofilms. The water now continues beyond the zones of biofouling but would have been changed through the events described above. These may be summarized as:

- Losses in the chemicals that could be nutrients for the biofouling microorganisms or be accumulated within the biofilms
- Losses in flow due to the plugging of voids within the porous media by biofilms and other accumulating materials
- Greater level of clustering within the water due to additional inputs of polymeric materials as the water passed around, and through, the plugging zone

Variable contents in microorganisms, chemicals and amount of total suspended substances (TSS) may occur depending upon the state of the various biofilms forming the plugging zone. It may be expected that the content loading may be lower when the biofilms are stable, and higher when the biofilms are unstable and, as a consequence, sloughing.

In traditional practice, little attention has been paid to the potential impact of a biofouling on the chemical composition of the product water (that has passed through that zone of biofouling and been affected by that interaction). The product water may be very atypical of the water entering the zone of biofouling and so cannot be easily used to determine the risk of biofouling (e.g., plugging). In order to compensate for these potentially serious anomalies, a sampling practice has to recognize this potential impact. To do this, the options range from taking a sequential series of samples over a time period that will allow a reasonable "picture" of the status of the biofouling (and its effect on the product water characteristics) to be obtained.

In the various samples listed below, attention will be paid to the sampling procedure since this step is crucial to the interpretation of the data gathered by various procedures. In all of the cases described, there is a fundamental premise that the microbial activity and impacts are clustered within community (consortial) growths that are dominated by various biofilms.

Water Well Plugging

Water wells often appear, with a cursory examination, to be nothing more than a hole in the ground from which water can be drawn. There is no easy way to view the often dense and widespread massive growths that are plugging up the well where this is indeed happening. Sometimes, these growths may be away from the well's screen back in the porous media and not visible when camera-logging the well. If these plugs could only be extracted and shown at the true size, then perhaps the operators of such installations would appreciate the challenges of water well rehabilitation. Often the growths, encrustations and tubercles observed inside the well and on the pumps and pipes are only just the "outer edge" of this much larger biofouling event which can extend for many feet back into the formation and be longer in length than the well screen itself.

There are three simple steps that should be used in the determination of biofouling. These are:

- Stress the biofouling so that the incumbent microorganisms, accumulates and polymers can be released into the water that will be sampled.
- Employ a sequential sampling technique that will bring water through the biofouling zones so that the size and form of the biofouling can be predicted.
- Determine the significant parameters thought to be relevant to the biofouling (e.g., plugging, corrosion, taste and odor).

There are therefore a number of decisions which need to be made concerning the methodology to be employed to recover a sequence of samples that would allow the size and the form of the biofouling in and around the water well to be determined. Consideration has to be given to (1) Stress, (2) Sample, and (3) Determine. Each of these steps is, to some extent, site specific. In other words, applying a standard protocol may not be applicable to every site.

Conversations with the well operators, examination of the performance records and a direct evaluation of the well site itself can go a long way to determine the form of stress to be applied. Usually the stress must be some sufficient divergence from the normal functioning of the well to cause the environment within which the biofouling is occurring, to change. This change should be enough to cause at least some part of the biofouling microorganisms to want to relocate (in other words, pack up their bags and leave). This act of relocation involves the organisms moving out of the attached biofilms (slimes, encrustations, tubercles etc.) and entering the flowing water. Once these organisms have entered the water, they can be recovered by pumping and sampling and subsequently detected through culturing.

Stress means that the environmental conditions in and around the well have to be changed. This can be done in many cases by simply turning off the pump and letting the well sit idle for at least twice as long as it would normally do. Ideally in these circumstances, the well should be shut down for seven days to really traumatize the biofouling microorganisms.

Sloughing from the slimes (biofilms, etc) occurs reasonably quickly as the conditions become more reductive. A net result is that the conditions become more reductive causing some of the microbes to slough off along with the colloidal slimes in which they have been growing. Normally the time factor for this sloughing can be measured in minutes and hours once the reductive conditions have been developed. Nominally, four hours of shut down would be an absolute minimum with twenty four hours more acceptable with the ideal being seven days. Naturally, many water well operators would be very annoyed by the thought of a one-week shut down and unhappy about even a one-day closure. Generally, reality dictates shut downs of at least eight hours if an observable trauma (releases of biofouling microorganisms) is to be achieved.

Sequential sampling of the water pumped out of the well after the shut down period has ended is used to determine the form and position of the biofouling zones in and around the well. High populations recovered immediately at the start of pumping have clearly originated close to the pump in side the well and adjacent well screen.

EXAMPLES, WELL PLUGGING

A variety of examples are presented of the different ways in which the form and management of plugging in water wells can be performed. The manner selected to do this is to provide examples from various studies of particular events that are important in the management of plugging and biofouling problems in water wells. Each example references a particular location. All examples using that location will be addressed at the same time.

Location: City of North Battleford, Saskatchewan, Canada

The City of North Battleford is in the western central region of Saskatchewan sited on the North Saskatchewan River. It has historically used a blend of ground water from wells sited along the river and surface waters from the river itself. The ground water offers the distinct advantage of lower treatment costs than the surface water. The location of the string of wells is such that much of the water comes through river recharge. The relatively high nutrient loadings associated with this recharge has led to relatively short life expectancies because plugging significantly reduces the specific capacity and yield. New wells have been installed periodically in a numeric sequence with the last ones being #16 and #17. Earlier wells have either been abandoned or are on stand by status. These observations are limited to wells #9, #15, #16 and #17 and cover relevant aspects of a joint Agri-Food Innovation Fund Program, Canada Agriculture (PFRA-TS) and Droycon Bioconcepts Inc.

The main research topic was related to the improvement of the specific capacities of the wells so that they could generate a greater (and more economical) part of the water delivered to the city. This project employed the ultra-acid-base (UAB™) treatment scenario to remove the plugging. This treatment uses a combination of heat and pH shifting along with the CB4 (ARCC Inc., Florida) to remove the plugging material and return the well closer to the original specific capacity.

Examples from the Battleford studies that took place from 1996 to 1998 include the diagnosis of the plugging using the BART water testers and the application of the UAB treatment on wells #15, and #16. Also evaluated were the potential long-term impacts of this treatment on the sustainability of the wells, and the economic aspects of maintaining the productivity of the wells.

City of North Battleford Ground Water Wells 15 and 16

Diagnosis of the Plugging Using the BART Water Testers. Wells 15 and 16 were selected as the test sites for the Ultra Acid Base (UAB) treatment process. This was done because of the well-documented treatment history that existed concerning the well field. In this field, the water bearing formation is only 18 meters (60 ft) below the surface. This close proximity of this water bearing formation to the surface allowed an economical extraction of the ground water. Also for the testing phase, the sand core samples (from the water bearing formation) were convenient. To do this, bore holes were drilled 3 and 6 meters from each well for detailed biological testing.

The BART water testers were used to evaluate the biological activity in the water bearing formation in the area around the screen and gravel pack. Water samples were collected during a standard two-hour pump test using sterile 250ml plastic containers. For sampling, the pump was activated at 4.1 Igpm/min with 500ml water samples being collected at increments of 1, 3, 5, 10, 15, 30, 60, 90 & 120 minutes. Prior to the start of any of these pump tests, the wells were shut down a minimum of 8 to12 hours (to cause sloughing of bacteria). This shutdown time was found to be a necessary step in order to allow enough time for the biofouling bacteria attached inside the biofilms to begin to slough into the water and allow their presence to be determined. This shut down therefore allows for a more accurate indication of the variations in, and aggressivity of, the various biological communities close to, and far away from the well.

This sloughing/detachment of bacteria from the biofilms on the surfaces will cause the water moving towards the well at the start of the pump test to have a higher bacterial population (because of the shearing and sloughing). The delay in the time for these (sloughing) bacteria to reach the well will give an indication of how far these bacteria traveled to reach the well. Using the changing time lag on samples taken during pumping, an idea can be gained as to where the more aggressive bacterial populations are around a well. When this slug of sloughing bacteria reach the well it would be easily recognized by a shortening in the time lag for the BART test which have reacted and, possibly, a shift in the reaction codes observed. Thus by knowing at what pumping time the elevated biological samples were observed, and knowing the pumping rate and porosity of the water bearing material, the distance to the various biological plugging events can be determined.

UAB Treatment of Well 15. This well was first subjected to the UAB treatment process in October 1997. This treatment caused a significant reduction in the biological activity (aggressivity). BART testing was conducted again in April 1998 and both the IRB and DN bacteria remained suppressed to a highly aggressive level from the extremely aggressive state observed prior to treatment. The SRB were also reduced by the treatment with the time lags being extended by one day longer (and therefore less aggressive) than before the treatment. The HAB had a low aggressivity immediately after treatment but by the following April, the time lag shortened to indicate a medium aggressivity. This may have been because the HAB further out in the formation may have moved in towards the well dominating the utilization of the organic debris left after the UAB treatment.

Biological results obtained from a pump test conducted 10 months after the treatment (August 1998), the HAB and SRB were at a medium aggressivity while the IRB, SLYM, and DN were all still highly aggressive. This gradual differential increase in the aggressivity of the nuisance bacteria means that a preventative maintenance program has to be put in place after a treatment. Because of the increasing aggressivity of the bacteria, there was a risk that the well would rapidly degenerate to the severely biofouled conditions that the well was in before the UAB treatment.

A second UAB treatment on well 15 was undertaken in August 1998 (10 months after the first treatment) to again control the biofouling. At this time, core samples were taken at distances of 3 and 6 meters from the well. The IRB were found to be clustered within 3 meters of the well while the SRB were recovered in the most aggressive state at the 6 meters of the well. While the IRB and SRB were clustered, the HAB were found to be highly aggressive all the way out from the well screen to 6 meters out. This study placed the IRB closer to the well than the

SRB while the HAB were found to be ubiquitous in both of these clustered communities.

UAB Treatment of Well 16. Based on the pump test water samples, this well was found to be highly aggressive with IRB, HAB and SLYM bacteria and medium aggressive with SRB. The cores showed that IRB, SRB, HAB and SLYM are all highly aggressive at 3 meters from the screen. At 6 meters from the screen, the HAB and IRB were only medium aggressive while the SRB was at a low aggressivity. The SLYM was highly aggressive at both the 3 m and 6 m cores. From this data, the well can be categorized as severely biofouled with the SLYM bacteria dominating particularly at the 6-meter distance from the well.

A very effective preventative maintenance program is an essential post-treatment strategy. To begin the development of a suitable strategy for these wells, an ongoing monitoring program was put into place. The objective here is to treat the wells before the biofouling became sufficiently serious that it would permanently damage the production potential for the well. Biological testing of both the piezometers and production well is now being conducted at regular intervals every three months along with the specific capacity of the well.

Sampling Protocols for the Wells. Water samples were collected from production wells 15, 16 and 17 during a standard two hour pump test. This test involved samples being collected at 1, 3, 5, 10, 15, 30, 60, 90 & 120 minutes. Water samples were also collected at all six core sampling sites. Water samples from the core holes were extracted from the top, middle, and bottom third of the water bearing formation (over the depth of 18.3 to 24.4m). Water samples were also collected at well 17 to determine the extent of biological activity surrounding a well as a control in which there had not yet been any significant loss in specific capacity. Wells 16 and 17 were constructed at the same time and are "virtual twins" with well 16 being 20 meters from the river and well 17 set 200 meters from the river.

Well 15: This well had been previously (UAB) treated in October 1997 at which time a doubling in the specific capacity was obtained. The iron, manganese and sulfate levels in water samples collected in August 1998 (10 months after treatment) remained lower than prior to the treatment application in 1997. However, the levels have been observed to be slowly rising. From laboratory follow-up studies, it appears that the caustic soda used in the first UAB treatment may not have reached the designed effectiveness at controlling the biofouling. This laboratory testing revealed that the caustic soda caused a swelling of the clay. It was therefore decided to re-treat this well with a new, reformulated UAB treatment process excluding the use of caustic soda to create an alkaline shift.

Well 16: This well had the lowest iron, manganese, sulfates and total hardness levels but had the highest nitrate and phosphorus levels. These high nitrate and phosphorus in well 16 could be expected due to its close (20m) proximity to the North Saskatchewan River.

Well 17: Unlike well 16, well 17 installed at the same time three years ago has not had any significant loss in specific capacity. However, the water chemistry showed high levels of manganese and total hardness that were approaching the levels observed in well 15 before the initial treatment in October 1997. Total nitrogen levels from well 17 were the highest of all the wells tested with 0.54 mgN/l (ppm). While the nitrogen level in well 17 was the highest, the total phosphorus level was the lowest at 0.03 mgP/l. This distorted N:P ratio of 18:1 and the low concentration of phosphorus may be one of the critical factors, which has restricted the biofouling in well 17. Particular care should be taken to monitor the ongoing shifts in the N:P ratio. Rising phosphorus levels could indicate increasing levels of bacterial aggressivity and declining specific capacities.

Application of the UAB treatment on Wells 15, and 16. The UAB treatment process uses a combination of chemicals and hot water to remove the plugging biofilms that are in the water bearing formations and gravel pack around the well. Heat (hot water) is used to facilitate a more effective destruction of the biofilms that would then allow the various treatment chemicals to more completely penetrate the regions of severe plugging surrounding the well intake area.

In the UAB treatment process, the water in, and around, the well is maintained during the treatment process at a temperature of at least 65°C (150°F). One of the major advantages of using relatively high treatment temperatures is that there are dramatic increases in the rates at which chemicals react. This advantage can be used to either reduce the amount of chemicals needed to achieve rehabilitation or extend the treated zone. The selected wetting agent (e.g., the surfactant CB-4) is applied to facilitate the penetration of heat and chemicals into the biofilms forming the plugging. This combination of heat and chemistry is used to break up the biofilms and keep the disrupted plugging materials in suspension so that they can be more easily dispersed.

Sequences in the Selected UAB Treatment Process. There are three phases in the process. The first phase involves a screen cleanup and shock phase using a suitable acid (e.g., hydrochloric acid, 1 to 15%) and a wetting agent (e.g., B-4, 1%) solution heated to 80°C (180°F). This phase is applied to remove screen incrustations and open up the screen slots and the voids in the adjacent gravel pack that has been either plugged or restricted with biological slimes. This first step opens up more pathways (doors) into the main plugging regions in, and around, the well ready for the main treatment process. To further encourage disruption particularly of plugging on the well screen, a wire or chemically resistant brush (chimney sweep) can be moved up and down the screen surface to assist in opening up the slot area.

The second phase is intended to complete the disruption of the impacted plugging zone. This is done by doing a "pH flip-flop" taking the pH up from 2.5 to 10 in the plugged formation in and around the well. Applying a pH shift of seven units over such a short period of time can cause severe disruption of the plugging formation and is lethal to many of the bacteria. This is accomplished in two steps using a suitable acid solution in the first step, and a suitable base in the second step. Both steps involve the use of a wetting agent, hot water, surging and pumping clean. An overnight contact time is usually required after the first (acid) step since the acid also helps to dissolve iron and manganese oxides that collect in the biofilms and encrustations.

The third phase is the disperse phase. This phase was designed to facilitate the dispersion and removal of the biofilms along with the other associated plugging material from the aquifer. Removal is achieved by surging (air or mechanical), bailing (bailer or air lifting), and pumping. The purpose of surging (re-development) the well is to more completely suspend the disrupted plugging material so it can be removed by air lifting (bailing) and pumping. As a final step, the treated water is pumped from the well until the water is clear and the pH has returned to its original (ambient) levels. Regularly changing the pumping rates can cause additional detachment of plugging material and improved rehabilitation.

Background information on the Wells . The selection of wells 15 and 16 for this intensive study was based on requirements from the City of North Battleford to examine alternative methods to improve the production from the well field. This selection was influenced by the extensive biological/chemical testing capability of DBI and the hydro-geological analytical capabilities possessed by PFRA-TS.

The well field derives its water principally through induced infiltration from the North Saskatchewan River. The ground water obtained from the aquifer is generally of good quality and is preferable to the city's surface water supply directly from the river. This prefer-ence for ground water is due to the lower treatment costs limited to the control of iron and manganese.

Well 15 was installed in July 1989 and was located 175 meters (575 ft) northeast of the North Saskatchewan River. The well has a double-walled gravel pack (12 inch total), with a 80/1000slot stainless steel wire wrapped screen interval located in a 0.29 porosity formation sand between 18.3 m (60 ft) and 24.4 m (80 ft) below ground surface.

Well 16 (approximately 300 meters down river from well 15) was installed in May 1995 and is located 20 m (65 ft) northeast of the North Saskatchewan River. This well has a single walled gravel pack (6 inch) with a 30/1000 slot stainless steel wire wrap screen interval of 9.14 meters (30 ft).

Well 15 had an original specific capacity (SC) of 18 Igpm/ft when it was installed in July 1989. However, the SC of this well has been declining since it was placed into service. The yield from the well has been declining dramatically because of the limited available drawdown. In 1993, the city began acid treatments on a yearly basis to address the declining SC and yield associated with well 15. These acid treatments ended in 1996 because they were shown to be an ineffective rehabilita-tion technique. After the four years of acid treatments, the specific capacity of well 15 had fallen to only 18% of the original.

At this time the City of North Battleford inquired about an experimental treatment process (UAB) being developed by Droycon Bioconcepts Inc. (DBI), and being tested as a part of a joint pilot project with Agriculture Canada's Prairie Farm Rehabilitation Administration-Technical Services (PFRA-TS) Division. During the discussions between the city, DBI and PFRA-TS during the fall of 1997, the specific capacity continued to decline to 1.96 Igpm/ft or 10% of the original SC.

Well 15 was treated (Table Fifty-Four) using the UAB treatment in October 1997 with the final outcome being a doubling in SC. Well 15 was selected to be treated again (in 1998) to determine if such a severely biofouled well could continue to improve with additional UAB treatments.

Table Fifty-Four

Influence of UAB Treatment on the Production Characteristics Of Wells 15, City of North Battleford

	Before	After UAB	Before UAB	After
Date (mn/yr)	10/97	10/97	08/98	08/98
SWL (m)	4.5	4.5	4.25	4.33
SWL (ft)	14.7	14.7	13.9	14.2
PWL (m)	15.4	11.0	12.9	12.6
PWL (ft)	50.6	36.0	42.3	41.2
DD (m)	10.9	6.5	8.7	8.2
DD (ft)	35.8	21.2	28.4	27.0
DD improvement (%)		41%	(19%)	5%
SC (Igpm/ft)	1.9	3.3	2.7	2.8
SC improvement (%)		68%	(23%)	5%
PR (Igpm)	70	70	75	75

Notes: SWL is the static water level, PWL is the pumped water level, DD is the drawdown, SC is the specific capacity and PR is the pump rate given in imperial gallons per minute (Igpm). The before treatments were tested on October 6, 1997 and August 10, 1998 respectively, and the after treatments were on October 10, 1997 and August 13, 1998. The original SC for well 5 was 20 Igpm/ft in July, 1989 and so the decline was by 83% to 17% of the original SC when the treatments started

Well 16 was selected for UAB treatment (Table Fifty-Five) in 1998 because of the dramatic loss in SC of 35% since its construction 3 years ago (May 1995), and its positive (aggressive) biological indicators indicating that biofouling was already very severe.

Essentially, the two wells selected from the well field were different in that one was nine years old, severely biofouled and had lost 90% of the original SC (well 15). The second well was only four years old, was biofouling rapidly and had lost 34% of its original SC (well 16). A comparison could be made between a well close to becoming totally plugged (i.e., well 15) with an almost total loss in SC and a well which was still relatively new (four years old) but was showing the symptoms of becoming rapidly plugged (i.e., well 16).

Table Fifty-Five

Influence of UAB Treatment on the Production Characteristics Of Well 16, City of North Battleford, August of 1998

	Before	After UAB
SWL (m)	3.8	3.8
SWL (ft)	12.4	12.5
PWL (m)	8.4	6.9
PWL (ft)	27.6	22.6
DD (m)	4.7	3.1
DD (ft)	15.3	10.1
DD improvement (%)		34%
SC (Igpm/ft)	13.1	21.1
SC improvement (%)		61%

Note: the original SC for the well when installed in May, 1995 was 20 Igpm/ft. The well had declined by 34% before the UAB treatment but was recovered to 6% better than the original SC by the treatment.

From the data presented in Tables Fifty-Four and Fifty-Five, the UAB when applied to the relatively young but rapidly biofouling well (16) allowed the plugging material to be removed so that the well recovered to its original SC. The older well (15) had already been subjected to treatment by acidization. This had been followed by the two UAB treatments in 1997 and 1998. These treatments did not cause a major recovery of the SC to the original levels but rather a limited recovery in the SC that declined again at a rate of 0.06 Igpm/ft/month. However, even a limited recovery in ground water production could be beneficial in lowering the overall water treatment costs to the community.

The relatively young well (16) did not have such a well entrenched plugging and the UAB treatment process was able to shock, disrupt and disperse the plugging materials and allow the well to return it its original production characteristics.

These treatments on wells 15 and 16 demonstrate that it is better to treat a well sooner rather than later. If treatment is applied to a biofouled well before the SC has dropped to below 40 to 60% from the original values, then there would appear to be a greater probability of returning that well to its original specific capacity (as was the case with well 16). However, once the well's specific capacity has fallen to well below 50% (severely plugged), then the probability of a full recovery is markedly reduced. With such severely plugged wells, the main objective would be to gain a modest but economically acceptable improvement in the SC that can validate the costs of rehabilitation and ongoing preventative maintenance being applied to that well.

Bacterial Aggressivity in the North Battleford Wells (summary)

Well 15 was subjected to the UAB treatment process in October 1997. A significant reduction in biological activity was observed after this treatment. BART testing conducted in April 1998 that showed that the IRB and DN bacteria had been reduced to a highly aggressive level (from extremely aggressive prior to treatment). The SRB aggressivity has been reduced even further since the treatment with an extended time lag of one day compared to immediately after treatment. Of the bacterial groups monitored, it was the HAB consortium that became more aggressive after treatment moving from a low to a medium aggressivity. It would appear that the impact of the treatment in October, 1997 on well 15 was to moderately suppress the DN and IRB, more radically suppress the SRB with their roles being partly replaced by the more aggressive HAB. Samples from the two cores taken at 3 and 6 meters from the well located the most aggressive IRB to be within 3 meters of the well while the most aggressive the SRB were between 3 and 6 meters of the well. Concurrently, the HAB were highly aggressive from the well screen all the way out to 6 meters showing that these bacteria dominated at that well site. From the subsequent data, it would appear that the HAB were the much more successful colonizers after the well had been treated.

Well 16, based on the pump test water samples, contained highly aggressive IRB, HAB and SLYM with a medium aggressivity for the SRB. The cores showed that all of the IRB,

SRB, HAB and SLYM were highly aggressive at 3 meters core sampling site from the screen. But at 6 meters from the screen, only the SLYM remained highly aggressive while the HAB and IRB had dropped to a medium aggressivity and the SRB to a low aggressivity. This three-year old well had, therefore, a different biofouling flora causing the biofouling (i.e., loss in SC) than the older well 15. Here, the dominant bacterial group was the SLYM that were very aggressive both at the well screen and out six meters into the surrounding formation. For the other three groups of bacteria, the SRB was the one most tightly clustered around the well while the IRB and the HAB remained clustered around the well itself. With this well, the early biofouling leading to the 34% loss in SC appears to have been primarily precipitated by a massive growth of SLYM, IRB and HAB at the well screen and extending outwards by >3, <3 and <3 meters respectively. The SRB appear to be clustered very tightly around the well screen and may be growing within, and protected by, the biofilms formed by the other bacteria groups.

Well 17, has not shown any loss in specific capacity in the last three years since construction. Based on the pump test water samples taken over the two hours sequence also employed for wells 15 and 16, the HAB, IRB, SLYM were found to have a medium aggressivity while the SRB had a low aggressivity. A core (C-90) was drilled 3.9 meters from the well. When sampled and tested, the IRB and SLYM were found to be highly aggressive while the HAB had a medium aggressivity while the SRB was not detected. This clustering of IRB and SLYM almost 4 meters from the well would suggest that these bacteria are forming a potentially plugging biomass away from the well. If these bacteria begin to shift towards the well, then it is probable that detectable plugging will occur with the early symptoms being a significant loss (i.e., >15%) in the SC and the occurrence of SRB. These observations support the need for a preventative maintenance (PM) plan so that the plugging (i.e., SC loss) would be controlled and so not seriously compromise the well's production.

There appears to be some correlation between the biological data taken from the pump tests and the biological data from the core samples.

These correlations lead to a better understanding of water well biofouling for wells 15, 16, and 17. As the aggressivity of the bacterial groups increases within the well and extends away from the well (as was the case with wells 15 and 16), then there is a high probability that significant biofouling is occurring in the well. When the aggressivity is found to peak in cores sampled away from the well, then it would appear that the plugging potential is there and will become a reality once these bacterial groups move in, and around, the well. Biological testing using a simple test system such as the BART tests can be helpful in determining the:

- rate of biofouling,
- type of bacteria that are biofouling the well,
- path that the water is taking surrounding the well,
- position and form of biological events involved within the zone of influence in the well field,
- nature of the plugging process.

Comparing these factors can give an insight into the treatment "train" that could be the most effective and beneficial for the optimum economical usage of the well. The core sampling has become a very important tool in determining the biological activity and the placement of biofouling surrounding a well.

Conclusion: Water wells that are likely to plug will reflect this with a degeneration in the SC of the well and increases in the biological aggressivity in the water. For the SC, the critical percentage losses of concern would be:

> <15%, not significant,
> 15 to 40%, significant and a PM program should be established,
> 40 to 60%, very significant and rehabilitation treatment should be undertaken to recover the well as close as possible to the original SC,
> >60%, it may be very difficult to recover the well to the original SC and the most that may be achievable will be half way back to the original SC.

Biological aggressivity would become serious when at least one of the bacterial groups become very aggressive in the well water samples during a pump test. The occurrence of very aggressive SRB is an additional concern since this would mean that the plugging has now matured and includes a potential corrosion/taste and odor problems as well as plugging.

Economic Aspects of Well Rehabilitation   Energy savings can be translated into economic savings after a successful water well rehabilitation. This is because of the decreased water level drawdown at a given pumping rate would correspond to decreases in power requirements and operating power costs. The energy savings for a given flow rate can be most easily calculated by subtracting the post-rehabilitation energy costs from the pre-rehabilitation energy cost for a given production volume. For well fields where the operating wells are coupled into grid systems, it may not be so easy to define the cost savings achieved through treating one of the wells within the field. This was true for the city of North Battleford well field and so the cost savings then have to be calculated based upon theoretical considerations.

Economics of Improving Well Efficiency   The economics of well performance and rehabilitation has been addressed (Helweg, O.J., 1982. Economics of Improving Well and Pump Efficiency. Groundwater, Vol. 20, No. 5). Two equations are involved. Equation 1 estimates the electrical power required to pump water from a particular depth. Equation 2 is now applied to determine the daily power costs of producing water from the well. For the wells being evaluated, the following assumptions are being made: Well 15 will continue to pump at 70 Igpm; Well 16 will continue to pump at 200 Igpm and that the pumps run 24 hours/day, 365 days/year.

$$KW = \frac{Q\ (s + SWL)(0.746)}{3956\ (e_o)} \qquad \text{(Equation 1)}$$

Where KW, power used in kilowatts; Q, discharge in GPM; s, drawdown in feet; SWL, static water level in feet; e_o overall efficiency "wire-to-water" of the pump and motor; e_m efficiency of the motor; e_p, efficiency of the pump, at present pumping rate; 0.746, converts horsepower to kilowatts; 3956 converts GPM-ft to horsepower; and e_o =$(e_m)(e_p)$.

The total cost (TC) of power used per day can now be calculated using equation 2:

$$TC = KW * C \qquad \text{(Equation 2)}$$

Where TC, Total Cost per day of electricity; KW, power used in kilowatts; C, cost of energy, in dollars/KWH.

Before well 15 was treated, the cost of pumping water from the well could be calculated using the following data: Q = 70 Igpm which would be 84 Usgpm; s, 35.83 ft; SWL, 14.73 ft; $e_o = (e_m)(e_p)$ or (0.90)(0.5) which equals 0.45; KW (calculated using equation 1) = 3.81 kwh/h; and TC (from equation. 2) = 3.81 kwh/h * 0.0327$/KWH = $0.125/h. This was equivalent to an annual cost of $1,092.

After well 15 had been treated, the drawdown (s) was reduced from 35.8ft to 21.2ft. This caused some power savings based upon the following data: Q= 70 Igpm or 84 Usgpm; s, 21.2ft; SWL, 14.73ft; $e_o = (e_m)(e_p)$ or (0.90)(0.5) = 0.45; KW (from equation 1) = 3.30 kwh/h; and TC (from equation 2) = $0.108/hr. This worked out to an annual cost equivalent to $945 per year with net savings to the operation of the well of $147 per annum. While the change in drawdown did not yield significant savings to the utility, the additional water generated by the improved SC led to more water becoming available from the ground water sources that subsequently required less water treatment than the surface waters which required more extensive and costly treatments.

Well 16 showed an additional power saving of only $50 due to the minimum 5ft shift in the drawdown. Here, as with well 15, the greater production resulting from the well returned to the original SC and production levels. This would also be a savings that would be achieved by now being able to lean more heavily on the ground water sources which require less treatment than the surface water.

There are clearly potentially large secondary economic benefits from the rehabilitation of plugging water wells. In the case of the well field serving the city of North Battleford, there were a number of assumptions involved in the operation of these wells:

- Wells will be operated so that drawdown will remain at pre-treatment levels.
- Cost of treating surface water is $0.387/m³ more that treating ground water.
- If all of the water had to come from the North Saskatchewan River, then the water treatment costs would be much higher than if the water had been generated from the wells.
- Distribution costs are unaffected.

For well 15, the effect of the treatment was to recover drawdown with a resultant improved SC. The drawdown shifted from 35.8ft (before) to 21.2ft (after), an equivalence of 14.6ft. The increase in flow (Q) after treatment was the equivalent of SC improvement of 3.29 multiplied by the recovered drawdown:

	Q	=	3.29 * 14.6
or	Q	=	48 Igpm
or	Q	=	25.2 million Imp. gallons per year (115,000 m³/year)

If this well was pumped at the same drawdown as before treatment, it would result in a 19.6% reduction in the overall (yearly) surface water requirements, and a 5% increase in the total water to the city for the year from the well field. In economic terms, the surface water treatment costs (C1996$) were 38.7 cents per cubic meter higher than for treating ground water. On an annual basis, the increased of 115,000 m³/year increase in available ground water from the wells would result in savings of $44,372 per year in reduced treatment costs.

This is based on the 1996 costs which for the surface water plant treatment amounted to 52.7 cents per cubic meter (220 Imp.gallons) while the well water plant treatment costs were commonly 14 cents per cubic meter.

The total savings per year, therefore, are the sum of power savings achieved through the reduced drawdown ($147) and the reduced treatment costs associated with the ground water ($44,372) for a total savings of $44,519.

Well 16 also showed an improvement through reduced drawdown of 5ft but the recovery in SC was more compelling at 21.1 Igpm/ft. which translated into 105.5 Igpm (55.5 million gallons or 252,000 m³ /year). Such an increase in available ground water has been calculated to have the potential for a 43% reduction in the city's requirement for surface water and an 11% increase in the total water available from the well field. Savings associated with this greater reliance on ground water would translate into a savings of $97,621 per year through the UAB treatment of well 16. Clearly, there are secondary spin-off benefits through the effective treatment of plugging water wells that extends beyond simply the power savings through having more available drawdown.

Conclusion: The effective treatment of plugging water wells generates spin-off benefits that extend beyond simply the recovery of available drawdown. It extends into greater production capacities, better water quality, and the extension in the longevity of the active life of the well to more assured production characteristics (where a PM program is effectively introduced). These secondary economic benefits have to be recognized in the evaluation of the cost effectiveness of any well treatment program.

Measurement of TSS using Laser Particle Counting - Laser particle counting is a valuable tool in the determination of the relative sizes and numbers of particles present in water over the size range of 0.4 to 120um (microns). Interference by the particles with standard cyclic pulses of laser light is used to detect particle numbers and sizes. When there is a sloughing event occurring from the plugging zones, this can be reflected in high particle counts in the water. Number and sizes are used to compute the total suspended solids (TSS) and the mean size of the particles gives some idea as to the form of these particles.

> *Well 15:* Prior to treatment in October 1997, the mean particle size of a representative water sample was 2.8um while after treatment these particles increased to a mean size of 5.2um as the void spaces were opened up during treatment, allowing larger particles to migrate into the well. On April 6, 1998 six months after treatment, the mean particle size is now 2.8um but 60% of the particles are in the 8 to 16um range compared to only 19% before the treatment on October 6, 1997. This indicates that the larger void spaces opened up during treatment still were allowing the larger physical and biological particulate matter to migrate into the well.

> *Wells16 & 17:* A comparison of the laser particle counts of these two wells were derived knowing that well 17 has a higher SC and less biofouling than well 16 (based on BART tests for aggressivity). Both wells were pumped at 200 Igpm to give a representative comparison. After two hours of pumping the following results were obtained.

> Well 16 had a mean particle size of 3.4um compared to well 17's slightly larger 3.6um. While the mean sizes were similar, the distribution of particles were for well 17 where 73% of its particulates were larger than 8um and for well 16, 639% of its

particulates larger than 8um in size. This would indicate that both wells were able to allow the larger particles to pass through the formation sand and pack signifying larger void spaces and less restriction (biofouling) around the wells were occurring than well 15 which was severely plugged.

Conclusion: Laser particle counting can give some indication of the throat size around the well. If only small particles are present in the water column, this may mean that the plugging is more restrictive and the throat size around the well has shrunk (a symptom of a rapidly plugging well).

<u>Well Treatment and Maintenance, the Use of Mesocosms</u>   For the selection of suitable treatment chemicals and procedures, model wells (mesocosms) can be used to develop a suitable chemical treatment train that would be effective on the biological problems at well 15 and 16. Laboratory mesocosms are small-scale replica of the conditions that would be experienced at and around the well screen in the well. The mesocosm is inoculated with bacteria from the well being evaluated and then fed with a nutrient feedstock to speed up the plugging process which is commonly measured by losses in permeability and drainable void spaces. In April of 1998, core (water bearing) material was collected at each well site and mesocosms constructed to simulate the actual well environment conditions in these two wells (15 and 16).

Preliminary tests comparing the Original Ultra Acid Base (UAB) treatment that was conducted in October (1997) and refined methods and procedures were completed on the model wells, prior to treating well 16, and retreating well 15. Results of model well testing showed that the caustic soda base used as a pH buffer in the October 1998 treatment of well 15 caused undesirable results. These included causing up to a 38% swelling in the clay particles which, in turn, restricted the water pathways to the screen reducing permeability.

Research has also shown that sodium hypochlorite and calcium hypochlorite (HTH) cause less clay swelling as a base pH and also disinfect at the same time. Also, it has been found that when using granular calcium hypochlorite, the crystals must be fully dissolved in a mixing container filled with clean water and allowed to settle for at least one hour. Then the solution must be decanted slowly so as not to allow the particulate matter to enter the well. If this does happen, there may be an increase risk of calcium encrustation on the well screens.

Conclusion: Laboratory mesocosms provide a potentially effective and rapid method for economical evaluation of a proposed treatment regime before the (expensive) application of the treatment scenario to the well.

Potential Preventative Maintenance (PM) Procedures   PM procedures are still being developed and tested in the laboratory. The focus is to be able to use as much *in situ* equipment as possible. This includes the use of submersible pumps with down hole check valves relocated to allow the pump to be used to lift and drop a volume of water continuously, thereby surging the well. Some of the PM procedures worth considering with surging are: standard bleach (6%), sodium hypochlorite (12.5%), calcium hypochlorite (65%), and CB-4 (0.75%). More extensive PM on problem wells may require airlifting along with a combination of acid and base pH "flip-flopping", CB-4 (an effective wetting agent/ penetrant), or higher concentrations of chlorine solutions.

Conclusion: It is absolutely critical that accurate records be kept on even the smallest details of the well after it has been treated. Any losses in SC or decrease in time lag difference (TLD) using the BART test system must be noted and acted upon in a timely fashion using the appropriate PM to ensure effective ongoing management of the well with a minimum of plugging. Overpumping of the well must also be avoided, as the increased entrance velocities created during overpumping will cause an increase in the biological activities, and accelerate the rate of biofouling close to the well intake area.

Location: Paper and Pulp Mill, Montana

A major paper and pulp mill in Montana was established near a river and the water was produced from a series of water wells set into the lower section of an aquifer which was recharged by snow melt from the Rocky Mountains. The well field consisted of a set of high yielding wells each capable of 5,000 gpm once developed. Unfortunately the combination of river recharge into the aquifer and leachate from the oxidation ponds set downstream from the aquifer led to rapid plugging. This resulted in the wells losing 80% or more of their production capacity. There are two findings from this research: (1) the variation in the form of the plugging that occurred across the well field, and (2) the size of the zone of plugging around the wells determined by coring and microbiological testing.

The Size of a Biofouling Event Around Well M15   A series of four satellite wells were drilled around well M15 extending with one well north and east with two to the south which was the direction from which the ground water flow was coming. These satellite wells were positioned 6' (north, N6), 13' (south, S13), 18' (east, E18) and 48' (south, S48). Core samples were taken to depths of 141', 138', 128' and 118' respectively since this represented the actively producing zone moving towards the screened area of the well. Differences were most apparent between the E18 and the S48 wells indicating that the biofouling extended beyond the 18'. For the purposes of this example, the comparison will be limited to the differences in the bacterial flora between these two wells. This is summarized in Table Fifty-Six.

Table Fifty-Six

Bacterial Populations (log. cfu/ml) in the E18 and S48 Satellite Wells

	E18				S48			
	118	128	138	141	118	128	138	141
PB	4.9	4.6	4.2	4.6	4.2	2.3	2.8	5.2
FP	3.9	3.3	1.2	0.0	1.3	1.2	1.2	3.6
AIRB	3.8	3.5	3.6	3.5	2.8	1.7	3.4	5.2
TIRB	4.7	4.9	4.6	4.8	2.2	1.7	2.2	4.6
HAB	4.9	>5.4	>5.4	4.9	4.2	2.2	2.8	5.2
ML	0.0	3.0	0.0	0.0	1.5	1.1	1.1	0.0

The five bacterial groups were: PB, pseudomonad bacteria; ML, molds; FP, fluorescing pseudomonads; AIRB, atypical iron-related bacteria in which the colonies did not accumulate iron (i.e., go brown); TIRB, typical iron-related bacteria which did generate typical brown (iron rich) colonies; and the HAB, heterotrophic aerobic bacteria. All of this data was obtained using standard extinction dilution spread agar plates and the standard recommended media. The four columns on the left under satellite well E18 represent the four core depth samples for that well while the final four columns are for well S48.

From the data, it is clear that there are much larger populations of bacteria at the E18 position at all of the four core depths included in Table Fifty-Six. For the pseudomonad bacteria, lower populations by greater than one order of magnitude were observed in the S48 128' and 138' samples indicating lower bacterial activity at these sites. However, the trend for the fluorescing pseudomonads (FP) differed in that at E18, there was a reducing bacterial population with depth but at the S48 site, the trend was reversed. Here, the highest FP population in S48 was recovered at the greatest depth (141'). Dominant bacteria in the E18 cores was a mixture of pseudomonad bacteria, typical iron-related bacteria and heterotrophic aerobic bacteria with the latter group dominating particularly at the 128' and 138' depths. At the S48 site, there was generally a lower bacterial activity than the E18 with the larger populations being recovered at the 118' and 141' depths.

There was a focus of heterotrophic bacterial activity in the E18 satellite well cores particularly dominating at the depths of 128' and 138'. This would suggest that there were much high bacterial populations in the voids at the 128' to 138' depth which might indicate that a greater movement of ground water (to the well #15) had occurred at those depths creating the greater amount of biological activity. For the S48 well, the largest populations were recovered at a depth of 141'. This may indicate that the ground water moved from a greater depth (of 141' or more) at a distance of 48' from the wells to a depth of approximately 132' when 16' from the well (based on the data from E18. Both N6 and S13 cores consistently gave very high populations (>5.4 log. cfu/ml) for the PB, TIRB and HAB which would indicate that the bacterially generated plugging was actually occurring must intensely within a radius of 18' of the well 15. It is therefore possible to project that the size of the bacterially generated plug could be envisage as being at least 141' in height with an average diameter of 30' with extensions radiating outwards over the 128' to 141' depths to a possible diameter of extension of at least 96'. Such "plug beasts" would create a panic if they could be hauled out for people on the surface to see but there, down in the ground, it remains a case of "out of sight, out of mind!"

Conclusion: The installation of satellite wells and the bacterial analysis of core samples from different depths can give an appreciation of the size of a plugging zone and the dominant nuisance microorganisms causing the problem.

<u>Variation in Plugging Events within a Contained Well Field</u>   Eight production wells were installed to meet the water needs for a paper and pulp operation. The wells all primarily received ground water from the south through aquifers being fed by ice and snow melt in the Rocky Mountains. Secondary recharges came from a river to the west of the

wells and some from the oxidation ponds to the north of the well field. Recharges could therefore come from three sources: south – from the mountains; west – from the river; and north – from the oxidation ponds. The wells were set in a paired arrangement 11 – 12, 14 – 13, 17 – 15 and 20 – 16 with the latter of each pair being closest to the river on the west. This combination of water sources could cause major influences on the bacterial groups causing the plugging which was being reported in all of the wells. This work was undertaken in 1988 and was one of the first full scale application of the BART water testers to determine the aggressivity of the various bacterial consortia in the wells.

Chemical and bacteriological determinations were undertaken on the well water from the field of eight wells. Of significance from this data are a number of observations that distinguish the wells as being influenced differently by the various ground water sources. River recharge influenced the oxygen concentration in the closer four wells to the river (12, 13, 15 and 16). The higher oxygen readings on the four wells closest to the river were +4.1, +2.8, +0.1 and +2.8 ppm respectively above the corresponding values for the more distant wells. This additional oxygen probably would arise from the more oxygenated state of the river water recharging into the well field. Iron was higher in the pair of wells 15 and 17 while manganese was high in the paired wells, 16 and 20. This would suggest that the wells were being subjected to separate incidents such as an old alluvial river-bed in which iron and manganese had become accumulated within different parts of the sediment. Nitrates appeared to be entering the well field from the north (where the oxidation ponds are located) reaching concentrations of greater than 20 ppm nitrate. Erratically high (>0.5 ppm P) were present in three of the wells with no clear pattern as to the cause.

The net effect of these differences in the chemistry of the well waters can be seen, to some extent, in the bacterial loadings determined by the BART water test system. IRB tended to dominate in the three pairs of wells to the north while the heterotrophs (HAB) dominated to the south in a similar pattern. SRB were detected in the second pair of wells from the north (13 and 14) and in well 16 but were not extremely aggressive.

Conclusion: Water wells within the same well field do not have to all possess the same characteristics in terms of chemistries or plugging. Each well should be treated separately and problems, management and potential rehabilitation methodologies should be addressed separately.

Over the time period of 1986 to 1989, an evaluation was made of the influence of age and depth of water wells on the loadings of various bacterial indicators in central and southern Saskatchewan. No significant correlations were observable for the 57 wells investigated with depths ranging from less than 5 to greater than 100 m. Fifty-eight of the wells had sufficient history that their ages could be established as ranging from new to greater than 20 years. Similarly, no correlations were observable from the 67 wells examined between turbidity (from 0 to greater than 20 F.I.U.) and the size of the bacterial populations recovered from the wells. Forty-four wells were examined weekly for eight weeks.

Traditional concepts of ground water management have tended to presume that the water is essentially sterile (Saskatchewan Agriculture Bulletins, 1981) and therefore not a tenable site for intense microbial activity. Associated with this was the concept that the bulk of the biological activity would occur at any recharge site where soil or sediment water was permeating downwards into the aquifer. This study was designed to determine the validity of these traditional concepts.

Table Fifty-Seven
Characterization of Well Field Suffering from Plugging

Wells away from river:	11	14	17	20	<recharge
Wells close to river	12	13	15	16	<recharge
pH	7.7	7.4	7.9	7.6	
	7.8	7.4	7.0	7.6	
Oxygen	2.9	2.4	4.8	4.0	
	7.0	5.2	4.9	6.8	
Total Fe 0.1	0.8	0.1			
	0.1	0.1	0.2	0.1	
Total Mn	0.02	0.03	0.03	0.3	
	0.0	0.1	0.0	0.2	
Total P	3.0	0.4	0.8	0.2	
	0.4	0.2	0.3	2.8	
Nitrate	25	2.5	2.0	1.8	
	35	25	0.8	2.5	
IRB (TL)	3.0	3.5	3.0	ND	
	6.0	3.0	3.5	ND	
SRB (TL)	ND	6.0	ND	ND	
	ND	6.0	ND	5.0	
HAB (TL)	3.0	4.0	4.0	5.0	
	ND	5.0	4.5	3.0	

Location: Central and Southern Saskatchewan, Canada

Of particular interest was the length of pumping time, depth of the well, age of the well, turbidity of the water, the interrelationship of various groups of bacteria which have been purported to commonly occur in wells and the impact of this on bioaccumulation of metals. It should be noted that over this period (1986 – 1990), the BART water testers were developed and parallel studies were conducted against the standard agar spread-plate techniques. These data and findings relate to the agar spread-plate technique and the populations recorded are consequently lower than those commonly found using the BART water testers because of higher sensitivity and broader spectrum of potentially culturable microorganisms.

Influence of Pumping Times The act of pumping a well is to cause the flow of water approaching the well screen to have an increased velocity followed by turbulence. If the pumping is sufficiently demanding, then a drawdown of the water column head may also occur. These factors disturb the incumbent microbial flora which may be sessile (attached to the pump, impellars, well screen, gravel pack and ground water media) or planktonic (suspended within the water well column). Gradual settling of these planktonic microorganisms to the bottom of the well has been observed when the water column is quiescent for a prolonged period. Frequently vertical profiles of the bacterial population of wells reveal much higher populations in the lower third of the well water column compared to the upper two-thirds. The act of pumping is to disturb both the planktonic and sessile flora causing sporadic passages of bacteria through the pump to the water distribution system.

A study of the rate of microbial releases from a well in Saskatchewan was conducted. Here, the well was taken out of production over night and then subjected to the period of pumping. Samples drawn aseptically at the well head were subjected to extinction dilution/spread-plate analysis using Brain Heart infusion (1/4 strength) agar and incubated aerobically at room temperature for 14 days. When well 1 was examined in 1986 and again twice in 1988, a baseline bacterial loading of 4.5 x 10^3, 2.2 x 10^4 and 1.6 x 10^4 cfu/ml were recorded. At different sampling times up to a maximum of 25 minutes, water samples were taken aseptically at the well head and recorded. In 1986, the microbial loading was reduced during the pumping period over a range from 22 to 53%.

When the same well was pumped the first time in 1988, the baseline microbial loading was found to be five times higher than 1986. There was also a greater variation in populations that were observed over the 20 minutes pumping time ranging from a low of 6.8% of the baseline data to a high of 590% of the mean. The highest reading occurred 15 minutes into the pump time. This indicated that there was an erratic sloughing of bacteria from the plugged zone.

When the second pump test was performed 20 days later, there was not the same "flushing" action of microorganisms out of the system due to sloughing (i.e., large standard deviation). There was an intense bacterial release at between 4 and 5 minutes into the pumping process. Similar fluctuations in the bacterial loadings for two other wells at the same facility as the test well confirmed that the process of pumping for a short period of time did not produce consistent bacteriological data in the product water.

Conclusion: Bacterial numbers will vary considerably when a well is first pumped after a period in which it is not producing. This is particularly the case when a treatment has been applied to a well to cause more bacteria to slough from the deeper regions of the aquifer.

Impact of Well Depth on the Bacterial Population in the Product Water

Fifty-seven wells were tested in Saskatchewan to determine what effect the depth of individual water wells would have on the bacteriological quality of the product water. For this examination, each well was pumped for 5 minutes after a period of at least one hour dormant period (Table Fifty-Eight).

Table Fifty-Eight
Influence of depth (meters) on Bacterial Loading (log. cfu/ml) on 57 Wells

Bacterial Group	Depth (meters)				
	<5	5-10	11-15	16-100	>100
HAB	4.2	3.5	3.5	2.6	3.7
	(2.6)	(1.0)	(1.6)	(1.5)	(0.7)
IRB	2.2	1.8	2.2	1.2	2.5
	(2.8)	(1.0)	(1.1)	(1.0)	(1.6)
TC	0.9	1.0	0.6	0.8	1.2
	(1.5)	(1.6)	(0.8)	(1.0)	(1.2)
FC	0	0.2	0.1	0.1	0.2
	(0)	(0.4)	(0.2)	(0.4)	(0.5)
Well Numbers:	5	13	6	22	11

Note: TC (total coliforms) and FC (fecal coliforms) were estimated using MF standard methods and the log cfu/100ml. Standard deviations are shown in parentheses below the mean values recorded for each group of bacteria investigated.

No statistical differences were noted between any of the groups of wells for the HAB and IRB and very wide variations between individual wells because of large standard deviations. Similarly, there appeared to be little difference for the fecal indicator bacteria although no FC were recorded for the wells of less than 5 meter depth this was a small sample size.

Conclusion: Bacterial activity and coliforms can be detected in water wells in Saskatchewan in depths up to greater than 100m.

The Relationship of Well Age to Bacterial Loadings One generally held concept is that bacterial incursions into a water well are likely to occur over a prolonged period of time which may usually be measured in years. Some practical experience has found that these bacteria can generate a heavy biofouling, commonly caused by iron-related bacteria. Symptoms of biofouling have been observed in a matter of weeks or months after the development of the well to create significant microbial populations quickly. The relationship for 58 wells is given in Table Fifty-Nine.

Table Fifty-Nine
Influence of age (years) on Bacterial Loadings (log. cfu/ml) in 58 Wells

Bacterial Group	Age (years)				
	≤ 5	5-10	11-15	15-20	>20
HAB	3.7	3.1	3.9	2.6	2.9
	(1.8)	(1.0)	(1.5)	(1.2)	(1.1)
FLOR	1.2	1.0	1.7	0.3	0.5
	(1.5)	(1.9)	(2.3)	(0.7)	(0.5)
IRB	1.7	1.7	1.9	1.1	1.7
	(2.9)	(1.3)	(1.7)	(1.0)	(1.0)
TC	0.3	0.7	1.0	1.4	1.3
	(0.4)	(0.9)	(1.5)	(1.8)	(1.2)
FC	0.1	0.1	0	1.0	0.5
	(0.2)	(0.1)	(0)	(0.3)	(0.7)
Well Numbers:	6	19	8	11	14

No difference was noted between the age of the well and the populations of HAB and IRB. There did, however, appear to be a trend for the older wells to exhibit lower populations of FLOR bacteria. Considerable variations were observed in the total coliform population with the highest populations being recorded in the wells aged 10 years and older.

Fifty-eight water wells in Saskatchewan of known age were included in a bacteriological survey to determine whether there was a higher incidence of microorganisms in the older well installations. The wells selected for this study ranged in age from less than 5 to greater than 20 years. Aseptic water samples were taken at the well head after 5 minutes of pumping following one hour of quiescence. The study wells were differentiated by age over periods of five years length and the grouped data subjected to statistical analysis with the means and standard deviation calculated.

Conclusions: Bacteria were recovered from wells of all ages and the only trend would appear to be reducing populations of fluorescing pseudomonads and elevating populations of total coliforms as the wells age.

The Relationship of Turbidity to Bacterial Populations in Water Wells

It is a generally held belief that crystal clear water is likely to contain a lower population of bacteria than turbid water. In this study, 78 wells were sampled and the turbidity determined by the standard method using nephelometric turbidity units (ntu). Concurrently, the bacterial population was determined by vortexed, extinction dilution/spread-plate analysis at room temperature using Brain Heart infusion (1/4 strength) agar for the heterotrophic aerobic bacteria. Sampling of each well was performed five minutes after the start of pumping. To ensure continuity, each well was shut down for one hour before sampling. No statistical differences were observed between the water based on turbidity and the incumbent bacterial population. Regression correlations (as R) ranged from -0.006 to -0.24 except for waters with ntu values of between 1 and 5. Here, there was a positive R value of 0.52. There was also no correlation between turbidity and the samples showing a positive detection of TC (48% positive) and FC (16% positive).

It was found that where relatively shallow water wells (up to 40 meters in depth) were examined for HAB and non-sheen total coliforms, water with coliform populations in excess of 330 coliforms/100 ml were recorded in seven incidents while the gross aerobic bacterial population was less than 0.4 million cfu/ml. No high coliform (i.e., >330 coliforms/ 100 ml) populations were recorded in waters containing higher gross aerobic populations (>5.3 log cfu/ml).

Conclusion: The turbidity of the water cannot be used as an index of the likelihood of bacterial population in well water samples.

Interrelationship of Different Bacterial Groups in Water Wells During the survey of the 44 wells in provincial parks in Saskatchewan in 1988, some interrelationships were seen between various groups of bacteria. Consistent differences were noted in the relationship between the HAB and IRB using the Brain Heart Infusion (1/4 strength) and Winogradsky Regina agar respectively, employing the agar spread-plate technique. For the data obtained over the eight week period for the 44 water wells, a trend was observed in which some wells consistently produced bacterial populations where these two types (IRB:HAB) were at approximate parity (1:1, group 1). In other wells, the IRB:HAB ratio occurred over a ratio of approximately 1:3 to 5 (group 2). In a third set of data, ratios were observed to be approximately 1:50 (group 3).

For the three groups, a linear regression analysis indicated that there was a close correlation between the population sizes for each grouped ratio. In group 1 (1:1), an R value of 0.99 was obtained. The regression equation was $y = 1.24x + 13,900$ where x is the population of IRB and y the population of HAB. For group 2 (1:3), the equation shifted to $y = 3.19x + 10,700$ with an R value of 0.82. For group 3 (1:50), the equation was $y = 18x + 14,905$ with a regression correlation of 0.84. There could therefore be a potential to undertake some diagnostic studies of a potentially biofouled well by evaluating the ratio of IRB to HAB populations. These ratios remained relatively stable for individual wells.

It could be postulated that a greater potential for iron-related bacterial biofouling would occur in group 1 since an equal number of IRB to HAB was observed. The risk of an iron bacterial type of biofouling would appear to be reduced in groups 2 and group 3 as the relative ratio of IRB to HAB declined.

Conclusion: The bacterial loading in a water well may reflect the potential of iron-related plugging with the group 1 IRB:HAB ratio (1:1) representing the greatest risk. The bacterial population within water wells may be relatively stable over time (this study was for 8 weeks for weekly sampling).

Location: Atlantic Region, Canada

Identification of Specific Dominant Bacteria Within Water Wells

Traditionally, emphasis has been directed towards the determination of hygienic risks by the application of a test procedure sensitive to fecal indicator bacteria such as the coliforms. There is growing concern that some of the indigenous bacteria within water wells, treatment and distribution systems could be nosocomial pathogens and present a significant risk to immunologically stressed individuals. To address this concern, a greater level of attention has to be directed towards the identification of the non-conform bacteria present within these systems. A study conducted in eastern Canada on five closely associated wells, all of which were less than 40 meters in depth and set in a close association (less than 200 meters apart from at least one of the other wells). This study was conducted by isolating the dominant colony types enumerated by extinction dilution/spread-plate techniques at room temperature and subjected to identification using the M.I.D.I. system (MIDI Inc., Newark, DE). A variety of dominant bacteria types were recorded from the wells during normal operation procedures (Table Sixty).

Table Sixty

Dominant Microbial Species Isolated from Five Wells In New Brunswick, Canada

Well 1	*Pseudomonas pickettii*
Well 2	*Pseudomonas fluorescens*
	Pseudomonas chloraphis
	Janthinobacterium lividum
Well 4	*Pseudomonas chloraphis*
	Micrococcus lylae
Well 5	*Janthinobacterium lividum*
	Rhodococcus rhodocrous
	*Leptothrix cholodnii**
	Euglena species A*
	Clorella species A*
	Micrococcus luteus
	Pseudomonas fluorescens
Well 6	*Pseudomonas fluorescens*
	Janthinobacterium lividum
	Chromobacterium violaceum

In addition to the bacteria identified by the MIDI technique, *Leptothrix cholodnii* was identified directly from the water (*) along with algal species (A) of *Euglena* and *Chlorella*. At the time of sampling, each of the wells displayed a different spectrum of dominant bacteria. For example, one well (#5) appeared to be dominated by *Janthinobacterium lividum* and *Rhodococcus rhodochrous* while, when subjected to pump tests, two days after the preliminary examination the bacterial flora was dominated by *Micrococcus lylae* and subsequently by *Pseudomonas fluorescens*. Gross aerobic bacteria displaying purple colony forms dominated another well (#6). Both *Janthinobacterium lividum* and *Chromobacterium violaceum* were identified. In the blended water from the wells entering the treatment place, *Pseudomonas fluorescens* and *Micrococcus lylae* dominated the water. While in the distribution system, *Staphylococcus hominis* and *Micrococcus roseus* were recovered although these were not evident at any of the previous sites along with *Chromobacterium violaceum*. When water was pumped through a greensand filter medium, the predominant bacteria isolated from the greensand filter included *Pseudomonas fluorescens* at the very top of the filter where considerable plugging was experienced; *Pseudomonas aureofasciens* was found to be dominant underneath the top surface. When the greensand was washed with water from the wells during the backwash cycle, the incumbent bacteria isolated from the washed core material as dominated by *Micrococcus roseus*, *Rhodococcus equi* and *Pseudomonas syringae*. In this particular water system, each of the wells appeared to be producing a distinctive bacterial flora that became mixed in the treatment plant and consequently penetrated into the distribution system. Such events need to be closely monitored given that there is a growing concern as to the role of different pseudomonad species as both community induced and hospital induced nosocomial pathogens. In consequence, a greater level of bacteriological evaluation needs to be applied to distribution waters to control the level of these more recently recognized hygienic risks.

Conclusion: Plugging can involve a variety of bacterial species. In these wells, species of Pseudomonas were the most often present.

Location: Kneehill M.D., Alberta, Canada

The Municipal District (M.D.) of Kneehill is in central Alberta. Most of the water used by the small communities and single users has come from two aquifer formations (the Horseshoe and the Paskapoo). The wells vary in depth from 30 to greater than 150 meters and draw water mostly from these formations.

The M.D. of Kneehill has had to face a number of problems with the water wells relating to plugging, taste and odor problems. As a result of this, the M.D. was selected as the first site for an intensive study of the factors influencing the failure of wells within that district. The study was jointly undertaken by the PFRA-TS of Agriculture Canada and Droycon Bioconcepts Inc. over the years 1995 to 1997. It was a four phased study involving:

(1) water well users questionnaire (360 replies received).
(2) microbiological survey of selected wells including both microbiological and chemical testing.
(3) detailed diagnostic testing of seven of these wells to provide further evidence of the extent of biofouling and the types of bacteria involved. These diagnostic tests also gave a well performance benchmark permitting calculation of the relative merit of any future well rehabilitation effort.
(4) field testing of the UAB prototype treatment process. Six of seven wells were subjected to the more detailed investigation.

From this study, a number of improvements were made in the understanding of the operating life of the water wells, the dominant forms of bacteria that were shortening the functional life of the well, and some of the elements necessary to implement preventative maintenance for these wells. Each of these aspects of the studies will be addressed separately below. It should be noted that the long term goal would be to create sustainable water wells capable of generating greater confidence in the production and significant extension of the active life span of the well.

Results of the Questionnaire Survey

A questionnaire survey was designed by the Prairie Farm Rehabilitation Administration (PFRA) to determine basic well information (age, depth, etc.), the type of water supply problems encountered by the well users, and the maintenance practices, if any, that have been used. The survey undertaken by staff members of the M.D. of Kneehill encompassed eight townships. Data analysis of the well owners' responses was carried out both by PFRA Earth Sciences Division and DBI. The data in this section was taken from the report, Groundwater and Water Wells in the Municipal District of Kneehill Alberta, published by the PFRA in 1997.

A summary of the concerns identified by well owners with active wells found that of the 275 water who responded to the responded to the questionnaire 220 were well owners. These 220 owners reported on a total of 251 wells that were currently being used. About 75% of the users (taking into account that 30 users had multiple wells) reported on a variety of water quantity or quality problems. These included the following:

- *49.4% had water quality problems only.* Of the 124 wells, 122 had an ongoing concern and 2 had seasonal concerns.
- *9.6% had water quantity problems only.* Of the 24 wells, 5 had an ongoing concern and 19 had seasonal concerns.
- *14.7% had BOTH water quality and quantity problems.* Water quantity was generally noted as a seasonal concern (35 wells) while water quality was mainly an ongoing concern (36 wells).
- *26.3% reported no concerns.*

Well users reported that water quality in the existing wells to be a bigger concern than water quantity (64% of owners reported quality problems while only 24% reported quantity problems). Some of this difference is likely due to the fact that slight changes in water quality are more immediately noticeable than small progressive reductions in well yield. It is not until the well yield drops to a small fraction of the original specific capacity that decline is observed. Only about 31% of the wells identified to have water quality or quantity problems were ever treated. Well owners possessing abandoned or standby wells also identified similar concerns. Although water quality was a significant concern, users appear to be willing to live with water quality concerns as long as a well is producing an adequate flow to meet essential needs.

A total of 68 survey respondents provided detailed information on 69 abandoned or standby wells. Of these 69 wells, 68 respondents reported water quality or quantity problems:

- *27.5% had water quantity problems only*
- *20.3% had water quality problems only*
- *43.5% had BOTH water quality and quantity problems*

Effects of Age on Well Status

The survey was evaluated to assess how the age of a well might affect its status. Analysis showed that older wells are more likely to be abandoned. Based on a survey of 153 wells of known ages, the median age of reported active wells was 14 years. For the standby or abandoned wells, the median age was 20 years. It was found that only 9% of wells with confirmed ages of less than 15 years old had been placed in a standby status or abandoned. In contrast, 17% of wells older than 15 years were standby or abandoned.

Symptoms of Biofouling Identified in the Well Owner Survey

The well owner survey data were reviewed from a biological perspective to estimate the percentage of wells in the study area which may be biofouled based on symptoms identified by well owners. In general, very few symptoms indicative of potential biofouling were reported in wells less than 5 years old. In wells older than 5 years, a majority of the water quality problems reported by well owners may be related to biofouling. Taste, odor, and black and red slimes were the symptoms indicative of biofouling most often reported by owners with water quality problems. Other symptoms reported included visible mineralization in the water, reduced well yield, and colored water.

Determination of Biofouling using the BART Water Test System

Using the BART water testing system, microbiological analyses were performed on water samples from 40% of the wells reported in the well owner survey. Water samples were collected, on a random basis, from these 134 wells and tested for the presence of IRB, SRB and TAB. Bacteria in the water samples were ranked by their aggressivity based on the test results obtained. For example, indications of very aggressive bacteria are: IRB reactions occurring in less than 3 days, SRB reactions occurring in less than 4 days or TAB reactions occurring in 1 day or less. Water samples found to contain these very aggressive bacterial populations were used to identify wells that are likely to be biofouled.

The results from the BART survey were used to asses the role of biofouling relative to concerns identified in the well owner survey and to determine the nature, rate, and extent of biofouling that is likely present in the study area. These data were also used to identify any factors that appear to encourage or promote biofouling (e.g. age of the well, geology in the completion zone, etc.). The data in this section was taken from the report written by Cullimore and Legault in 1997, ("Microbiological Investigations of Water Wells in the M.D. of Kneehill, Alberta") and published by the PFRA.

Results of the BART tests indicated that 68% of the 134 sampled wells contained at least one type of very aggressive bacteria. This would mean that 68% of the wells sampled are likely to have become subjected to severe biofouling. This test data also showed that the wells in this area were particularly prone to SRB infestation since these dominated many of the wells recognized to be at a higher risk of severe biofouling.

The dominance of SRB in this region was at odds with the observations from other areas in the Canadian prairies where the IRB tend to dominate. One possible reason for this SRB dominance could be related to the fact that this region is partially underlain by gas bearing formations.   Consequently, the SRB could be using the influent permeating methane as an organic carbon nutrient source. At the same time, the lineaments in that area could be providing pathways for this methane to more readily reach the near surface wells and the oxidative regions associated with these installations.

Comparison of the BART Test Results to the Well Owner Survey

A comparison of the BART test results to the well owner survey confirmed that many of the water quality and water quantity problems reported by the owners may be related to biofouling. Comments from the well owner survey were matched to 118 of the 134 wells sampled where the data from the owners had been adequate to include the information. These comments showed that 73 of the sampled wells were experiencing water quality problems according to the owners while 35 were reported having water quantity (production) problems. The remaining ten wells were believed by the owners to have no problems.

Of the wells reportedly experiencing water quality problems, 61% contained very aggressive bacteria. In addition, 74% of the sampled wells reportedly experiencing water quantity problems contained very aggressive bacteria. This would suggest a somewhat closer link between bacterial aggressivity and production problems rather than water quality issues.

Influence of Well Age on Well Deterioration Caused by Biofouling

Results from the BART testing and the owner survey indicated that, in general, symptoms of biofouling can show up within five years but as wells age beyond five years there is a significant increase in the percentage of wells containing very aggressive bacteria. At the same time, the number of owner reported symptoms related to biofouling increases. For example, nine of the sampled wells were known to be under five years of age and only three (33%) of these contained very aggressive bacteria. In contrast, of the eleven sampled wells that were between 6 and 10 years of age, seven (64%) were found to contain very aggressive bacteria. The replacement of wells became the most significant for wells over 15 years old.

Influence of Geological Formation Type on Biofouling

A review of the well completion data was carried out to identify sampled wells that drew ground water solely from either the Horseshoe

Canyon or Paskapoo Formations. In this study, a total of 43 wells were identified with 13 wells draw water from the Horseshoe Canyon Formation while the remaining 30 wells draw water from the Paskapoo Formation. Where it exists, the Paskapoo Formation overlies the Horseshoe Canyon Formation and is coarse grained with some thick sandstone lenses. In contrast, the Horseshoe Canyon Formation is generally fine-grained and has a high bentonite content that leads to a low water supply potential.

Biofouling, for all three bacteria types (SRB, IRB, and HAB), was more prevalent in the Horseshoe Canyon Formation. In both cases, SRB were the dominant type of bacteria present. All the wells completed in the Horseshoe Canyon Formation contained very aggressive bacteria. In the Paskapoo formation only 66% of the wells contained very aggressive bacteria.

Conclusions: Bacteria were found in all of the wells tested and that supports the concept that bacteria are ubiquitous within ground water. There were links established between wells experiencing user-recognized problems and the aggressivity of the bacteria recovered from the wells. The dominant bacteria found to be the most aggressive was the SRB. Testing for nuisance bacteria should be a standard part of procedures to determine the form and cause of biofouling in water wells. Sequenced sampling from pump tests provides a means to determine the nature and extent of the biofouling around a well.

Evaluation of the Rehabilitation of Selected Wells

In 1996, seven wells within the study area were selected for more detailed well diagnostic tests. These tests were used to establish the degree of biofouling and to determine aquifer and well performance characteristics. Diagnostic procedures performed on each well in 1996 included:

    (1) gathering background data ,
    (2) an analyses of water chemistry,
    (3) well logging,
    (4) down hole camera inspections,
    (5) pump tests,
    (6) microbiological tests for aggressivity.

Upon completion of these tests, each well received a shock chlorination treatment. In July and August 1997, PFRA undertook a joint venture with DBI to field test the UAB treatment technology within the M. D. of Kneehill, Alberta. Well treatments were applied to six of the seven wells that received detailed diagnostic testing during the summer of 1996. Field-testing provided an opportunity to test the operational constraints of the UAB treatment equipment. It also provides a means to examine the impact of the treatment on the microbiological activity and subsequent impacts within actual water well environments. Results of the field tests were used to evaluate how effectively the treatment process could be used to rehabilitate a biofouled well. If successful, this should now allow the establishment of effective treatment procedures that may improve the confidence level in operating water wells.

Prior to the treatment applications in 1997, pump tests and down hole camera inspections were repeated in order to establish pre-treatment conditions and also to determine if there had been any changes in well performance over the past year. These tests were performed again shortly after treatment to evaluate the effectiveness of the treatment process on well performance and to examine the impact of the treatment on the levels of microbiological aggressivity within the well.

To evaluate the long-term effectiveness of the treatment, pump tests, camera inspections and microbiological tests were repeated in May, 1998 (nine months after the treatment).

Pump Test Results. In order to evaluate the long-term effects of biofouling on well performance, pump tests were performed on the six wells selected for treatment in 1996 and in 1997 before treatment was applied. During each test, water was pumped from the well for two hours at a constant rate and the water level recorded at regular time intervals to provide a drawdown curve. The two-hour specific capacity of each well was then calculated as the pumping rate (Igpm) over the total drawdown (ft). These tests were run again in 1997 immediately after treatment to evaluate the effectiveness of the treatment applied and again in 1998 approximately 9 months after treatment for comparison purposes.

The diagnostic pump tests performed before treatment, in 1996 and 1997, were used as a benchmark for any changes in well production with respect to time. Specific capacity results (Table Sixty-One) from these tests were compared to provide some quantitative measure of the effects of biofouling on well production. The following table shows pre-treatment specific capacity results from 1996 and 1997.

Table Sixty-One

Changes in Specific Capacity Following Treatment,
Six Test Wells in Kneehill M.D.

Well Number	Post-treatment Specific capacity (Igpm/ft)		
	1997	1998	%change
1	1.152	1.270	+10.0%
2	0.216	0.195	-9.7%
4	0.113	0.110	-2.6%
5	0.021	0.019	-9.5%
6	0.678	0.508	-25.0%
7	0.214	0.236	+10%

Pump tests conducted in May 1998 indicated that the specific capacity had decreased in four of the six study wells. The remaining two wells showed slight increases in specific capacity. Losses in specific capacity ranged from 2.6% to 25% of the 1997 post-treatment specific capacities. Some losses in specific capacity are to be expected over time as the bacteria re-establish plugging formations in, and around, the wells following treatment. Sudden large losses in specific capacity could indicate that areas of biofouling in the well environment were not adequately penetrated by the treatment chemicals and grew back quickly following the treatment. It is possible that some of the study wells had become so severely biofouled that they would require a second treatment to remove all the plugging material.

The gradual losses in specific capacity progressing over the nine month period following treatment underscores the importance of regular well maintenance. For the four wells showing reductions in the specific capacity, the losses over the nine-month period were variable. Annual losses for wells 2, 4, 5 and 6 were 12.9%, 3.5%, 12.6% and 33% which would translate into a life span 7.7, 28.6, 7.9, and 4.0 years respectively assuming a linear rate of specific capacity decline.

Regular preventative maintenance conducted on plugging wells would not only slow the development of (plugging) biofilms in a well but also extend the life span of the well. Newly installed wells or wells that have been effectively treated would therefore benefit from regular monitoring of bacterial activity as a convenient indicator of the status of plugging at that well. When increases in activity are noted, the well owner can apply appropriate well treatments to effectively extend the life span of the well.

Conclusions: The UAB treatment was effective in the recovery of significant specific capacities in five of the six wells tested. The well that did not show recoveries was possibly being subjected to factors not related to biofouling (this was reflected in the difficulty in obtaining precise data from the well). It was possible to extrapolate the potential longevity of the treated wells through repeated measurement of the specific capacity on an annual basis. In the five wells successfully treated, the recovery in specific capacity and production was not short-lived but rather durable with the improvements still being evident after nine months. Preventative maintenance is an essential part in the ongoing management of water wells if they are to be considered sustainable rather than disposable.

Location: Plugging mesocosms, Canada Agriculture, Regina

It has been estimated that there are around 200,000 water wells operating in 1998 in the Canadian prairies. If it were to be assumed that the average cost of well installation was C$5,000, then the minimum investment in the present operating wells would amount to C$1,000,000,000! If the average operable life span for a well was 15 years, then this would mean that each year there would have to be an investment of $66,000,000 just to maintain the number of operable wells. Canada Agriculture through the Prairie Farm Rehabilitation Administration –Technical Service (PFRA-TS) recognized this cost and initiated a joint project with Droycon Bioconcepts Inc. to determine more effective management strategies to extend the useful life span of the wells. This became known as the Sustainable Water Well Initiative (SWWI).

As a part of the SWWI, a series of experimental mesocosms (1L) and one megacosm (100 L) were constructed both at the University of Regina and at PFRA-TS. These were simulations of aerobic (oxidative) porous media (fractured sand or gravel) set around a slotted well screen and through which a nutrient feedstock was passed to encourage biofouling. The location of the aeration, sequence for pumping, and the manipulation of the redox front (primarily by shutting off aeration or nitrogen sparging) were used to assess biofouling. Plugging was measured commonly by the drainable volume in the mesocosm during the plugging process. As the biofouling increased, it occupied the void volume thus reducing the free (drainable) volume. Alternative measurement included the rate at which the free drainable volume was released from the mesocosm. In a pristine (unbiofouled) mesocosm, there is very little resistance to flow during free drainage discharge. However, once the voids become "coated" with biofilms and the plugging processes have started, there is a much slower rate of "free" discharge reflecting the resistance created by the plugged voids. In general, when the void volume occupied by plugging reaches between 40 and 80% (depending on the media porosity), this resistance can become sufficient to effectively prevent drainage. The mesocosm thus becomes plugged completely even though there may be some free interstitial water left within the plugged medium. There are a number of stages in the plugging process for a mesocosm that would reflect those stages that occur in water wells. These include:

1. Diminishing total drainable volume (this volume loss would reflect the volume of the voids occupied by the plugging)
2. Slower rates of draining reflecting the increasing resistance to water flow created by the occupation of the voids by the plugging process
3. Lengthening times to complete drainage of water from the mesocosm due to (2) above
4. Reductions in the quality of the discharge due to sloughing of colloidal elements from the plugging material during the drainage process.

It was through research using mesocosms biofouled with IRB and/or SRB that the various improvements in water well rehabilitation have been, and continue to be, developed. The next part of this example describes a typical mesocosm experiment in which an IRB generated plugging is developed.

Conclusion: Valuable information concerning the rate and form of various biofouling activities can be monitored using laboratory scale mesocosms. This information can aid in the development of effective treatment strategies, suitable materials for construction and appropriate management practices.

COMMENTS ON BIOFOULING IN OTHER ENVIRONMENTS

Plugging and biofouling is not an event limited to just the ground water environment and water wells but occurs widely. The following section of examples briefly covers parallel examples of biofouling and/or plugging that may have relevance to particular circumstances associated with wells. A brief overview of the potential for biofouling in other environments is listed below.

Surface water is often subjected to biofouling that initially changes the clarity of the water followed by the degeneration in color and taste and odor problems. Since water interfaces commonly with the sediment underneath, there are often interactions between the water and the sediment. Where there is a turbulence in the water, this can drive sediments up into the water to reduce quality. When the water is relatively still, then there is the potential for the sediments to release gases entrapped within the sediments usually as occasional streams of bubbles. The common gas is methane being generated by the methanogenic bacteria.

Organic pollution of a surface water body clearly has the potential to: (1) generate microbial aggressivity; (2) remove oxygen from the water as a result of microbial respiration; and (3) reduce the water quality to unacceptable levels. Frequently, a dominant factor with organic pollution of surface waters is the potential hygiene risk that would be associable with the detection of fecal coliforms (e.g., Escherichia coli). This natural preoccupation with hygiene risk potentials rather than the global impacts on the microbial ecosystem has led to an inadequate understanding of the systems involved.

Biofouling is most commonly recognized by the senses of sight and smell. If the water looks cloudy or colored (by algal or bacterial growth) or has an odd odor ("rotten" egg, septic and earthy musty are three common ones), then there is probably a serious microbial biofouling occurring. At this time, the idea of surface water as being subject to a "biofouling" has not been accepted even though it is an almost inevitable follow-on to an organic pollution event. Inorganic pollution is often not associable with biofouling in surface waters because of the often very aggressive nature of the polluting chemicals (e.g., toxic, carcinogenic). What is frequently not appreciated is that these chemicals will often become entrapped (accumulated) in the slime (EPS) matrices around the cells. Toxic chemicals may accumulate at a distant from the cells and so not impact directly on the activity of the cell.

The nature of surface waters is such that many of the elements of biofouling may be suspended in a colloidal matrix within the more tightly clustered water. Generally, biofouling will result in very aggressive populations of nuisance bacteria that will reflect the nature of the impact. For example, very short time lags in the following BART tests would reflect very aggressive bacteria and a probability of a biofouling event (Bart type, time lag in days, comment):

SRB <2 BT – anaerobic conditions, severe organic pollution
SRB <1.5 BB – risk "rotten" egg odor high, anaerobic conditions
SLYM <3 BA – heavy organic pollution, hygiene risk may exist
SLYM <1 CL – extreme organic pollution, BOD may be high
IRB <2 GC – aggressive pseudomonads, aerobic biodegradation
IRB <1.5 FO – anaerobic conditions
HAB <1 UP – aggressive aerobic bacterial activity.

Soil is a porous medium like the media around water wells in the sense that it has a porosity and permeability and also generates redox fronts. As in water wells, the establishment of a redox front (usually laterally) in the soil forms a focus site for microbial activity. Iron pans in soils are a classic example of the product of a stable redox front in which there had been an ongoing bioaccumulation of iron (in much the same way as it bioaccumulates around wells!).

Within the soil, there are natural lateral stratifications of different microbial groups as the redox shifts from oxidative to reductive and the voids become saturated with water. Similar stratifications can also occur around plant roots where there is an ongoing exchange of organics, nutrients, water and gases. Testing soils for the microbial aggressivity is therefore challenging. Generally, this is resolved by making composite samples from soils taken to a depth of 10cm or 4inches. Because of the complex nature of the community structures in the soil, there is almost inevitably a large level of variability that can weaken the "scientific" validation of findings. This somewhat parallels the inaccuracies in ground water investigations because of the variable rates of sloughing from the biofilms that dramatically affects the microbial populations in water samples.

Biofouling in soils can lead to a variety of impacts. The most obvious of these involves the generation of slime layers that compromise the permeability of the soil. Commonly, these slimes harden through the accumulation of iron and/or carbonates and fill the void spaces commonly at the redox front. This front is often associated at the level with the top of the water table. Other impacts, less recognized, include losses in the water holding capacity (due to the destruction of the slimes containing EPS), deposition of salt crusts (as a result of the "wicking" of water in biofilms up to the surface where it evaporates leaving salt deposits), and losses in soil texture and structure (due to the destruction of mycelial structures and EPS in and on the soil particles).

The generation of impermeability in a soil can seriously compromise the ability of that soil to allow recharge of water back into an aquifer beneath (e.g., through rainfall). Losses in water retention and texture would lead to a soil not being able to retain moisture for long enough time periods to support plant growth. Relatively little is known of the subtle relationships between the negative aspects of microbial growth causing biofouling and the positive aspects enhancing the fertility of the soil to support plant growth. An example is given in the next section of the experiences that have shown the impact of biofouling in high sand content greens. Here, lateral black plug layers have caused losses in permeability and severe die back in the turfgrass.

BART testing of soils requires a different strategy to water. The easiest manner to test is to follow the following sequence of testing events. This assumes that you have a composite soil sample that is ready to be tested. The sequence would be:

1. Weigh out 0.1 g soil
2. Unscrew the double cap of the BART and turn the two caps inside facing upwards
3. Lift out the inner tube
4. Carefully turn the tube over in a manner that causes the ball to roll into the up facing inner cap
5. Turn the inner tube back upright and add the 0.1g soil sample so that it drops into the base of the tube over the dried medium
6. Return the ball to the inner tube
7. Fill the inner tube with sterilized distilled or deionized water up to the fill line (15ml). The ball will float up as the water is added and the soil should disperse in the water
8. Cap the tubes and place in a secure upright place at room temperature out of direct sunlight. Observe daily for the recognized reactions and record reaction codes and time lags on the standard BART™ date interpretation sheet.

Black Plug Layering (BPL) in golf greens. Golf has become a major recreational industry with severe pressures by the users to generate turfgrass greens that are perfectly manicured, uniform, and provide an excellent playing surface. As a part of the industry strategy for meeting these demands, a high sand content green has been developed which involves the application of a regime that is not only ideal for the turfgrass but also suitable for the growth of many of the microorganisms in the sand green. Essentially, a "war" develops between the turfgrass roots and the microbes for space, water, nutrients and oxygen. While the turfgrass is "winning" nobody cares, but when the microorganisms "win" the outcome is a form of biofouling. This form of biofouling is most commonly seen as the black plug layer (BPL) which develops laterally just under the surface of the green. The BPL competes with the turfgrass for living space and the grass dies back leaving chlorotic patches of grass, bare patches of green and black globules oozing up through the green.

The BPL is an anaerobic complex consortium of bacteria, fungi and algae with the majority of the slime being dominated by SRB. The growth is complex including lateral slime plates and vertical columns all overlayered with perched water in which algae grow copiously. Controlling the BPL, once it has established itself, is very difficult and the traditional approach has been to attempt aerification and radical watering practices. If these fail, then the greens are rebuilt completely. Radical approaches include foliar feeding the turfgrass (to "starve" the soil microflora), aerification with high pressure oxygenated water injection and the use of surfactants.

Detecting the presence of a BPL in a green can most easily be achieved by coring the green and finding black slime patches in the sand. The position may vary with the maturity of the BPL and, in severe cases, the BPL is found between 4 to 100mm beneath the surface of the green. Of the BART tests, the SRB will commonly give a BA reaction in less than three days and an IRB will give an FO reaction in less than one day.

Natural Attenuation and Biodegradation, in the last decade there has now developed a growing recognition that Mother Nature is actually quite capable of biodegradation of even recalcitrant organic compounds. The primary event that allows this is the adaptation of the natural microflora at-site to perform these additional degradation functions. Essentially, management of these processes is akin to a farmer growing a new crop. The "crop" in this case is a complex consortium of microorganisms that are adapting to degrade the chemicals of concern. What has happened is that there has been a natural attenuation of the normal microflora to fulfill these biodegradative functions. This approach to biodegradation is particularly appealing because the science involved is extending the processes already initiated by Nature without the need to introduce any alien species "engineered" to specifically meet the perceived need.

Aerobic biodegradation functions often involve a dominance of pseudomonad bacteria in saturated media and fungi in semi-saturated conditions. These bacteria can become very aggressive and are often most easily determined using the FLOR or the SYM BART testers using the standard protocol recommended for soils (see above). Generally, a PB or GY reaction code will follow a CL reaction. Where the natural attenuation is proceeding rapidly, time lags can be as short as four to eighteen hours to a CL reaction with the PB or GY reaction following within a day. The time lag can be used to gauge the rate of biodegradation with the shorter the time lag, the faster the rate of degradation. By manipulating such strategies as aeration rates, nutrient application rates and timing, and pH amendment, it is possible to use the pH to control the rate of degradation.

Under anaerobic conditions, natural attenuation may function using a variety of dominant consortia such as the SRB, DN and IRB. Time lag interpretations would commonly be slower (x2) those of aerobic degradation with critical reaction being:

- SRB – BB, predominantly anaerobic biofilm; and BT where there remains an aggressive aerobic component
- DN - copious FO in one day would indicate that a dominant denitrification is occurring
- IRB – FO, predominantly anaerobic biofilm with IRB involved; BL, intense enteric and pseudomonad bacterial consortium dominating; BC, IRB dominating; RC, enteric bacteria dominating.

By monitoring the bacterial populations and their composition and aggressivity, a closer control can be applied to assure the optimized performance of a natural attenuation process.

Die back around oil and gas wells. One recent problem that has caused concern has been the die back of vegetation around oil and gas wells. Concerns relate to the possibility of some negative impacts on local vegetation from the operations of the wells. The causes of the die back are a natural focus for the wells where this occurs, particularly where a zero tolerance is being practiced to control such environmental aspects.

Microbiological aspects of the die back relate to parallel events to those observed in the die back of turfgrass in golf greens. The probable cause would be the organic compounds moving up, particularly from wells that are leaking methane gas and/or volatile hydrocarbons. These compounds move up through the ground water to the redox fringe where they become bioassimilated into the biofilms growing at those sites. This acts as a primary factor expanding the size and aggressivity of the plugging zones. One net result of this is that there is an acute competition with the roots of the local vegetation. Where the biofilms dominate to form copious plugs there is no longer enough water, nutrients and space to support the plants and these die back leaving soil which has now become compromised and has reduced permeability.

Coring the impact zones will frequently reveal colored layers of plugging growth dominated by the shades of black (SRB) and brown (IRB). At these sites the permeability becomes reduced two or three orders of magnitude. BART tests reveal very aggressive SRB in the black layers going to a BA in less than three days. In the brown layers, the IRB dominate commonly with a very aggressive FO reaction in less than two days. This culminates in BC and BR reactions occasionally ending with a BL. Throughout the impacted die back zone, the heterotrophic bacteria are very aggressive and non-pigmented zones registering low permeability often have very aggressive HAB populations which may give either UP or DO reactions commonly in less than one day.

What appears to be happening microbiologically at these impacted sites is that the nutrients moving up from the well reach the redox fringe commonly at the water table and stimulate biofilm growths (resembling a slime "umbrella" around the well). The growths now dominate pore spaces and restrict the root penetration causing the plants to die back.

Plugging in Oil Wells. There has never been a common acceptance that microbial biofouling could occur down oil wells beyond causing corrosion (through the activities of SRB). The mindset dictated that oil was not a suitable substrate for the microbial growth and fouling activities. An examination of the losses in production in an oil well parallels the form production losses in water wells. Traditionally, for oil wells, these events were thought to relate to physical plugging with fines and clays rather than microbial events. Microbiological testing using the BART tests revealed that there were highly aggressive HAB populations. Dispersing the black clogging paraffin and anthracene (P/A) deposits revealed considerable elements indicating that these deposits were microbiologically integrated. Within the black P/A deposits, elements of sheaths, filaments, EPS strands and globular elements were recovered. This may mean that the microbial fouling was stimulated through the microbes actually mining water from the oil in sufficient amounts to establish P/A dominated plugging. BART testing revealed that HAB (DO, time lag <2 days) and SRB (BT or BA, time lag <3 days).

Conclusion

Microbial activities are universal on and in planet Earth and the management of the planet has to involve an understanding of the various, often subtle, impacts that microorganisms can have on the functioning of the planet. It has to be remembered that microbial activities extend into a greater part of the planet than either the plants or the animals.

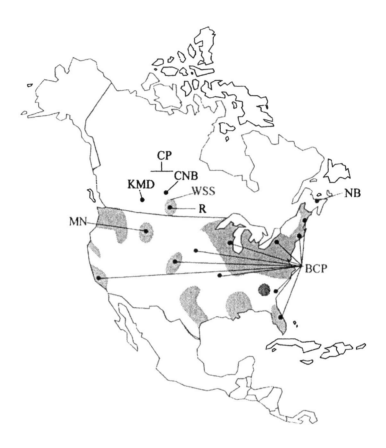

Figure Fifty-Nine, Map of North America showing the sites at which the studies discussed in Chapter Seven are located. Much of the studies were within the Canadian prairies (CP). BART testers were employed across North America (shaded areas) in many of experiences discussed in this book. A prime site for studies was the City of North Battleford (CNB) and the Kneehill Municipal District, Alberta (KMD). Mesocosm simulation of plugging events in the field were conducted in Regina (R) and two major surveys of water wells were undertaken in southern Saskatchewan (WSS). Intensive bacteriological studies were included in the studies in New Brunswick (NB) and the consortial nature of plugging was first located at water well fields in Montana (MN). BCHT projects (BCP) commonly also determined plugging in the various projects over the last decade using the BART tests.

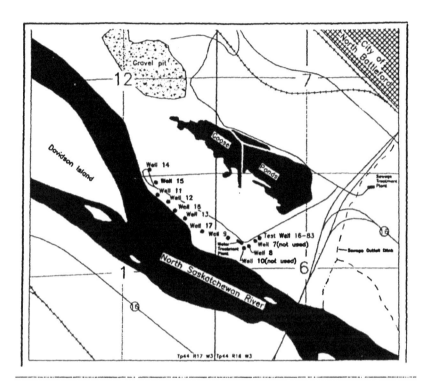

Figure Sixty, Location of the water wells in the field utilized by the City of North Battleford. These wells are located along the shore of the North Saskatchewan River and receive at least some of the water as recharge from the river. Wells in use currently are primarily #11, #15, #16 and #17.

BEFORE TREATMENT W
C86   C85   #15

AFTER UAB   TREATMENT W
C92   C91   #15

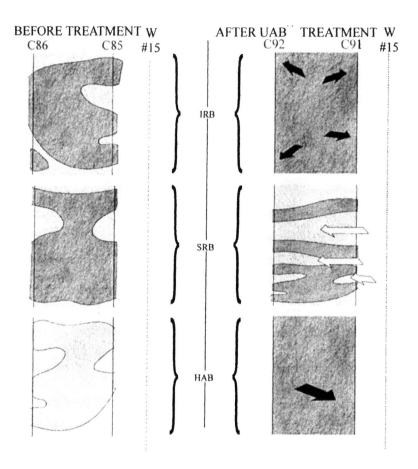

Figure Sixty-One, Profile of bacterial consortial biofouling of four cores sampled at various levels within the range of the well #15 well screen (18 to 25 meters) before (left, C86 - 6m away, C85 - 3m away) and after (right, C92 - 6m away, C91 - 3m away) treatment rehabilitation using UAB. The location of the highly and moderately aggressive (by shade intensity) IRB (upper), SRB (center), and HAB (lower) are shown for the cores sampled before (left) and after (right) treatment. The effect of the UAB treatment (right) on the bacterial plugging of the cores is shown by open arrow where the treatment reduces the bacterial aggressivity and by a black arrow where there is an increase in bacterial aggressivity. Note that well #15 was not included in the bacterial study because cores could not be safely taken so close to the well.

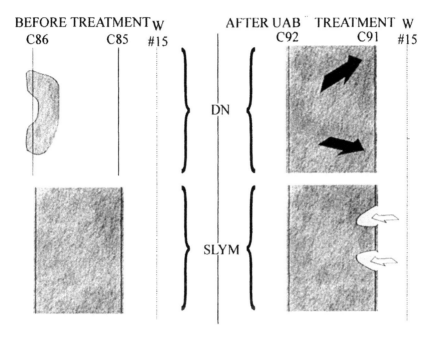

Figure Sixty-Two, Profile of bacterial consortial biofouling the comparative cores sampled at various levels within the range of the well #15 well screen (18 to 25 meters) before (left, C86 - 6m away, C85 - 3m away) and after (right, C92 - 6m away, C91 - 3m away) treatment rehabilitation using UAB. The location of the highly and moderately aggressive (by shade intensity) SLYM (lower), and DN (upper) are shown. The effect of the UAB treatment (right) on the bacterial plugging of the cores is shown by open arrow where the treatment reduces the bacterial aggressivity and by a black arrow where there is an increase in bacterial aggressivity Note that well #15 was not included in the bacterial study because cores could not be safely taken so close to the well.

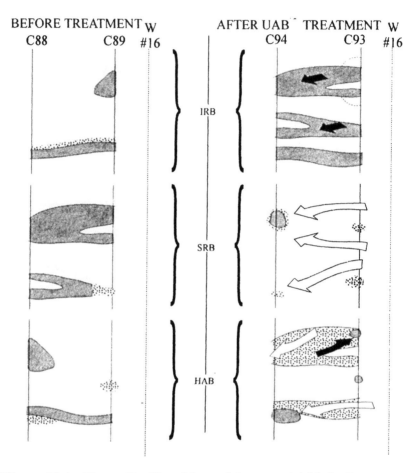

Figure Sixty-Three, Profile of bacterial consortial biofouling cores sampled at various levels within the range of the well #16 well screen (8 to 19 meters) before (left, C88 - 6m away, C89 - 3m away) and after (right, C94 - 6m away, C93 - 3m away) treatment rehabilitation using UAB. The highly and moderately aggressive (by shade intensity) IRB (upper), SRB (center), and HAB (lower) are shown in this diagram. The effect of the UAB treatment (right) on the bacterial plugging of the cores is shown by open arrow where the treatment reduces the bacterial aggressivity and by a black arrow where there is an increase in bacterial aggressivity. Note well #16 was not included in the bacterial study because cores could not be safely taken so close to the well.

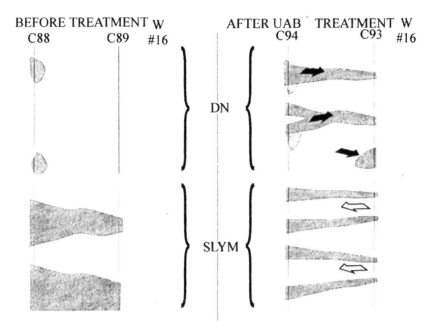

Figure Sixty-Four, Profile of bacterial consortial biofouling four cores sampled at various levels within the range of the well #16 well screen (8 to 19 meters) before (left, C88 - 6m away, C89 - 3m away) and after (right, C94 - 6m away, C93 - 3m away) treatment rehabilitation using UAB. The location of the highly and moderately aggressive (by shade intensity) DN (upper), and SLYM (lower) are show both before (left) and after (right) the treatment. The effect of the UAB treatment (right) on the bacterial plugging of the cores is shown by open arrow where the treatment reduces the bacterial aggressivity and by a black arrow where there is an increase in bacterial aggressivity Note well #16 was not included in the bacterial study because cores could not be safely taken so close to the well.

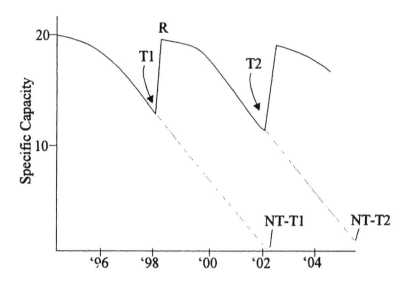

Figure Sixty-Five, Impact of the UAB treatment on the specific capacity of well #16 (T1). Also given is the long term impact on the specific capacity (y axis) over the next two decades (x axis) had no UAB (ie: T1 not applied) been applied at all. If there is no future UAB (T2) or equivalent treatments applied to control plugging (NT-T2. dotted line), then the specific capacity is shown as NT-T2.

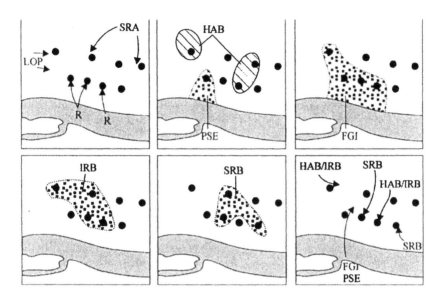

Figure Sixty-Six,   Map view of the well field in Montana that contained eight large producing wells (upper left) set alongside a river. This field was   subjected to mixtures of ground water from a snow recharged aquifer (SRA), the river (R) and from leakages from the oxidation ponds (LOP). The physical and chemical characteristics of the wells was affected by these water sources. In addition, there was a distinct clustering of the bacterial consortia around various wells. This is shown for the heterotrophic aerobic bacteria (HAB, upper center), pseudomonads (PSE, upper center), Fungi (FGI, upper right), IRB (lower left), SRB (lower center) and a summation of the creations of the dominant plugging groups (lower left).

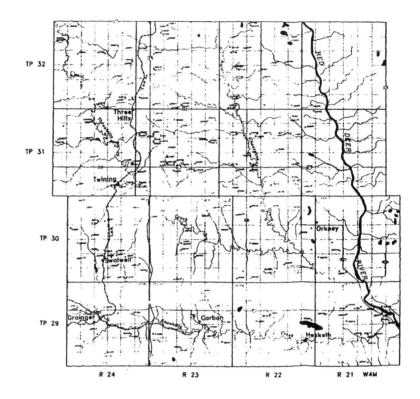

Figure Sixty-Seven, Location of 134 wells included in a microbiologi-cal survey of water wells in the municipal district of Kneehill, Alberta, Canada. These wells were selected by a well survey of 275 well owners to gather basic information on the wells for the major findings.

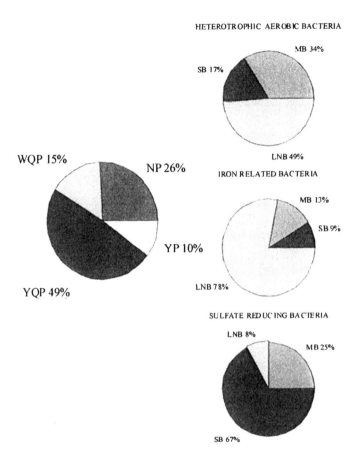

Figure Sixty-Eight, Summarized results of the microbiological study at Kneehill M.D. The pie chart (left) shows percentages of wells with no problems (NP), yield problems (YP), yield and quality problems (YQP), and water quality problems (WQP).   The other three pie charts (right) display the percentage of well severely biofouled (SB), moder-ately biofouled (MB), and low or no detectable biofouling (LNB).

8

MAJOR CONSIDERATIONS IN WELL BIOFOULING

There are a number of important considerations that are applicable to all water wells. These are summarized below and represent some of the major factors affecting the sustainability of water wells. These are listed in point form and are not any specific order relating to importance but all should be considered as relevant to the effective management of water wells.

1. There is no such thing as a sterile water well. Microorganisms will always be present and the level of biofouling will relate to the position and form of the microbial activities that occur.

2. Biofouling will commonly begin during the development of the water well. Once the well goes into production, the biofouling will accelerate and begin to impact on the specific capacity and water quality. Significant biofouling may be occurring even when a well has only lost 5 to 10% of the original specific capacity. It is not appropriate to consider that a well is not suffering from until after the specific capacity has declined by 40% or more. By this time the well may be so severely plugged that rehabilitation back to the original specific capacity may not be achievable.

3. Preventative maintenance should be practiced from the time that the well goes into production. Using the BART biodetectors, the time lag (as days of delay) and the reaction patterns signatures observed can be used to determine whether the microbes causing the biofouling are becoming more aggressive (i.e., shorter time lag) or changing in dominant microbial fouling (i.e., shifting in the reaction pattern signatures) can be tracked by routine testing (commonly monthly with repeat testing when there is evidence of increased aggressivity). Where treatments are applied, the success of the treatment may be determined through improvements in the specific capacity and water quality along with lengthening time lags and shifting reaction pattern signatures.

4. Biofouling does not have to occur just on the well screen and be visible when a camera logging is performed down the bore hole. The absence of any fouling on the screens does not mean that there is no biofouling in the well, it simply means that there is no observable biofouling on the well screen. There are many ways in which the biofouling can occur around a water well. These can range from a tight plugging forming in some of the regions close to the well screen and causing variable water quality as water enters the well from different formations to various structures sited at different distances from the well. The microbial activity will tend to focus around the sites of maximum water flow towards the well and where turbulence occurs. As plugging develops at this site, the water flow patterns change. Even (laminar) flow down the length of the well screen can reduce this focusing of microbial activity. Instead, the plugging occurs more evenly over the length of the well screen to have a less dramatic impact on water quality and specific capacity. On some occasions the microbial fouling can move towards the well along a redox front created by ground water recharge from surface water sources. This type of fouling can be insidious and cause sudden dramatic losses in water quality and specific capacity once the growths begin to encapsulate the well itself.

5. Biofouling is an all embracing term which can relate the losses in specific capacity (plugging), accelerated corrosion (often due to the activities of the SRB), losses in water quality (e.g., increasing turbidity due to microbial growth and biofilm sloughing, taste and odor problems resulting from microbial activities in the biofouled zones), and equipment failure due to encrustations and/or corrosion processes. When a biofouling event is evident, the BART™ biodetectors can be used to test the product water and confirm the type of biofouling that has been occurring.

6. Little attention has been paid to the average life of a water well in active production before the installation is reduced to a stand-by status or abandoned. While each well has to be treated as unique (Water wells even very close to each other can often exhibit totally different characteristics), there are some general observations. In the Canadian prairies, it appears that the average life span of a water well is fifteen years with the range commonly between five and thirty years. Given the cost of installing a new well, there needs to be more attention paid to increasing the life span of each water well so that it can become more sustainable.

Increasing the life span of the well would involve an active preventative maintenance program and effective treatment program which would control the level of biofouling in the well.

7. Treatment of a water well to remove biofouling is inherently difficult since the biofouled zones are commonly in the void spaces and difficult to access due to the buffering action of the surrounding media. In selecting a particular treatment strategy it should be remembered that no one size fits all, and careful consideration has to be given to the form and the location of the zones of biofouling in and around a well. It is not appropriate to claim that a specific treatment has been successful simply because the treatment product was located one or several hundreds of feet away from the well being treated. Where this happens, the only valid observation can be that there was a conduit (channel) through which the treatment fluids moved away from the targeted well. This conduit may have, at least in part, been formed by the biofouling of the surrounding porous media around a fracture or water lens within the media. The consequence of this would be that the treatment would impact on that surrounding mass of biofilms (a.k.a. slime tube) but not on the fouling occurring away from that (those passages). While such treatment would impact on the localized growths around the channels through which the treatment fluids are passing, it is less likely that there is an overall impact on the remaining biozones. Evidence of recovery has to relate to the recovery of specific capacity, a reduction of the biological burden in the water and a return of the water quality parameters to the original levels for that well.

8. While it is obviously in the interest of promoters of particular treatment practices to down play the involvement of microbes in plugging, corrosion and other biologically induced problems, it has to be remembered that a water well can never be completely "sterilized" by the common treatment practices employed in the industry. When a well has been treated, there will inevitably be some debris associated with the disrupted biofilms. This will include nutrients and viable cells. These living survivors now become "cannibals" living off the dead cells and disrupted organics and nutrients within and around the zone that was treated. This activity will rapidly generate a "rebound" of biological activity and restart the processes of biofouling. Promoters of specific treatment processes that devalue the

biofouling to a "minor" or "easily controllable" event do not take into account the inevitability of a "rebound" biofouling that will follow treatment unless there has been scrupulous care to remove all of the fall out debris from the treatment.

9. One of the major difficulties in assessing the success of a treatment from the microbiological viewpoint is when to test the water well to determine that the treatment has been effective. One effect of a treatment, successful or not, is that there is a "kill" zone in the immediate environment of the effective treatment zone. Beyond that zone is a secondary impact region where the biofilms and associated encrustations would have been disrupted with a variable kill rate. Here, the surviving microbes begin to cannibalize the dead cells and utilize the disrupted organic material. This leads to localized "blooms" of growth. When the water is first pumped after such a treatment it can be expected that the initial product water has a very high particulate loading with a low level of microbial aggressivity (while the killed cells predominate in the water). As the water from the secondary impact zone is pumped out, the characteristics of the water change as the particulate material is flushed out (e.g., turbidity and chemical parameters drop) and the aggressivity increases (as the "cannibal" microbes grow within the debris created by the treatment). Frequently during this latter event the aggressivity recorded using the BART biodetectors exceeds by one or two orders of magnitude the background levels observed prior to treatment. It is, therefore, recommended that the determination of the effectiveness of an applied treatment be delayed until the biological activity in the impacted zone has stabilized and the debris has been essentially flushed out of the system. The recommended time delay for conducting this assessment for the effectiveness of a treatment is a minimum of six weeks. The bottom line is recovery of the specific capacity back to its original levels by an effective rehabilitation treatment. Remember that the more the specific capacity has been lost by biofouling (particularly when it has dropped by greater than 40%), the less there is a probability that the well can be rehabilitated completely.

10. It is uncommon for a water well to have a single localized site of biofouling around the well. More often, there are complex communities both layered one on top of another and as concentric circles of growth around the well. In the latter event, each of the

concentric biozones may have a very different bacterial community structure. Usually the shift in community structure moving away from the well is from an oxidative (+Eh) to a reductive form (-Eh). Aerobic bacteria are more aggressive closer to the well in the oxidative zone with IRB tending to dominate. However, away from the well in the reductive zones, anaerobic bacteria tend to dominate including the SRB and the methanogens (biogas producing bacteria). Between the oxidative and reductive regions lies the redox front where a variety of facultatively anaerobic bacteria can dominate. These will include enteric bacteria and also those aerobic bacteria able to utilize alternative respiratory substrates such as nitrates. Pumping water from a well having these various biozones means that the product water will contain first the aerobic bacteria often dominated by IRB, then the nitrate respiring aerobes and facultative anaerobes followed by the anaerobic SRB and the methanogens. Sequential sampling of the product periodically for at least two hours with at least three samples becomes critical to a more comprehensive determination of the extent of the biofouling.

11. The presence of anaerobic bacteria (SRB and/or methanogenic) can have direct and serious implications. For the SRB, this can mean the development of "rotten" egg odor, generation of black water and slimes and the initiation of corrosive processes. Where the methanogenic bacteria dominate deep in the reductive parts of the formation, methane will be produced. This methane gas may lock into the voids in bubbles (foam blockage) or move towards the oxidative regions within and around the well. Where this happens, methane may enter the water well column and even vent off from the well with the associated risk of ignition. Methane may also be degraded within the oxidative regions by methane degrading bacteria (methanotrophs) and this can seriously increase the amount of plugging. The presence of methane in a water well has sometimes been thought to be totally associated with leakages of natural gas from neighboring gas and oil fields, but on some occasions this gas could be locally generated by methanogenic bacteria in the immediate vicinity of the well. The detection of aggressive methanogenic bacteria in water samples from a well would indicate that there would be a potential for locally generated methane production.

12. Since most of the bacteria associated with the biofouling events occurring within and around a well are residing within biofilms,

encrustations, nodules and tubercles, they would not necessarily be expected to be always present in the product water. Casual sampling of water from a water well contains intrinsic errors since the sample cannot be representative of the various biological events going on within and around the well in the ground water. To induce the bacteria to detach from the biofilms, encrustations, nodules and tubercles, the environment within the well has to be changed from that normally experienced by the resident microbes in the well. Commonly, wells follow a routine schedule of pump and rest normally on a twenty four hour schedule. To break that routine the easiest thing to do is to shut the well down for long enough to break the routine cycle and stress the incumbent bacteria. That stress is created primarily by the shifting redox front (moving towards the well) and the loss of hydraulic flows (that would bring nutrients towards the biozones and biofilms). Ideally, the well should be shut down for a week prior to sampling but very often a producing well cannot be taken out of service for that long. Practical experience would dictate that a twenty-four hour shut would normally be effective with an absolute minimum of eight hours. For wells being run almost continuously, it is often difficult to get the user to shut the well down at all. In these cases, it has been found that a shut-down of two hours will at least begin to cause releases of the bacteria to occur from the biofouled regions. Sequential sampling once the pump has been turned back on allows the location of the fouling to be predicted (the later the sample in which the bacteria is first detected, the further away from the well these bacteria are infesting).

13. Many specialists still down-play biofouling events as being either "of no major significance" or "easily controlled (not a problem)" using a preferred treatment procedure. In reality, the covert nature of the bacterial infestations in and around the water well often makes them difficult to detect and/or easily discounted. Microbiologists over decades have concentrated on trying to identify individual species apparently involved in the biofouling event rather than looking at the community (consortia) of bacteria that are involved. Given that these bacteria may infest different sites in various community structures, the dominant communities can often be used to project the likely source of nutrient factors stimulating the biofouling event. For the BART™ testers, the relative aggressivity of the different bacterial groups can be used

to predict the source(s) of the biofouling. For example, aggressive IRB are likely to be found in the earlier samples taken during a sequential pump test (see 12 above) and would mean that there was a high iron and possibly manganese concentration in the ground waters entering the well and that the conditions were fairly oxidative. Aggressive SRB, on the other hand, are associated very much with reductive ground waters with a high sulfate and/or organic loading. Field experiences have found that frequently the SRB can become exceptionally aggressive when the well is being impacted with methane generated biogenically local to the well or arriving at the well through passages from natural gas formations or leaking gas and oil wells. Heterotrophic aerobic bacteria (HAB) tend to dominate where there is a high organic input in the ground water and the conditions are more oxidative. These microbes also can be stimulated through their ability to respire nitrate when there is no oxygen present in the water. SLYM (slime-forming bacteria) produce very thick, often sticky, slime coatings that often form at redox fronts and discharge outlet ports. These bacteria also thrive in high organic ground waters but do not bioaccumulate iron and manganese to the same extent as the IRB. Denitrifying (DN) bacteria become very aggressive in ground waters impacted by sewage and high nitrogen organic wastes that have passed through an oxidative environment. When this happens, oxidative bacterial nitrification causes the production of nitrate as a terminal product from the oxidation of ammonium. When the ground water returns to a more reductive environment, the DN begin to reduce the nitrate partially to nitrite and then completely to dinitrogen gas. The presence of aggressive DN indicates a heightened potential for enteric bacteria and sewage to be present. It is prudent to check waters with aggressive DN for the presence of total and fecal coliforms as a measure of the potential health risk.

14. Indirect tests for the presence of bacteria include the use of the ATP as a gauge for the number of viable cells within a given water sample. The premise here is that the greater the level of activity, then the higher the amounts of detectable ATP. It is generally considered that living microbial cells possess 0.1% of the dried cell mass as ATP. This translates to one bacterial cell possessing 5×10^{-16} grams. The advantage of the ATP test is that is does relate to the active microbial cells only and so reflects the aggressivity of the microbes rather than simply recording the

numbers of cells whether they be active or passive within the natural waters and biofouled interfaces. Another indirect test is the use of the laser particle counter which is able to detect and size particles in the water. The range detectable is commonly from 0.4 to 120 microns and so covers the full range of both cell sizes and colloidal structures that may be associable with the microbial population. Often the data is presented as total suspended solids (TSS) along with the mean size of the particles detected and distribution of the cell volumes. Generally, bacterial activity is most often associated with biocolloidal structures ranging in size from 6 to 32 microns. On some occasions, the size range for these structures can be quite narrow (e.g., 8 to 11 microns) and, on other occasions, may "castellate" with separate peaks every 4 to 6 microns. This is a common occurrence when filamentous growths or stalked bacteria (e.g., *Gallionella*) are present. Direct microscopic examination is often employed but tends to bias the observer to that which is the most obvious rather than that which is dominant. A good example of this is the detection of *Gallionella* based on the observation of the stalks that can sometimes go right across the field of view when observing a slide. It has to be remembered that the stalk of the *Gallionella* is dead and contains no ATP. It is essentially an artifact used by the bacteria to dispose of the surplus ferric iron. Each stalk came from a single cell that had been growing much more slowly than the masses of slime (biocolloidal particles) that are often richly populated with consortia of bacteria some of which would have sheared from biofilms. The basic lesson here is that ground water samples from wells may often give very false impressions of the form, extent, function and nature of the biofouling that is occurring in and around a well. It is necessary now to move from an implicit denial of the role of microbes in ground water to an acceptance recognizing that there are many challenges still to be addressed if the true role of microbes in wells and aquifers is to be determined.

15. Over the whole history of water well rehabilitation there has been a reliance on methods that are applied cold to the well (i.e., without the deliberate application of heat). The reason for this has at least, in part, been due to the untested belief that water wells cannot be heated successfully. The reason for applying heat to a well during treatment is two-fold. First, it is well known that the application of heat speeds up chemical reactions and, therefore,

would lead to faster removal chemical encrusta-tions and alien structures from the well. Second, once the temperature is elevated greater than 40C° above the normal well temperatures, it can be expected that there would be massive reductions in living microbial cells ranging from a minimum of one to a maximum of five orders of magnitude lethality. In the last two decades, progress has been made in heating up wells using such techniques as the patented blended chemical heat treatment (BCHT·, ARCC Inc., Daytona Beach, FL). This technique has been subjected to rigorous trials with the U.S. Army's Corps of Engineers and has been found to rehabilitate badly plugged wells. Recovery to original specific capacities has been found to be achievable where the losses have been less than 40% (i.e., the well still retained a specific capacity of >60%). However, as the specific capacity of the well falls over a critical range from 60 down to 20% of the original specific capacity then so the probability of full recovery becomes reduced although a partial and economi-cally viable recovery may still be fully achievable. Here, the main message is conduct preventative maintenance from the day the well goes into production and never let the well sink to below 60% of its original specific capacity. If these rules are followed, then a sustainable water well becomes a reality.

9

USING THE WELL FOULING INDEX (WFI)

With contributions by K.S.Singh and B. Keevill

Introduction

Over the course of the last century, ground water has moved from a readily available limitless resource to one that, like surface waters, can become limited. This limitation can be generated by a combination of over-use, abuse and lack of concern for addressing the factors causing a well to fail. In essence, quality control and quality assurance are minimalized to those essential to the end user rather than to the maintenance of the production from the well.

In the last three decades, there has developed a growing understanding of the physical, chemical and biological factors that can impact on a well's production. These factors can be crudely grouped into the physical and chemical factors that may be considered to be geochemical and could lead to a clogging of the well, and the biological factors which could lead to a plugging of the well. Both of these groups reflect potential fouling pathways that could restrict the production of water from the well in a variety of ways such as losses of production and degenerating water quality.

Since 1985, advances have been made in many areas relating to the management of water wells. Of particular interest to the fouling of water wells has been the establishment of the specific capacity as a primary measurement for the production capability of a well. A second aspect developed over the last ten years is the application of the BART testers to determine the level of aggressivity of the various groups of bacteria in the water well's environment.

To develop a sensitive method for relating the impact of biological plugging and geochemical clogging on the fouling of a water well, a strategy has been developed in which a well fouling index (WFI) is generated based upon the specific capacity and the aggressivity of various groups of bacteria using the BART testers. Equal weighting is given to the historical trend analysis for specific capacity in the well being examined and the historical aggressivity of four selected BART testers. The proposed formulation is essentially based upon an equal weighting of the specific capacity decline and increases in aggressivity (through the shortening of time lag shifts). Where the decrease in time lag is faster than the decreases in the specific capacity, then that event may be considered to be more related to biological plugging factors (BPF). Should there be no such relationship (due to the time lags remaining relatively long and the aggressivity low), it may be considered that the dominant factors influencing the losses in specific capacity are not biological (i.e., associated with the BPF) but rather affected by geochemical clogging factors (GCF). In generating this analysis, the key computation is in the calculation of the general well fouling index (WFI) that uses an equally blended mixture of the specific capacity data and the time lag recorded using selected BART testers. Where the time lag shortening precedes the drop in specific capacity then the dominance may be considered to be biological and the BPF is supported as the probable cause of the fouling. If it is found that the decline only occurs in the specific capacity but not in the aggressivity (i.e., shortening time lags), then the dominant impact may be considered physical and/or chemical and the GCF is supported.

The well fouling index (WFI) therefore indicates the extent of well fouling and their type such as biological (due to bacterial aggressivity) or hydro-geochemical (caused by geological and chemical characteristics of formations around the wells). This index relates the major fouling events such as the losses in specific capacity (SC) and increase in bacterial aggressivity associated with a decrease in the time lag (TL). WFI can be represented as the function of SC and TL at a particular time.

$$WFI = f(SC_t, TL_t) \quad \text{Equation 1}$$

Where, SC_t equals the loss in the specific capacity at a particular time and TL_t equals the change in the time lag after a particular operating period.

The WFI risk factor would be considered as low, medium, and high depending upon the changes as the well ages in the specific capacity (Table Sixty-Two) and time lag as an indication of bacterial aggressivity (Table Sixty-Three). This WFI was developed on the basis of robust concepts generated through experiences gained from historical and experimental data. Prime cornerstones of this WFI concept are that: (1) bacterial aggressivity (BA) is inversely proportional to the TL and can be determined by BART testers; and (2) changes in the specific capacity of a well can be related to the current time production capability of that well. TL is taken to be that time lag prior to the first observation of a positive reaction and it is this TL that gives an index of the aggressivity for the particular group of bacteria being examined. It should be noted that the different BART testers do have different significance to aggressivity for the time lags recorded.

Table Sixty-Two

Relationship between the WFI Risk Factor and SC_t

WFI Risk Factor	SC_t
Low	$SC_t > 0.85\ SC_o$
Medium	$0.85 SC_o > SC_t > 0.6\ SC_o$
High	$SC_t < 0.6\ SC_o$

SC_o, original specific capacity

Table Sixty-Three

Relationship between the WFI Risk Factor and TL_t

WFI Risk Factor	TL_t (d) IRB	SRB	HAB	SLYM
Low	> 8	>8	>5	>6
Med.	5-7	5-7	3-4	4-5
High	< 4	< 4	< 2	< 3

TL_t (d) refers to the time lag (in days) at the time of testing (not original)

In general, the relationship between WFI, TL_t, BA, and SC_t can be represented as follows:

$$WFI \propto (1/TL_t) \propto BA \propto (1/_{Sct})\quad \text{Equation 2}$$

WFI can be considered as an important monitoring parameter in the development and management of a rehabilitation program for water wells. The rehabilitation-success potential would be dependent upon the WFI generated and the robustness of the data used in its generation. Relationships between the rehabilitation potential and the WFI can be expressed as follows:

$$\text{Rehabilitation Potential} \propto (1/\text{WFI}) \quad \text{Equation 3}$$

The above relationship (3) suggests that the monitoring of WFI on a regular basis (preferably monthly and not less than quarterly) after production has started should be mandatory to obtain a reactive sustainable operation of water wells using this factor as a "turn key" stage in the process of applying a preventative maintenance protocol to these wells.

Evaluation of WFI

WFI can be determined by a cumulative weighted evaluation of following parameters:

$$\text{Factorial decline (FD)} = (SC_o - SC_t) / SC_o \quad \text{Equation 4}$$

$$\text{Shift in time lag } (Tl_s) = (Tl_o - Tl_t) / TL_o \quad \text{Equation 5}$$

$$\text{Mean shifts in time lags } (TL_m) = (TL_{sx} + TL_{sy} + TL_{sz}) / N_b$$
$$\text{Equation 6}$$

$$\text{WFI*} = (FD/2) + (_{Tlm}/2) \quad \text{Equation 7}$$

where,

FD	= factorial decline in the well's specific capacity
SC_o	= original specific capacity of the well
SC_t	= specific capacity of the well at a particular time
TL_s	= shift in time lag for a specific BART type after a certain time period
TL_o	= average original time lag for a specific BART type
TL_t	= average time lag for a specific BART type after certain time
TL_m	= mean of shifts in the time lags for all of the BART types used
$TL_{s(x,y,z)}$	= time lag shifts for BART types x, y, and z
N_b	= number of BART types used in the test

*It should be noted that the WFI is based upon an equal weighting being given to the specific capacity and also to the mean shifts in the time lags using the BART testers. This determination of the WFI has been designed for use on new wells going into production for the first time or wells in which both the specific capacity and time lag data (using the BART testers) has been gathered routinely from the beginning of production.

Wells that do not have data extending back to the original start up of production cannot generate a meaningful WFI since there may have been significant changes in the specific capacity and/or the time lag data generated by the BART testers prior to the start of testing. Consequently, correction factors would be required to compensate for this lack of confidence due to an inadequate data base for an aged well. This will be addressed as a separate issue later in the chapter.

Considerations in the WFI Monitoring Protocol

There are seven major considerations that have to be taken into account in the generation of a WFI. These are listed below and may affect the frequency and form in which the data is collected and interpreted.

1. It is recommended that four types of BART testers be employed for the determination of original time lags (TL_o). These would consist of: IRB-BART, SRB-BART, SLYM-BART, and HAB-BART. The number of BART types could be reduced to only those types which did detect aggressive bacteria (i.e., generated a reaction pattern after a certain time lag, see Table Sixty-Three).

2. BART test types that routinely do not generate a time lag through the detection of any reaction patterns may be eliminated from the WFI evaluation protocol. Elimination would involve no reactions being detected within ten days of the start of each testing stage. Elimination would involve the removal of such data as may have been generated for that specific group of bacteria and the TL_m would be generated from those bacterial groups that are detected using the BART testers.

3. In case of all of the BART types failing to exhibit any type of reaction patterns in the original well samples, then a time lag value of 10 days can be assumed for TL_o and recommended testing should be limited to the SLYM and SRB-BART.

4. It is recommended that the water sample for BART™ testing should be collected after 30 minutes of pumping at the maximum rate. This would allow the sample to reflect any potential

plugging problems in the formation around the well rather than localized bacterial growths infesting the well water column itself.

5. Specific capacity and the original set of time lags (TL_o) using BART testers should be monitored at monthly intervals after the well has gone into production. If the well was not used during any monthly period, a pre-sampling period should include a twelve hour continuous pumping from the well followed by a twelve hour shut down period after which the sampling sequence (after 30 minutes of maximum discharge) can be conducted.

6. In the event of a sudden dramatic reduction in the time lag (TL_{t2}) from the previous TL_{t1} for a particular BART type, it is recommended that the test be repeated using duplicate BART testers of that particular type. The mean time lag for that particular BART should then be calculated using the three time lag values obtained (one from the original test, and two from the duplicate repeat test).

Application of the WFI in the Sustainable Water Well Management

WFI can be applied effectively in the water well management program to predict the extent of fouling and its impact on the production capacity of the wells. Figure Sixty-Nine gives a conceptual presentation of the stages of decline in the specific capacity that may occur. The stages are described below.

In the first phase (A) of operation, the well continues to operate at the original specific capacity with no physical symptoms of fouling. This does not preclude the possibility of bacterially induced fouling that has not yet affected the specific capacity (Figure Seventy). Once fouling begins to physically affect the hydraulic characteristics of the well for whatever reason, there is a slow but accelerating degeneration in the specific capacity (phase B). At some point, the degeneration of the specific specific capacity declines in a linear (arithmetic or logarithmic) manner (C) although there would probably be variations resulting from localized events within the well. This decline will continue until the plugging and/or clogging effectively seals off the well (phase D). At this time the specific capacity effectively is reduced to zero (phase E) and the well is no longer capable of producing water without successful treatment.

This forms a conceptual representation of the application of WFI in the water well management but does not attribute the cause to either clogging (inanimate) or plugging (animate). To determine this, the form of the fouling is classified as being generated by one or both of the two factors included in the WFI calculation: (a) predominantly a geochemical clogging (GC), a form of fouling commonly associated with the hydraulic and geochemical inanimate characteristics of aquifer formations but not directly through localized biofouling in, and around, the well, and (b) animate biological plugging (BP) that results from plugging infestation zones (called biozones) that in some manner impede the movement of water into the well.

Because two factors are recognized as being potentially significant, the analysis of the WFI is a two-stage process. The first stage assesses the current WFI based on historical data (for specific capacity and aggressivity using the BART testers). This gives a factorial estimate of the current status of any fouling around the well. It is a composite estimate based upon shifts in the specific capacity (Equation 4) and time lags observed from the BART testers (Equations 5 and 6). The WFI can be finally computed by assuming an equal weighting of the GC and BP in the determination of the changes in the SC and TL (Equation 7). The WFI calculation uses Equation 7 as the basic WFI. This factorial simply indicates the potential for a fouling event to be occurring as a result of changes in the specific capacity and the time lags over the period since the well was new. It is also possible to predict, through factoring in the relevant contributions of the GC and BP, the overall fouling event (as measured by WFI). This makes the assumption that the GC is essentially a physical and/or chemical inanimate phenomenon that does not involve any major biological activity, hence the term "clogging" is used to reflect the inanimate nature of this event. A second assumption is that the BP is essentially a biologically driven event with physical/chemical reaction components being secondary. This is measured by the increases in bacterial aggressivity and, where this dominates, the term "plugging" is applied. The relative importance of GC and BP can be apportioned by the following equations:

$$GC = FD_{(SC)} / WFI \quad \text{Equation 8}$$
$$BP = FD_{(TL)} / WFI \quad \text{Equation 9}$$
$$GCF = (1 - (GC / (GC + BP)) * 100 \quad \text{Equation 10}$$
$$BPF = (1 - (BP / (GC + BP)) * 100 \quad \text{Equation 11}$$

Where $FD_{(SC)}$ is the factorial decline in the specific capacity; $FD_{(TL)}$ is the factorial decline in the time lags; GCF is the geochemical clogging factor; and BPF is the biological plugging factor.

The relative weighting of the GC and BP are calculated in Equations 8 and 9, respectively. This weight is apportioned by calculating the factorial component in the WFI that can be used to generate a risk assessment for fouling (Table Sixty-Four) that can be related to the GC and the BP. The relative importance of the inanimate fouling (GC) to the animate fouling (BP) can be determined by Equations 10 and 11 where geochemical clogging factor (GCF) is the factorial presentation of the potential for inanimate clogging (Table Sixty-Five) and biological plugging factor (BPF) is the factorial presentation for the animate plugging (Table Sixty-Six). Where the factor is given as a positive percentile, then it is the dominant factor in that particular fouled well. The attribution of importance is given by the scale of the positive percentile for either the GCF or the BPF (Table Sixty-Four). The form of the fouling that would occur in a plugging is given in Figure Seventy where the time lag falls earlier and faster than the specific capacity. Where a well is being clogged then the time lags will not fall until after the well has begun to significantly lose its specific capacity (Figure Seventy-One).

In practice, wells being compromised by biological plugging (BPF) events are likely to exhibit more deviations from the linear reductions in the specific capacity over time than a physical and/or chemical clogging (GCF). This would cause reduced confidence in the linear regression analysis expressed through the correlation coefficient (R^2). The potential for a clogging (GC) or a plugging (BP) event to be dominating a fouling well is given in Table Sixty-Seven. Here, a high R^2 value (greater than 0.9) for an arithmetic linear regression analysis of the loss in specific capacity over time would imply the fouling to be of a clogging type. If the R^2 value is less than 0.8 the increased variation would support the probability that the well is being subjected to a plugging (BP) event.

Table Sixty-Four

Significance of Calculated WFI Values to Fouling Risks

Calculated WFI Value	Fouling Risk Significance	Potential for Recovery
< 0.1	Not Significant	N/A
0.1 - 0.3	Significant	VG
0.4 - 0.6	Serious	RPS
> 0.7	Very Serious	LPS

N/A, not applicable; VG, very good; RPS, reasonable probability of success; LPS, low probability of success.

Table Sixty-Five

Significance of Calculated GCF Values to Cause of Fouling

Calculated GCF % Value	Potential Cause of the Fouling
< -21%	BP dominant
> -20% to < +20%	Both BP and GC active
> +21%	GC dominant

GCF, geochemical clogging factor; BP, biological plugging; GC, geochemical clogging

Table Sixty-Six

Significance of Calculated BPF Values to Cause of Fouling

Calculated BPF % Value	Potential Cause of the Fouling
< -21%	GC dominant
> -20% to < +20%	Both BP and GC active
> +21%	BP dominant

BPF, biological plugging factor; GC, geochemical clogging; BP, biological plugging

Table Sixty-Seven

Potential Significance of the Regression Correlation (R^2) Analysis of Well Age (x) to Specific Capacity(y)

Regression Correlation (R^2)	Potential Cause of Fouling
> 0.9	Clogging dominant
<0.9 to > 0.8	Clogging and plugging
< 0.8	Plugging dominant

From the tables above (Tables Sixty-Four to Sixty-Seven), an understanding of the potential origin of the decline in specific capacity can be determined. The formulations rely upon the logging of the well's specific capacity and bacterial aggressivity in a routine manner from the time the well enters production (i.e., has completed development). Under ideal circumstances, both the specific capacity and the bacterial aggressivity (using the BART testers) testing should be performed every month and a trend analysis undertaken to determine shifts in the WFI, GCF and the BPF along with a regression analysis of the specific capacity to obtain the regression correlation coefficient. For wells exhibiting a slow rate of aging (i.e., very little differences in the data on a monthly schedule) then the monitoring could be pulled back to a quarterly schedule.

Rehabilitation of a well included in this monitoring program should cause radical changes in the WFI (declines), specific capacity (increases), GCF (approaches zero) and the BPF (should initially go more positive and then approach zero depending upon the time of the testing relative to the rehabilitation treatment). This anomalous behavior is often due to the impact of the treatment which has killed and disrupted the growths in and around the well. The debris and cleaned surfaces left behind become natural sites for colonization after the completion of the treatment. It can be expected that, following treatment, there will be shifts in the parameters associable with the destabilization of the fouling around the well. If the monthly monitoring is continued through the post-treatment period, it can be expected that these perturbations in the data would continue for one to three months. This means that reliable data (for predicting the reoccurrence of fouling) may not be obtained until four months after the treatment.

Monitoring the WFI on Aged Wells

The premise for the generation of the WFI is that there is a relationship between the fall in the specific capacity (as an indicator of fouling) and the fall in time lags as the bacteria associated with the fouling become more aggressive. To achieve a WFI with some level of confidence, data should have been gathered from the onset of its production life. Failure to do this means that the essential background information for the new (pristine) well would be absent and the SC_0 and various TL_0 could not be incorporated into the calculations rendering it impossible to calculate the WFI, GCF and the BPF. A system of compensation has therefore to be applied to correct for this short fall in the available data.

There are a number of presumptions that have to be made with respect to an aged well that is now to be included in a WFI assessment. These are:

1. The well is already likely to have been subjected to some level of plugging and/or clogging that may, or may not have, already impaired the specific capacity of the well

2. The specific capacity, if declining, will do so (phases B and C, Figure Sixty-Nine) at an erratic but fundamentally linear rate

3. Time lags may have already shortened due to the presence of - aggressive bacteria in the event of plugging but may not necessarily have done so in the event of a clogging.

Given these observations with respect to an aged well, there are a number of strategies that can be employed to generate the WFI, GCF and BPF for that well. Since it is not possible to generate the original data set that forms the basis for the generation of the WFI, GCF and BPF, a trend analysis approach has to be taken.

Trend analysis means that a set of readings for both the specific capacity and the time lag to aggressivity for the four standard recommended BART™ testers would have to be undertaken on a monthly basis. After five months of monitoring and six sets of data, a pattern can now be observed. The methods for interpreting the data are given below.

Specific capacity data can be analyzed by linear regression analysis. If there is an increase in the slope towards the x axis, then this would indicate that some form of fouling was likely to be occurring. The steepness of the slope can be used to extrapolate the rate of decline, and the time of intercept with the x axis (time) would allow a theoretical prediction of the length of time before the well suffers from failure. This does not take into account the sudden collapse in hydraulic conductivity (see phase D, Figure Sixty-Nine) which would shorten that length of time before failure.

The potential for the well to be suffering from plugging and/or clogging can be first determined by the correlation coefficient (R^2) generated by the regression analysis. Using Table Sixty-Seven, the relationship between R^2 value and the likelihood of clogging and/or plugging occurring can be inferred. Using the BART™ testers, the shifts in the time lags recorded over the testing period can be used to determine the aggressivity of the bacteria. Here, the sequence of time lag shifts for the four recommended BART™ types (see Table Sixty-Three) can be used directly to assess the WFI risk as being low, medium or high for each of the four bacterial groups for which aggressivity is being determined.

Where a new well is being placed into an existing well field, there is naturally an interest in using the original data from that well to generate WFI, GCF and BPF values assuming that the new well would have had similar characteristics to the other wells in the field when they were new. This may not be appropriate since it assumes that all of the wells in the field are essentially the same and will follow the same production life cycle and suffer from the same form of fouling. Since this is often not the case, such cross-application of data may yield incorrect interpretations. It should also be remembered that this concept assumes that the ground water supply (and water table) do not change since clearly, losses in specific capacity could also be caused by a falling in the available ground water supplies around the water well.

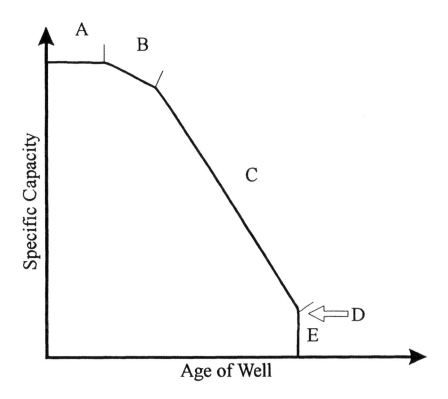

Figure Sixty-Nine, Theoretical relationship between the age of the well (x axis) and the specific capacity (y axis) where there has been a loss in capacity due to fouling. This is shown as a trend analysis going through five stages. These are: (A) stable production no loss in capacity due to fouling; (B) slow degeneration in specific capacity as fouling begins; (C) is an erratic linear loss in the specific capacity due to fouling; (D) is the point at which there is a collapse in the hydraulic conductivity which leads to (E) abandonment of the well due to total loss of production.

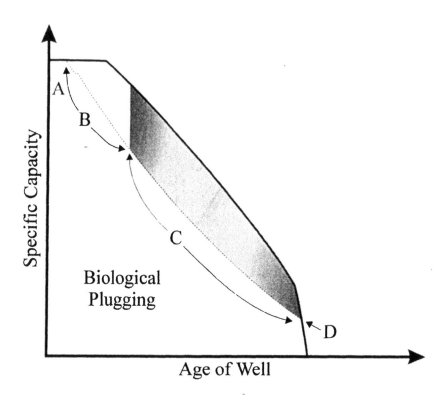

Figure Seventy, Theoretical relationship between the losses in specific capacity (see Figure Sixty-Nine) and biological plugging. A typical format for the development of the plugging is shown in four stages: (A) time lags begin to fall as aggressive bacteria associated with the initial formation of the biofilm infest the well; (B) time lags shorten further as the aggressivity increases and there is now evidence of fouling through a falling specific capacity (phase B, Figure Sixty-Nine); (C) time lags continue to fall and fouling has a major effect on the specific capacity which declines further; (D) very aggressive bacteria cause a radical degeneration in the specific capacity leading to well failure.

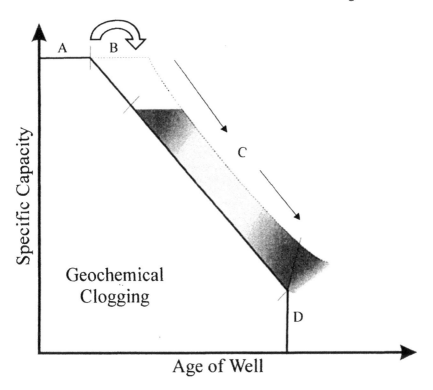

Figure Seventy-One, Theoretical relationship between the losses in specific capacity (see Figure Sixty-Nine) and clogging. A typical format for the development of the clogging is shown in four stages: (A) specific capacity begins to fall (phase B, Figure Sixty-Nine) prior to any reductions in the time lags since aggressive bacteria associated with the initial formation of the biofilm have not yet infested the well; (B) time lags now begin to shorten but only after there has been a decline in the specific capacity; (C) time lags now begin to drop and show a decline having similar, but delayed, characteristics to the specific capacity; (D) Time lags now become short but the radical degeneration in the specific capacity of the well has already led to well failure. In this event, the losses in specific capacity can be considered to be due to predominantly inanimate clogging of the well.

APPENDIX

This appendix is composed of items of interest to the reader and is subdivided into two sections:

Conversion tables
Culture Media Formulations

Conversion Tables

This book has been written predominantly in metric and the following tables are provided for those readers who would wish to use the more traditional units.

Temperature conversion, Centigrade to Fahrenheit:

°C	°F
-17.8	0
-11.1	12
-4.4	24
0	32
5.6	42
7.8	46
11.1	52
15.6	60
20	68
24.4	76
28.9	84
33.3	92
35.6	96
37.8	100
43.0	110
49.0	120
60.2	140
71.4	160
82.6	180

Metric-English Equivalents

Metric unit	Multiplied by	= English Unit
m	3.279	ft
L	0.2642	gal
cm	0.394	in
kg	2.203	lb
g	0.0353	oz
kPa	0.145	psi
km	0.62137	mile
ha	2.47	acres
km^2	0.386	sq. mile
m^3	1.308	cu. yds

Culture Media Formulations

Most of the microbiological culture media in common use in the analysis of waters and wastewaters is described in the *Standard Methods*. Additional formulations are available in the various manuals published by the manufacturers. A selection of these manuals in common use is listed below:

DIFCO Manual (Difco Laboratories, Detroit, Michigan 48232 USA)

BBL Manual of Products and Laboratory Procedures (BBL, Division of Becton, Dickinson and Company, Cockeysville, Maryland 21030 USA)

Microbiology Manual (E. Merck, Darmstadt, Germany).

It should be noted that other companies also publish lists of their culture media products and various appropriate medthodolgies applicable to their product lines and may be an equally appropriate source of information.

Two media which may not appear in the above manuals are listed below (amounts given are per liter):

AA agar medium,

Trypticase peptone	10g
Mannitol	10g
NaCl	1.0g
Nitrofurantoin	0.35g
Crystal violet	0.02g
Agar	14g

WR agar medium,

Ferric ammonium citrate	10g
$MgSO_4$	2.4g
NH_4NO_3	0.5g
K_2HPO_4	0.5g
$NaNO_3$	0.5g
$CaCl_2$	0.15g
Agar	18g

Note: correct pH to 7.4 with 1 N NaOH.

SELECTED BIBLIOGRAPHY

Abiola, A.T., Cullimore, D.R., Mnushkin, M. and Mansuy, N., Studies on the Utilization of Turbidity and Particulate Loadings in Groundwater as Correlatable Factors with the Bacterial Population, Proc. of the Microbiology of the Deep Subsurface, WSRC Information services, 1990.

Alford, G., Mansuy, N. and Cullimore, D.R., The Utilization of the Blended Chemical Heat Treatment (BCHT™) Process to Restore Production Capacities to Biofouled Water Wells, Proc. of the Third Outdoor Action Conference on Aquifer Restoration, Ground Water Monitoring and Geophysical Methods. Association of Ground Water Scientists and Engineers, Orlando, Florida. May, 1989.

Alford, G. and Cullimore, D. R., *The Applications of Heat and Chemicals in the Control of Biofouling Events in Wells*, Sustainable Water Well Series, D. R. Cullimore, series ed., Lewis Publishers, Boca Raton, FL, 1999.

Andorfer, G. and Elisco, D., producer, Discovery Channel, "Titanic: Anatomy of a Disaster." Stardust Visuals, Pittsburgh in cooperation with RMS Titanic Inc., New York. 105 minutes, VHS format, 1997.

Anon., Nuclear Fuel Waste Disposal Review: Microbiology Tutorial sponsored by the Scientific Review Group, Mississauga, Ontario, December 13, 1993, Federal Environment Assessment Review Office, Environment Canada., 1993.

Anon., Problem Organisms in Water: Identification and Treatment, AWWA Manual, M7, American Water Works Association, 1995.

Anon., User Quality Control Manual in Support of the BART™ Biodetection Technologies, Droycon Bioconcepts Inc., Regina, SK, Canada, 1996.

Anon., " BART™s Take a Dive." News and Notes for the Analyst 21(2) June, 1997.

Anon., *Biofouling and Water Wells in the M.D. of Kneehill*, PFRA and M.D. 40 Kneehill, Alberta, Canada, 1997.

Anon., *Field Testing of Ultra Acid-Base (UAB™) Water Well Treatment Technology in the Municipal District of Kneehill, Alberta, Canada*, PFRA-TS, Agriculture and Agri-Food Canada, Regina, Saskatchewan 1998.

Atlas, R. M., *Handbook of Microbiological Media*, CRC Press, Boca Raton, FL , 1997.

Atlas, R.M. and Bartha, R., *Microbial Ecology: Fundamentals and Applications*, Addison-Wesley, Menlo Park, California, 1981.

Barbic, F. F., Dragomir, M., Bracilovic, M. V., Djindjic, S. M., Djorelijevski, J. S., and Zivkovic, B. V., Iron and Manganese Bacteria in Ranney Wells, *Water Research*, 8,1974

Baresi, L., Mah, R.A., Ward, D.M. and Kaplan, I.R., Methanogenesis from Acetate: Enrichment Studies, *J. Appl. Environ. Microbiol.* 36, 1978.

Caldwell, D.R., *Microbial Physiology and Metabolism*, Wm. C Brown Publishers, Dubuque, IA., 1995.

Cullimore, D.R., Bacteriological Aspects of Iron and Manganese in Water, Proc. 39th Ann Conv. of the Western Canada Water and Wastewater Association, Saskatoon, October, 1987.

Cullimore, D.R., Black Plug Layer Research, A Compendium of Technical and Scientific Papers, Regina Water Research Institute Report 71, 1991.

Cullimore D.R., Microbes in Civil Engineering Environments: Biofilms and Biofouling , *Microbiology in Civil Engineering*, P. Howsam, Ed., E & F.N.Spon, London, 1990.

Cullimore, D.R., An Evaluation of the Risk of Microbial Clogging and Corrosion in Boreholes, *Water Wells Monitoring, Maintenance, Rehabilitation*, P. Howsam, Ed., E & F.N.Spon, London, 1990.

Cullimore, D.R., *The Practical Manual of Groundwater Microbiology*, Lewis Publishing, Chelsea, MI, 1993.

Cullimore, D.R., *Practical Manual of Bacterial Identification*, RWRI, University of Regina, Canada, 1996. (Note: second edition to be published by Lewis Publishers, Boca Raton, FL., 1999.)

Cullimore, D.R. and Alford, G.A., Method and Apparatus Producing Analytic Culture, U.S. Patent 4,906,566,1990.

Cullimore, D.R. and Alford, G.A., Method and Apparatus for the Determination of Fermentative, Analytic Cultured Activities, U.S. Patent 5,187,072, 1993.

Cullimore, D.R., Abiola, A., Reihl, J. and Mnushkin, M., *In Situ* Intercedent Biological Barriers for the Containment and Remediation of Contaminated Groundwater, Droycon Bioconcepts Inc. Report 14, 1990.

Cullimore, D.R. and Abiola, A.T., Novel Technologies in Monitoring Biofouling, Pro. 42nd Annual Convention of the Western Canada Water and Wastewater Association, Regina, SK, Canada,1990.

Cullimore, D.R., Abiola, A.T. and Reihl, J., Preliminary Bacteriological Investigation of the Rocanville Water Distribution System, Regina Water Research Institute Report 62,1990.

Cullimore, D.R., Abiola, A.T., Mnushkin, M. and Reihl, J., Evaluation of the Potential Biofouling of Water Wells, Newcastle, New Brunswick, Regina Water Research Institute Report 64c, 1990.

Cullimore, D.R., Abiola, A.T., Reihl, J. and Mnushkin, M., Evaluation of the Potential Biofouling of Water Wells, Newcastle, New Brunswick, Regina Water Research Institute Report 64a, 1990.

Cullimore, D.R., Lebeden, J. and Legault, T., Investigation of Bacterially Driven Biofouling of Water Wells in Central Alberta, Canada, Pro. AGWSE National Education Program "Biological Aspects of Ground Water" Las Vegas, September, 1997.

Cullimore, D.R. and Legault, T., Microbiological Investigations of Water Wells in the Municipal District of Kneehill, Alberta, Rural Water Development Program, PFRA, Canada Agriculture, 1997.

Cullimore, D.R., Legault, T, Keevil, B., and Alford, G., The Laboratory Synthesis of Iron Related Bacterial Clogging in Well Screen and Gravel Pack, Pro. AGWSE National, 1997.

Culllimore, D. R., and McCann, A. E., The Identification, Cultivation and Control of Iron Bacteria in Ground Water, In: *Aquatic Microbiology*, Skinner, F.A. and J. M. Shewan, Eds., Academic Press, New York, 1978.

Cullimore, D.R., Mansuy, N and Reihl, J., To Conduct a Microbiological Assessment of the Potential for Enhanced Biofouling from a Gasoline Spill in Selected Water Wells in Missoula, MT, Regina Water Research Institute Report 57, 1988.

Cullimore, D.R., Mansuy, N. and Alford, G., Studies on the utilization of a Floating Intercedent Device (FID) in the Biological Activity Test (BATS) to Identify the Presence and Activity of Various Bacterial Groups in Groundwater, Proc. First Int. Symp. on Microbiology of the Deep Subsurface, USDOE, Orlando, FL,1990.

Cullimore, D.R., Ornowski, S. and McLean, W., Microbiological Assessment of Slime and Sludge Samples obtained from a Sanitary Landfill Microcosm, Regina Water Research Institute Report 76, 1995.

Cullimore, D.R., Ornowski, S., Reihl, J. and Ostryzniuk, E., Utilization of Various Straws as Substrates to Retard the Growth of Micro-Algae within Confined Water Bodies, Regina Water Research Institute Report 75, 1994.

Davis, D. G., Palmer, M. V., and Palmer, A. N., Extraordinary Subaqueous Speleothems in Lechuguilla Cave, New Mexico, *NSS Bulletin* 52, 1990.

Gehrels, J, and Alford, G., Application of Physico-Chemical Treatment Techniques to a Severely Biofouled Community Well in Ontario, Canada., *Water Wells Monitoring, Maintenance, Rehabilitation,* P. Howsam, Ed. E & F.N.Spon, London, 1990.

Geldreich, E. E., *Microbial Quality of Water Supply in Distribution Systems*, CRC Press, Boca Raton, FL, 1996.

Hattori, T., *The Viable Count, Quantitative and Environmental Aspects*, Science Tech. Publishers, Madison, WI, 1988.

Helweg, O.J., Economics of Improving Well and Pump Efficiency, *Groundwater*, 20 (5), 1982.

Howsam, P., Microbiology in Civil Engineering, FEMS Symposium No. 59.E. & F.N. Spon, London, 1990.

Howsam, P., *Water Wells, Monitoring, Maintenance, Rehabilitation*, E. & F.N. Spon. London,1990.

Jeter, R.M. and Ingrahm, J.L., The Denitrifying Bacteria in *The Prokaryotes,* Starr, M.P; Stolp,H.; Truper, H.G; Balows, A. and H.G. Schedel, Eds. Springer-Verlag, New York, 1981, chap 73.

Legault, T., The Ultra Acid-Base (UAB™) Water Well Treatment Technology, 4th Year Work Term Report, Faculty of Engineering, University of Regina,1997.

Mah, R.A. and Smith, M.R., The Methanogenic Bacteria, In:The Procaryotes Ed., Starr, M.P., Stolp, H., Truper, H.G., Balows, A. and H.G. Schlegel, Springer-Verlag, NY, 1981, chap. 76.

Mann, H., Tazaki, K., Fyfe, W.S. and Kerrich, R., Microbial Accumualtion of Iron and Manganese in Different Aquatic Environments: An Electron Optical Study. Catena (Suppl. 21), Cremlingen, 1992.

Mansuy, N., Nuzman, C. and Cullimore, D.R., Well Problem Identification and its Importance in Well Rehabilitation, in *Water Wells Monitoring, Maintenance, Rehabilitation,* P. Howsam, Ed. E & F.N.Spon, London, 1990.

Mansuy, N. and Layne Geosciences Inc., *Water Well Rehabilitation, A Practical Guide to Understanding Well Problems and Solutions,* Monogram in the Sustainable Water Well Series, D. R. Cullimore, series Ed. Lewis Publishers, Boca Raton, FL, 1999.

Morris, R., Grabow, W. O. K., and Jofre, J., editors of Health-Related Water Microbiology 1996, *Water Science and Technology* 35 (11-12), Pergamon Press, Oxford, 1997.

Pellegrino, C. and Cullimore, D.R.,The Rebirth of the RMS Titanic, A study of the bioarcheology of a physically disrupted sunken vessel. *Voyage* 25, 1997 .

Pipes, W. O., *Bacterial Indicators of Pollution.* CRC Press, Boca Raton, FL, 1982.

Prescott, L.M., Harley, J.P. and Klein, D.A., *Microbiology,* Wm. C Brown, Dubuque, IA, 1993.

Rowe, K., Fleming, I., Cullimore, D.R., Kosaric, N., and Quigley, R.M., A Research Study of Clogging and Encrustation in Leachate Collection Systems in Municipal Solid Waste Landfills, Report conducted for the Interim Waste Authority Ltd., Ontario by Geotechnical Research Center, University of Western Ontario, June, 1995.

Rowe, K., Hrapovic, L., Kosaric, N., Cullimore, D.R. and Quigley, R.M., A Laboratory Investigation into the Degradation of Dichloromethane, Draft Final Report. Report, conducted for the Interim Waste Authority Ltd., Ontario by Geotechnical Research Center, University of Western Ontario, May, 1994.

Schuler, M.L. and Kargi, F., Bioprocess Engineering, Basic Concepts. Prentice Hall PTR, Englewood Cliffs, NJ, 1992.

Singh, B., Harris, P. J., and Wilson, M. J., Geochemistry of Acid Mine Waters and the Role of Micro-Organisms in Such Environments: A Review. *Advances in GeoEcology*, Reiskiirchem 30, 1997.

Smith, S. A., Case Histories of Iron Bacteria and Other Biofouling, *Water Well Journal*, 41 (2), 1987.

Smith, S. A., Methods for Monitoring Iron and Manganese Biofouling in Water Supply Wells, American Water Works Association, Denver, 1992.

Smith, S. A., *Monitoring and Remediation Wells: Problem Prevention, Maintenance, and Rehabilitation*, CRC Press, Boca Raton, FL., 1995.

Sommerfeld, E.O. et al., Iron and Manganese Removal Manual, Saskatchewan Environment and Resources, 2nd Ed., May, 1997.

Stoffyni-Egli, P. and Buckley, D.E., *The Titanic 80 Years Later: Initial Observations on the Microstructure and Biogeochemistry of Corrosion Products*. Poster, S. Geological Survey, Canada, Bedford Institute of Oceanography, Nova Scotia, 1992.

Wells, W. and Mann, H., Microbiology and Formation of Rusticles from the RMS Titanic, *Resource and Environmental Biotechnology*, 1997.

INDEX

O

P

Q

R